Biosemiotics

AF828948

Volume 23

Combining research approaches from biology, semiotics, philosophy and linguistics, the field of biosemiotics studies semiotic processes as they occur in and among living systems. This has important implications and applications for issues ranging from natural selection to animal behaviour and human psychology, leaving biosemiotics at the cutting edge of the research on the fundamentals of life.

The Springer book series *Biosemiotics* draws together contributions from leading scholars in international biosemiotics, producing an unparalleled series that will appeal to all those interested in the origins and evolution of life, including molecular and evolutionary biologists, ecologists, anthropologists, psychologists, philosophers and historians of science, linguists, semioticians and researchers in artificial life, information theory and communication technology.

Kalevi Kull, Professor in biosemiotics, University of Tartu
Alexei Sharov, National Institute of Aging, Baltimore

More information about this series at http://www.springer.com/series/7710

Filip Jaroš • Jiří Klouda
Editors

Adolf Portmann

A Thinker of Self-Expressive Life

Springer

Editors
Filip Jaroš
Department of Philosophy and Social Sciences
University of Hradec Králové
Hradec Králové, Czech Republic

Jiří Klouda
Department of Philosophy and Social Sciences
University of Hradec Králové
Hradec Králové, Czech Republic

ISSN 1875-4651 ISSN 1875-466X (electronic)
Biosemiotics
ISBN 978-3-030-67809-8 ISBN 978-3-030-67810-4 (eBook)
https://doi.org/10.1007/978-3-030-67810-4

This Springer imprint is published by the registered company Springer Nature Switzerland AG
The registered company address is: Gewerbestrasse 11, 6330 Cham, Switzerland

This book was written with the support of the Grant Agency of the Czech Republic (Grantová agentura České republiky – GAČR) for the project
"Adolf Portmann – a pioneer of eidetic and semiotic approach in the philosophy of the life sciences." (GAČR 19-11571S)

Acknowledgments

It is both our duty and pleasure to thank all who took part in the formation of this book. This includes, foremost, the contributing authors. We further extend our gratitude to the Department of Manuscripts and Old Prints at the University Library Basel, particularly to Lorenz Heiligensetzer and Monika Studer (among other things, for allowing us to publish the rare photographs of Portmann printed on p. v and p. 70), and to the leading authority on Portmann's work, Roger A. Stamm, to whom we are grateful for his all-around advice and assistance. We thank the Philosophical Faculty at the University of Hradec Králové for the quality research environment it has provided us; our publisher, Springer, for their benevolent patience; and Kalevi Kull and Alexei Sharov for accepting our book in the Biosemiotics series.

Special acknowledgment is due to Andrew G. Christensen for translating the introduction and some of the chapters and for language editing and copy-editing the rest. And to Michal Hubálek for typesetting and layout of the book.

Contents

Chapter 1
Introduction

Filip Jaroš and Jiří Klouda

Forty years after his death, the work of Swiss biologist Adolf Portmann (1897–1982) still presents a challenge to several disciplines. A brief indication of the significance and complexity of his legacy is the fact that this zoologist, long regarded as a specialist on several species of marine fauna and on the embryonic development of vertebrates, achieved fame with a book about the special position of humans in the animal kingdom and thus became one of the founding figures of philosophical anthropology. In contrast, the culmination of his zoological research – a project to reform morphology based on the display of living beings, on the idea that "appearing" belongs to the basic function of living forms and is just as fundamental as preservation of the individual or the species – attracted (and continues to attract) the attention of philosophers and aestheticians but did not make a large impact on his fellow zoologists. The causes of this differential and somewhat contradictory reception are not primarily rooted in Portmann's work itself, as its creator did not see a radical divide between the individual areas of his research. Furthermore, he formulated the basic principles of both his anthropology and his phenomenal biology roughly at the same time, in the early 1940s.

If we are to provisionally identify the common denominator of both approaches, we can very generally define it as a programmatic *contextualization* and *anthropologization* of scientific knowledge. We must not, however, represent this merely as a critical problematization of seemingly objective biological knowledge or as a deliberate "anti-scientific" devaluation of exact research. Among his great inspirations Portmann counted J. von Uexküll's concept of *umwelt* – the idea that the existence of living beings always takes place in a specifically structured world and that it is necessary to understand both their particular anatomic structure and the totality of

F. Jaroš (✉) · J. Klouda
University of Hradec Králové, Hradec Králové, Czech Republic
e-mail: filip.jaros@uhk.cz; jiri.kl@centrum.cz

F. Jaroš, J. Klouda (eds.), *Adolf Portmann*, Biosemiotics 23,
https://doi.org/10.1007/978-3-030-67810-4_1

their life activities in the widest sense.[1] This idea – that the life of living beings carries a disposition to structure meaning, construct a world, and orient oneself within it – then provides guidelines for special research conducted by exact means. Physical and chemical analysis, which began an unprecedented path of development for biology in the nineteenth century, led during Portmann's lifetime to the revolution of the ability to read DNA, which many thought would reveal the final mysteries of life. Confronted with the enormous amounts of scientific knowledge being produced, however, it became even more clear that although biologists employ physical and chemical methods, biology itself is not (only) physics and chemistry – these exact methods must be *applied* according to what has *meaning* for living beings. The work of the biologist does not begin and end with exact measurement – this must be preceded and followed by contextual interpretation, which is also part of science. In Uexküllian terms: if biology is to give relevant accounts of life, it must orient its research more or less according to what is significant in the umwelt (and its corresponding organization) of the living being under consideration. To get a sense of the specificity of Portmann's method of work, we need only shift the accent in this otherwise apparent and commonplace claim: when biologists produce scientific knowledge, they must in some way *operate* with it, and with their science as a whole, in which case this "utilization of science" is itself no longer exact science; it takes place on another level – the very level that biologists themselves aim to investigate, the level of living beings who are trying to interpret their world.

As opposed to the primarily physiologically oriented Uexküll, Portmann relied on a wider phenomenal-morphological foundation which allowed him to go beyond the model of a closed "functional circle" (*Funktionskreis*) advanced by his great predecessor. The correlation between the particular distinctiveness of appearance of a living being (its expressiveness) and the complexity of its relation to the world (richness of experiencing the world) led Portmann to the idea of a certain hierarchy of living beings (*Rang*) that could be better incorporated into the general theory of the evolution of life (and thereby overcome the limitations of Uexküll's anti-evolutionary skepticism). With the acceptance of evolutionary ideas, the "reading of meaning" becomes a dynamic, ever-changing process, just like the "umwelt" in which it takes place. Thus, a living being must in some way *interact* with the bearers of meaning in its umwelt, not merely react, quasi-mechanically, as it would to explicit signals. A living form interprets its world (similarly as a biologist does factual observations), and this interpretation then speaks back to its world (it "manifests" through the behavior of interpreting beings).

A particularity of Portmann's position is that the contextualization of biological knowledge sketched above developed systematically from both ends, so to speak. One side of his project can be designated a methodology of biological cognizance, a philosophical reflection on the way in which humans relate to living nature, the general assumptions behind this relationship, and what this kind of cognition means for them. On the other (only seemingly opposite) side, there is his more empirically

[1] Cf. Portmann 1956.

oriented project of phenomenal morphology – an investigation into how living beings appear and relate to the world. An assumption behind this research into the autonomous behavior of living beings is that we ourselves are living beings. It is primarily our experience of ourselves as living, corporeal beings that allows us to comprehend how other living beings understand and navigate their world, and only standing on this foundation can we work with the methods, instruments, and results of modern science. In the specific autonomy of living beings (which Portmann calls "inwardness" or "centricity") lies the keystone holding together both aims of his research – we can only inquire about the nature of life because we who ask are living beings and thus mysteries to ourselves as well. Thus, Portmann's contextualization of biology is also an anthropologization (without, of course, succumbing to a naïve and uncritical anthropomorphism).

It will likely come as no surprise that this type of thinking met widespread acceptance specifically on the side of anthropology, and the biologist Portmann typically registered greater success in fields that utilize biological findings than in biology itself. His concepts of "secondary altriciality" and the "social womb" came precisely from this "contextualizing" method, when exact measurements of various factors of the prenatal and postnatal development of diverse species of mammals were compared and interpreted with consideration of the particularity of human cultural existence.[2] The fact that the analogically based "phenomenal morphology" did not receive similar acceptance can be accounted for by the ideal of a strict and objective science that had long presided over biology.[3]

The situation here has also changed significantly since Portmann's death, however. A range of explicitly contextual biological disciplines has gained in importance, particularly evolutionary development and ecology. Thanks to them, interpretive activity is no longer considered a behavioral trait only of higher animals (or exclusively of humans), but also a way in which an organism works with its genetic material – even at a level, that is, which was until recently considered an unambiguous and definitive level of exact description. It is no coincidence that many contemporary authors who capitalize on Portmannian theoretical impulses are associated with biosemiotics, a young field lying on the borders of several biological disciplines which relies on a similar understanding of science, or that this volume is published in the *Biosemiotics* book series.[4]

The aim of this introduction is not to provide an exhaustive, synthesizing overview of Portmann's research efforts. The multifaceted nature of his legacy and its

[2] J. Fischer (2009 and elsewhere) characterizes philosophical anthropology as an independent branch of Continental thought, for which philosophical utilization of findings from natural sciences is characteristic.

[3] This situation is echoed by the monumental history of biology edited by I. Jahn, where the bibliography for Portmann completely ignores his books and studies on the project of phenomenal morphology (Jahn 2000: 928).

[4] Some of the chapters published here – or rather, working versions of them – were presented at an international conference, "Adolf Portmann: a pioneer of the eidetic and semiotic approach in the life sciences," held on 6–7 June 2019 in Hradec Králové, Czech Republic.

potential application is better demonstrated by the fact that biologists, anthropologists, philosophers, and historically oriented researchers all come together over his work. This is also seen in the structure of this volume, though it is also true that the division into disciplinary sections is of relative appropriateness only, just as Portmann's biological, anthropological, and philosophical interests did not exclude but permeated each other, culminating in an original and inspiring theoretical approach. The fact that some chapters here are rather of a Portmann-inspired nature, or develop Portmannian ideas in new contexts, is by no means a betrayal of his legacy, as Portmann himself always emphasized the schematic nature of his texts and the provisional state of the presented conclusions of his research.

***

The volume we present our readers opens with two texts from the mid-1960s meant to familiarize them with the basic outlines of the foundations and goals of Portmann's biology, as well as the most important contexts in which it was developed and received. **Adolf Portmann** is represented by a translation of one of his theoretical texts, in which he presents the main principles of his conception of biology to a wider public. In the 1965 essay, "New Fronts of Biological Work," we find not only Portmann's original concepts of inwardness, relation to the world, self-presentation, and unaddressed phenomena, but also considerations on the conditions under which these could become topics of biological research. The author alerts us that these fundamental properties of living organisms have escaped the attention of biology mainly because its knowledge is methodologically oriented primarily according to the Cartesian model. Its goal is *general* findings; what is relevant is that which is valid as a common law for the widest possible range of living beings. Biology is able to find this at a macromolecular level, as clearly demonstrated by advances in the analysis of genetic code. This search of the general, however, must be complemented by research of the *particular* – what is specific to individual forms of life. Portmann's reform of morphology, therefore, is an attempt to open a new front for biology in which the particular would become a focus of specialized expertise.

The epistemological lines running through Portmann's reflections were further elaborated by **Marjorie Glicksman Grene** (1910–2009). This American philosopher of natural science was among the first interpreters of Portmann to familiarize the Anglo-American academic public with his thought. While her well known book *The Understanding of Nature*[5] contains an extensive introduction to Portmann's biology, we present here a previously unpublished text in which she follows up and philosophically develops Portmann's concept of living nature. "The Language of Nature Reread" is a lecture (delivered sometime in 1964–65) in which Grene tries to imagine an alternative to the Galilean (and Cartesian) conception of nature. Her main idea can be summarized as an attempt explicitly to work out an "ontology of the living," which implicitly lies at the foundation of Portmann's new biology. The

[5] See Grene 1974: 254–293, and 1965 (an earlier article on Portmann).

path Grene takes toward this ambitious goal is that of a non-Cartesian epistemology utilizing the model of "tacit knowing," adopted from M. Polanyi. Grene interprets the two horizons of biological knowledge outlined by Portmann – the Cartesian, impersonal cognizance of general laws and, conversely, the always personal cognition of phenomenally given specifics and particulars (which in the philosophically oriented chapters below will be further explained as primary and secondary, Ptolemaic and Copernican stances toward the world) – as akin to Polanyi's complementary components of human cognition. This always has two interdependent and indispensable components which continually restructure each other: a central and focal, though partial, component and, conversely, a peripheral and subsidiary, though complex, component.

The editors of this volume are honored that the first contribution in the section dedicated to biology and anthropology is by **Roger Alfred Stamm**. His text, "Adolf Portmann: The Researcher and the Teacher," uniquely brings together the perspectives of a former student of Portmann, a university biologist and expert, and the scientific steward of Portmann's papers in the archive of the University Library Basel. Stamm outlines the gradual genesis of Portmann's work and introduces his main research themes, books, and ideas. He familiarizes us with Portmann's research and teaching methods and also includes his extensive and multifaceted public activities. A critical evaluation of Portmann's works, taking up approximately the final third of the chapter, points to certain weaknesses and problematic moments (given the state of biology at the time and necessarily human limitations), which can nevertheless also be considered as providing potential material and perspectives for further development of the great researcher's heritage – indeed, many of the following chapters proceed along such lines. The appendices contain an overview of basic biographical data and a chronological annotated bibliography of Portmann's most important articles and books, making the chapter a comprehensive introduction to Portmann's works and ideas – something which until now has not been available in English.

In her chapter titled "Novel Ecological Mismatches in the Light of Jakob von Uexküll's and Adolf Portmann's Works," **Riin Magnus** looks for a connection between the two thinkers' theories relating to the current problem of human-induced rapid environmental change. The perceptual and behavioral characteristics of organisms are fine-tuned to the environments they have long inhabited. If these environments undergo rapid change, there arises a mismatch between innate reactions (or sometimes those learned during personal development) and their biological function. Thus, the relative rigidity of perceptual schemes, emphasized by Uexküll ("search image") and Portmann ("archetypal image"), can in some cases of rapid environmental change threaten the survival of a given species. Mechanisms that lead to a change of meaning in the framework of the umwelt of a given organism are no less important from a biosemiotic perspective. Here it is possible to apply both Uexküll's approach linking the significance of a perceived object with the tuning (*Stimmung*) of the animal subject and Portmann's theory of unaddressed phenomena, whose interpretation is not bound to a predetermined biological function.

In "Revisiting Basal Anthropology: A Developmental Approach to Human Evolution and Sociality," **Andres Kurismaa** demonstrates that Portmann's phylogenetic perspective on the early ontogenesis of birds and mammals is highly relevant to current research in the field of evolutionary development. In the framework of distinguishing between so-called precocial and altricial types of young, it is shown that humans are, like other great apes, primarily precocial but during their evolution (apparently already in the *Homo erectus* phase) secondarily acquired some altricial characteristics (especially limited locomotion following birth). Kurismaa shows that although the designation of humans as a secondarily altricial species is widely acknowledged, the morphological-structural approach characteristic of Portmann's view of living organisms has conceded to functional explanation in the style of Darwinian reductionism. Recent findings from neurocognitive and neurodevelopmental research, however, have confirmed the basic postulate of Portmann's basal anthropology, which relates the physiological early birth of human infants to the "openness" of our life mode, which in the phylogenesis of our species served as a preadaptation for dynamic cultural evolution.

Filip Jaroš deals with Portmann's anthropology in his chapter, "Portmann's View on Anthropological Difference," primarily with a view to determining the difference between animals and humans. Portmann's basal anthropology has a common goal with the tradition of German philosophical anthropology (M. Scheler, H. Plessner, A. Gehlen): to demonstrate through the results of biological sciences that humans have a special status among other living beings. In the framework of his concept of the "extra-uterine year," Portmann shows that humans start to acquire their typical form only towards the end of their first year of life, when they take on an upright posture and the foundations of speech and rational thought. Portmann's conviction about the different mechanisms of biological (i.e. on the level of hereditary changes) and cultural evolution seems to have been challenged in recent decades by the discovery of so-called animal cultures. It remains an open question whether the postulate of Portmann's basal anthropology (humans as a secondarily altricial species) signifies a qualitative difference, or whether it is more appropriate to emphasize the similarities in the early development of human and great ape offspring (a strong bond with the mother, considerable cognitive plasticity).

Compared to the previous section, the section devoted to philosophical problems in Portmann's work shows greater concentration on a few central problems (outlined in M. Grene's text). In "Portmann, Goethe and Modern Biology: Two and a Half Ways of Looking at Nature," **Markus Wild** presents Portmann's conception of biology in distinction to modern, mechanistic causation science, on the one hand, and to Goethe's conception of science on the other. The latter's notion of science deserves particular attention, as Portmann formulated a number of his ideas in texts on Goethe's method of research. The attempt to grasp a living being in its autonomous manifestations, irreducible to mere physical and chemical causal processes, led Portmann to formulate the concept of "authentic phenomena." This is presented and discussed in detail by Wild (partially in debate with M. Grene's interpretation), as it is through authentic phenomena that one of the fundamental properties of living organisms is realized – their "self-presentation." It was on these foundations that

Portmann built his phenomenological approach to living organisms, which, with its dynamic concept of the appearing form (*Gestalt*), does not deny its Goethean inspiration but avoids its explicitly non-biological, metaphysical assumptions.

Pietro Conte examines the position of Portmann's morphology among biology, aesthetics, and philosophy in his chapter, "Playing Appearances: On Some Aspects of Portmann's Contribution to Philosophical Aesthetics." He focuses here on the pivotal work *Animal Forms and Patterns* (Die Tiergestalt) and analyzes the shift in Portmann's thinking between 1948 and 1960, when the first and second editions were published. From the view of theoretical biology, Portmann's morphological thinking is characterized by a rejection of Darwinian functionalism, expressed in the motto "form follows function." A phenomenal form (*Gestalt*) is the result of the species-specific self-expression of each organism, going beyond considerations of mere self-preservation. The affinity of this idea with phenomenology led some authors (H. Arendt, M. Merleau-Ponty) to use Portmann's morphology in support of the thesis that a phenomenon is conceivable only in connection to a perceiver. This, though, is in contradiction with the concept of "unaddressed phenomena" – that is, morphological structures that at the time of their origin (and long after) had no addressee among living beings. As expressions of the creative power of evolution, these phenomena are, in accordance with certain directions of modern philosophy, interpreted as non-purposive, unmotivated, "aestheticizing" play.

Oreste Tolone, in "Ptolemaic Man and Copernican Man: In Favour of 'Depth Anthropology,'" focuses on the philosophical aspects of Portmann's basal anthropology, particularly the relation of the imaginative and intellectual approaches to the world. Portmann assigns the different ways in which humans inhabit the world to the poetically named categories of Ptolemaic man and Copernican man. The first type can be approximately characterized as being sensorily and bodily situated in the world, which is perceived through the senses, emotion, fantasy, and myth; the second type lives in the intellectually constructed world of modern science, in which rationality presents itself as a means of modeling and control. Portmann's main thesis is that modern Western civilization is headed toward a profound hypertrophy of the rational function. Tolone develops this warning in connection with the critiques of "technomorphic thinking" by K. Lorenz and E. Morin. The goal of the individual should be a balance between the Ptolemaic and Copernican modes of inhabiting the world, although an understanding of life derives from deep sources typical of the first approach, hence the designation of this anthropological program as depth anthropology.

Jiří Klouda also returns to the problem of unaddressed authentic phenomena in his chapter, "'The New Morphology' Between Biology and Philosophy: The Hermeneutic Dimension of Portmann's Thought." Considering a range of texts in which Portmann deals with the possibilities of the human cognitive stance toward nature, and which complement his theoretical manifestos on the reform of biology, Klouda suggests that central Portmannian concepts such as "authentic phenomena" or "self-presentation of a living form" are not standard objective and empirical concepts with which science could work under normal circumstances. Nevertheless, these concepts are not fundamentally dubious or mistaken: the conceptual space of the "new morphology" is a borderland where science and extra-scientific experience converge; its sense is one of critical and regulatory mediation between these two spheres. On a traditionally philosophical level, this corresponds to the border between the sensory and rational components of human cognition.

The section dedicated to historical contexts of Portmann's work begins with a study by **Karel Stibral**, "The Beauty of Organisms: Biological Aesthetics Between Darwin and Portmann." His comparison of the stances of both great biologists on the question of beauty comes to a conclusion that will surprise many: that it is Darwin who reserves a greater place for beauty in the animal kingdom, primarily by postulating a "sense of beauty" in many animal species. This sense of beauty comes into play in the selection of sexual partners as a significant, independent factor, which often works against natural selection in a purely utilitarian understanding: it does not raise the chances of self-preservation for its bearer. Attractiveness, according to Darwin, is not an indication of good physical condition, for example. Later neo-Darwinism almost completely suppressed this element of its founder's theory. Portmann, too, stands in the tradition of modern aesthetics, which places beauty in opposition to usefulness. He can still refer to "self-manifestation" as a basic property of living forms, but at the same time he must take care that beauty set in opposition to functional purpose does not again, paradoxically, become just an isolated "aesthetic function."

Ivana Ryška Vajdová charts one of the most interesting areas of Portmann's public activity – his long-term engagement in the Eranos group – in her chapter, "On the Brink of the Expressible: Adolf Portmann Meets Carl Gustav Jung on Eranos Ground." Her attention is focused mainly on the controversy between Portmann and C. G. Jung that arose from a session in 1950. Portmann delivered a lecture on "The Problem of the Primordial Image in Biological Perspective" in which he politely but decisively indicated the extent to which Jung's approach to archetypes is influenced by outdated concepts of instinct that biology had in the meantime fundamentally corrected. Jung's dismissive reaction reveals, among other things, the principle difference in understanding between the two authors of "unconscious patterns of behavior." Biological research approaches the human psyche differently than therapeutically oriented analytical depth psychology.

Our book culminates with a contribution from **Stanislav Komárek**, "The Czech Reception of Portmann." The phenomenon to which the author dedicates his attention is the advent of a unique school of "Czech Portmannism." While Portmann's philosophical anthropology achieved widespread international recognition in the 1960s, his conception of "phenomenological biology" was not, as a whole, accepted or developed in academia, with the exception of the Czech biologist and philosopher Zdeněk Neubauer (1941–2016). Thanks to him and his friends, Portmann became one of the key authors of intellectual dissent in communist Czechoslovakia, and following the 1989 revolution, Neubauer led an academic experiment to establish Portmannian theory as a paradigm of biological thought, education, and research. At Charles University in Prague there thus arose in the 1990s the original phenomenon of a Portmannian school, which continues to this day. It is a testament to the continuing power and inspiration of this enterprise that the author of this contribution and those of the previous two chapters number among the direct or indirect students of Neubauer.

***

As the reader of the contributions introduced here will discover, it was not an ambition of the editors to promote a unified or definitive "Portmannian" terminology in English, nor did Portmann himself endeavor to do this in his native language. Rather, we gave the authors significant freedom in translating both basic terms and titles of Portmann's articles and books – the only rule we have insisted on is that the interested reader always be able to easily find the original German or French terms and original publication titles in any given case. Likewise, we left it up to individual authors whether to use existing English translations, as we are aware that older English translations of Portmann's books, in particular, can be problematic. Here too, however, we have taken care that our volume should be fully understandable to those who are able to read only in English.

References

Grene, M. (1965). Beyond Darwinism: Portmann's thought. *Commentary, 40*(5), 31–38.

Grene, M. (1974). *The understanding of nature. Essays in the philosophy of biology*. Dordrecht: D. Reidel.

Fischer, J. (2009). Exploring the core identity of philosophical anthropology through the works of Max Scheler, Helmuth Plessner, and Arnold Gehlen. *Iris, 1(1)*, 153–170.

Jahn, I. (Ed.). (2000). *Geschichte der Biologie. Theorien, Methoden, Institutionen, Kurzbiographien*. Heidelberg/Berlin: Spektrum Akademischer Verlag.

Portmann, A. (1956). Vorwort. Ein Wegbereiter der neuen Biologie. In J. von Uexküll & G. Kriszat (Eds.), *Streifzüge durch die Umwelten von Tieren und Menschen: Ein Bilderbuch unsichtbarer Welten. Bedeutungslehre* (pp. 7–17). Hamburg: Rowohlt.

Part I
Paving a Path to the New Biology

Chapter 2
New Fronts of Biological Work

Adolf Portmann

If you consider the fields of contemporary biology, you will find most of its scientific production in areas that owe their growth to physico-chemical research. Over the last few decades, this area of life research has participated in the worldwide expansion and boom of technically oriented natural sciences; thanks to its practical implications, it belongs to the most important fields of research. Biotechnology has arisen and is growing continuously, and such research generates enormous industrial power.

The growth of this field shall continue – and if we, spellbound by the power, see only one aspect of life research, it is clear that the intellectual world that created these achievements shall determine biological research also in the future. From such a point of view, today's innovations are nothing more than an extension and consequence of the foundation that has determined the dominant attitude of research since the time of Descartes.

Such an image, however, is deceitful. For quite some time, a different type of work has been confronting this well-established approach, working in secret, in the shadow of the giant achievements of the physico-chemical technique, and therefore having a difficult time asserting itself. Yet an important change in the basic attitude toward the research object of biology – toward the organism – is happening there.

If I were asked which one of these two types of life research has a broader worldview and a more comprehensive framework of the basics, I would not hesitate a moment to ascribe the wider worldview to the hidden one. I would like to emphasize this with all clarity. The exceptional achievements of the first biological approach do not refute this, and I accentuate my thesis fully aware that the expansion of the biotechnological, physico-chemical work shall continue.

Adolf Portmann (1897–1982)

The chapter was translated by Lucie Nováková and Andrew G. Christensen.

A. Portmann

F. Jaroš, J. Klouda (eds.), *Adolf Portmann*, Biosemiotics 23,
https://doi.org/10.1007/978-3-030-67810-4_2

***

Were I to characterize a direction in biology that enables, in my opinion, a change in the image of nature, I would call it above all a qualitative life research.

Its foundation lies in the great early achievements of morphology at the beginning of the nineteenth century. Johannes Müller, who worked until 1858, developed it far beyond the focus on form into a general concept of living beings. His successor was the great Baltic biologist Karl Ernst von Baer. But the spiritual development of the second half of the nineteenth century removed this broad-minded view from the stage of the dominant biology, and only in this century was the value of this approach rediscovered.

When I speak about a qualitative life research, it is necessary to emphasize that its goal is not a mathematical formalization. This does not mean that it does not respect mathematical methods – it uses them as a supportive tool.

The core of this change, which in my opinion bears crucial importance for a complex biology, is the attempt to find a new approach to our concept of the organism, an approach that surmounts the separation of body and mind. It is necessary to overcome the separation of the psychic aspect of reality that, 300 years ago, under the influence of Descartes, divorced the exploration of matter from experience and inwardness. A new concept of living reality is being born, with a promise of great change.

The attempt to overcome the fatal separation of body and mind focuses on a reality of life in which these two areas, too often radically separated, work together – biology explores behaviour that is based on the influence of matter as well as on subtle mental processes.

The crucial innovation is the recognition of the fact that living beings appear in the world as subjects, and that they settle into, and intervene in, environments as relatively autonomous centres of action. This applies to plants as well as animals, and we would like to stress that this subjectivity should not be too closely identified with the emergence of consciousness. Saying that organisms have a "relationship to the world" does not imply anything about their "experience of the world".

The recognition of this subject-nature of organisms is a fact of great consequence, as well as an occasion for potential error. We have therefore to stress in advance that this new biological research never implicates anthropomorphism. It is not a question of ascribing human life experience to other living beings, nor of attaching even a tiny portion of our own life experience to lower and lower levels of life. It is about a serious acceptance of inwardness as a special mode of existence among living organisms. We recognize this special aspect of life primarily in ourselves, but however vigorously we fight against the projection of human feeling and experience, we should not forget that inwardness is natural to all living beings, even if it incomprehensible to us.

Organisms are forms of being that integrate their experience into their mode of existence and process their experience independently of conscious processes. Such integration of experience into molecular events that are not themselves based on experience makes a living being into a constantly changing entity.

It is therefore useful to illustrate this integration of experience with few examples. I can expose plants as well as unicellular beings or animals, polyps or higher organisms, to a changed rhythm of day and night or light and darkness. In a laboratory I can also expose sea organisms to an unusual change of low and high tide. These can be organisms without proper nervous systems as well as organisms with complex nerve centres. In all cases, the same thing happens in response to the new situation. The changed rhythm of light and darkness, low and high tide, is registered in the organisms through processes that are unknown to us; however, the organism responds as a whole: it modifies its old rhythm under the influence of the new time relations. The extent of the modification is not important to us at the moment. What matters is that the organism tends to respond to the time schedule in a very precise way; it therefore somehow perceives time. While the modification is faster in higher organisms aided by brains, what is significant here is that a response to a time change, setting the internal clock according to new circumstances, is possible at all levels of life. The newly acquired rhythm is maintained. If I change it by further experiments, the organism again adapts itself in accordance with the new circumstances.

I could cite many other examples, but it is important to focus on the main point. Organisms perceive events in their environment. The perception is meaningfully integrated into the existing structure that mediates contact with the environment. The living being is changed by the new experience, even if the change is a tiny one. We would like to consider above all those living forms that lack perception organs common for humans – eyes, ears or brain. Take, for example, a plant that alters its metabolic processes in response to a change in its environment and starts to bloom, or one that, under new circumstances, stops blooming and falls back into a vegetative period. Even such a being, in its relative autonomy, is a constantly changing subject. In using the word *subject*, however, we should not understand a conscious self with special attributes.

***

In the case of the internal clock, we still know very little about the actual mechanisms. The extensive use of these time-based processes in organisms with and without neural organs shows us that they must belong to the very basic processes of plasmatic organization. Among life's secrets, apparently, is the specific way that responding to time is integrated into the molecular order. Since in a biological assessment of this specific organismic characteristic we provisionally refrain from any statement concerning consciousness and limit ourselves to objectively comprehensible modes of action, we are free to operate with these modes of action as we would with physico-chemical facts. This type of approach extends the possibilities of biological work tremendously. The behaviour of subjects, previously considered inaccessible to research, becomes an object of natural science, but without the danger of anthropomorphism. In just a few decades, behavioural research has become an important field of inquiry.

In most cases it is impossible to say to what extent the integrated experience is conscious for the organism. It is important, however, that the dispute over causal or final research becomes obsolete in this approach. In life research, the processes and relationships explored through causal research take place in an orderly whole that is constantly changing under the influence of experiences and that, in turn, influences future experience. Researchers of the subjective must therefore accept and search for special structural possibilities in organisms that, not being completely closed, allow for change based on experience. The new behavioural research shows us more and more that such openness is an attribute of all living beings; while this is often limited to narrow areas of coping with life, in some cases there are more and greater possibilities for openness. Through their increased ability for learning, higher organisms attest that their central nervous systems have especially rich organizations of this kind. The differences can be great even among closely related organisms: some birds that learn to speak can incorporate new word formations into the structure of their nervous system throughout their lifetime, while other species of the same animal class have the capacity of acoustic learning for only a limited, sensitive period.

The knowledge of this openness, gained in recent decades, invalidates old formulations. Only a few years ago, one could persuasively, if simplistically, describe a dichotomy of human and animal experience in which humans were considered open to the world (*weltoffen*) and animals were bound to their environment (*umweltgebunden*). Even nowadays, few would deny the general truth of this distinction. Nevertheless, from now on, the concept of a world-open human and environment-bound animal will need a more careful characterization. We can observe openness to the world in animals as they process and integrate their experience, and we likewise see environmental constraints on humans' experience of the world! Our openness to the world is shown by the fact that we are able to recognize our environmental bonds and distance ourselves from them. In this possibility of a special "positionality", pointed out by Plessner already in 1928, lies the decisive characteristic of the human relationship to the world.

The insights I am talking about refresh concepts about life that were formulated more than a century ago. They remind us that the Greek word *psyche* originally signified the full character of living structure. The recognition in our time of the special inherent value of life is an important factor of the changes happening in life research. The study of the subjectivity of living beings leads to even further insight. The goal is to see the specific character of a life form in contrast to the unique character of another form. Here we demand something that deviates from the objective of most fields of current biology. According to the prevailing understanding, the real task of biology is to grasp the generally valid laws that characterize life processes.

Contrary to this is the requirement to observe and research also the big differences that distinguish the life form of a crab from that of a bee, or birds from mammals. Since the task we have in mind is of particular importance, we must reflect on how much the study of general laws has influenced biological research in recent decades. Only through an appreciation of this research can we truly understand the need for change and the importance of this new research goal.

We want therefore to remember that biochemical and genetic research has proved the existence of molecular structures such as nucleic acids in the whole range of life forms, even in cases of forms that lack the structure of cell nuclei. I also point to evidence that the sexuality of both plants and animals is based on the same basic processes: the mixture of two genomes and the transmission of a constant number of hereditary factors. Anyone who wants to capture the meaning of general laws must also consider the proof of the great unity of metabolic processes and the correlated work of enzymes in the construction and breakdown of substances in animals and plants.

The evidence of these and many other correlations has become the centre of biology to such an extent that a "general biology" could emerge on this base and become the centre of life research even for practical applications, for biotechnology. The imposing unity of this structure sometimes makes us forget that all these processes always serve an autonomous, special form of life. This peculiarity, which distinguishes larger groups, but also significant differences among individual species within groups, must remain the central point of research. Observance of this second aspect of separating multiplicity in full knowledge of the similarity of basic life processes seems to me an aspect of a new front in biology heralded by many signs.

In many areas of cutting-edge research today, peculiarity stands out in the search for general laws. For example, genetics primarily tries to find the general effects of hereditary factors and the general laws of the game of inheritance. But in this search it constantly comes across the specificity of the species with its unique combinations of genetic material. This is how genetics, despite its declared search of general law, enters the field of specificities of separate life forms.

There is a movement in the same direction in evolutionary research and evolutionary history, as well as palaeontology, the research of extinct life forms, however much some researchers here declare the tracing of the general as their ultimate goal. In these cases, the following questions prevail: What is a mutation? What are the general rules of evolutionary processes? How are the tasks defined and distributed in a germ? How are the individual areas of the germ being defined? But these questions also reveal something that was marginalised in the search for general results: the mighty specificity, the uniqueness of every species, genus or class, and, in the case of higher organisms, of every individual being.

Physiologists experience the same. We have already spoken about the ability of plants and animals to perceive processes in time and to sensibly include this experience in their life rhythms. We have also already mentioned the research of very general molecular processes enabling the management of time. But all these studies include some evidence demonstrating that these basic processes, which are likely very similar, are implemented very differently in various life forms. For one bird, the inner clock enables it to make nocturnal flights; in another, it helps with long-distance flights according to the position of the sun.

A special biology of plants and animals is not just a matter of museum classification, of mere order and overview, as many people might think; it is a part of a completely different research task, a necessary extension of the contemporary prevailing search for general laws. As soon as we turn for a moment away from correlation, from equivalence of form, development and function of various organisms, we come across the great puzzle of the peculiarity of every living form and the strangeness separating different ways of relating to the world.

If we look at a significant part of this relation to the world, at possibilities of eyesight, we can see the contrasts among living beings. The research of forms shows us several higher types of eyes that are able to mediate visual experience. Morphological research strives to trace the relationship of individual organs back to simpler, basic forms in order to understand their possible evolution. However, the deeper we penetrate into the function of these visual organs, the greater the differences in relationship to the world that are mediated by such different visual organs. Physiological analysis of the eyesight of insects brings us to a completely different world of visual sense in which the spectrum of qualitatively experienced colours is different from ours, so that in the realm of bees, for example, the colour red goes missing, but, on the other hand, ultraviolet emerges. Research on insects leads to a world where the perception of shapes differs very significantly from that of vertebrates. Physiologists teach us that the perception and memory of shapes in insects is significantly conditioned by the complexity of contours perceived, contrary to what we usually imagine if we think about visual perception in the realm of lower animals. For example, we would assume that simple geometrical shapes like a circle or a triangle can be perceived and remembered more easily than the complicated outlines of flowers. On the base of experiments, however, we have to change our opinion because significant shapes such as the outlines of flowers and nectar guides can be remembered by insects more easily thanks to hereditary processes. But finally we have to abandon this view as well and become acquainted with the idea that the possibility of mastering shape impressions depends mainly on the complexity of the contours and not so much on an inherited relationship to specific forms. We would also like to consider for a moment that green vegetation reflects significant amounts of ultraviolet light. While the human eye cannot perceive this, its meaning as a background in the experience of bees is an important component that can change our idea of the colour green. The strangeness increases even further when I imagine that the colour circle of a bee's eye does not contain red and purple between blue and yellow, but an unknown colour that we, with no possibility of experiencing such a colour impression, call bee purple. We have to accept that in the world of a bee, poppies are not red and many flowers have ultraviolet markings invisible to us. When I think that this same eye perceives the polarized skylight and, incorporating it into everyday life, uses this polarized skylight for orientation, the strangeness of this world grows even further.

Let us consider another way of life, for example the world of bats, which, as experiments have demonstrated, includes a method of orientation in space based on a special sound system in which the emitting of supersonic waves facilitates a very special kind of spatial experience. This capacity was called a "sonar system", after our own radar technology. It may be good to remember that we could discover the peculiarity of this world experience only after our technology had constructed the system of radar orientation for airplanes. Only thanks to general analogy could we understand the behaviour of the bats more clearly, without knowing anything about the real character of their experience. In this context, I might also recall that birds have a hereditary ability to orient themselves in space by perceiving the position of the sun and simultaneously registering the passage of time, and that this combination enables worldwide travel to birds migrating during the day. Many young birds flying without their parents travel thus with no previous experience.

Most recently, it has been rightly observed that nocturnal migratory birds have an inherited orientation system based on the position of the stars in the night sky and a perception of time and that this orientation system allows them to travel from the northern to the southern hemisphere and back. A debate is currently underway about the hereditary components enabling such orientation. The positions are still unclear, but the fact of dealing with time and its processing with impressions of the outside world is now beyond doubt.

***

We are only at the very beginning of a differentiation of life forms. The often very strong equivalents of structure of sensory organs have to be set against the differences in integration of the sensory impressions and differences in behaviour, as Jakob von Uexküll has tried to demonstrate in his book *Umwelt und Innenwelt*. Who knows what world of light is mediated by the sophisticated eye of the cuttlefish, of the eight-armed octopus or of a squid of the high seas? Who is able to understand the situation of a migratory bird who seems to fit best into the realm of light but during migration suddenly plunges into the completely different world of night for its great trip around the world, flying hundreds of kilometres without pausing. What enables that little aviator? What is going on in him when this nightly restlessness overcomes him?

We too readily think about the possibilities of animals' perception of the world through a kind of subtraction from our world, often giving the animal only crumbs that fall from the table of the rich.

Research will need to be able to explore special kinds of living diversity with new intellectual tools. It took us a long time to perceive, at least to some extent, the similarities and correspondences of animal types; we therefore concede that it will take a long time before we penetrate deeper into the peculiarity of this special world, if and in so far as the expressiveness of human language can accomplish this. It is already certain today that experimental research develops objective approaches to the subjective, to group-typical characteristics of living beings.

Once we explore special kinds of perception of the world, much will suddenly come into focus that was neglected or almost forgotten by the predominant goal of a general biology. The object of biology shall now be the whole life form, a full being whose forms and functions are interesting not only as an expression of general laws but as an independent whole, in appearance, behaviour and experience, studied in its relationship to the world in so far as we find objective access to it.

This will also bring a fresh perspective to peculiarities of appearance. The exterior of organisms that is accessible to the senses is often reduced in contemporary research to parts necessary for preserving the individual or the species: adaptations to the conditions of the environment, characteristics necessary for reproduction or development, differentiating sexes, or care of the brood, phenomena that should warn by colours or hide the animal from enemy eyes – the characteristics of appearance received attention as members supporting the vital roles!

I must try to explain more clearly this change of focus. The surface of the animal that is visible to eyes – the skin with all its shapes, colours and patterns – has previously been interpreted above all as a boundary, a border protecting the "interior" from the outer world and therefore securing metabolic processes, but on the other hand enabling a relationship to this outer world through sensory organs and all sorts of tools that help the animal develop an experienced environment and ensure its preservation.

Largely disregarded and almost completely forgotten, however, is another role of the surface: the innumerable patterns of animal skin structures that have no vital function and play no role in any important life-preserving task, and are neither adaptation, signal, nor camouflage colouring. Research has quite often banished them to the storage room of the labs – in many cases they are deemed to be only systematic characteristics enabling us to give a name to a plant or an animal.

In reality, these ignored parts of the living forms are much more significant – however with the disadvantage of showing no use and playing no immediate role in preservation. We call these characteristics unaddressed appearances (*unadressierten Erscheinungen*) because they do not address a specific recipient, because they are not among the characteristics we prefer (those whose meaning we can easily understand), and because they play no life-supporting role.

The streamlined shape of fish, camouflage patterns, protective colours, ostentatious warning patterns like those of wasps – these are easily comprehensible, textbook examples. There exist, however, uncounted patterns and skin organs that cannot rise to this honour; they are neglected or relegated to a realm of aesthetics that is generally considered to have nothing in common with biology.

The development of a new front in research transforms this realm of unaddressed appearances into a new problem. Such appearances are the instrument of a special manifestation of life, the self-presentation (*Selbstdarstellung*) of the organism, expressing special and species-specific features of an organism in the language of the senses. The main field of this demonstration of intrinsic value is the world of the eyes, the realm of light. Light, as the most important medium of appearance, becomes again a problem for the biologist, asking new questions to him as well as to the physicist.

So far, this research field of "Life and Light" has primarily been understood as a part of preservation – recall the role of light in the metabolic processes of green plants; appearance in the light field has also been examined in the category of utility. Thanks to the new twist, the light field is now being explored as an essential place of self-expression.

This shows that many of the previously most frequently observed features are actually results of more original appearances, transformed structures of self-expression initially playing no role in the preservation of the species, generally considered purposeless – though "purposeless" does not necessarily mean "meaningless". The study of animal appearance, particularly in so-called "lower" life forms (polyps, jellyfish, worms, echinoderms, etc.), reveals characteristics of appearance that play no role in the game to preserve species – indifferent markings, or "unaddressed characteristics". It can be demonstrated that many of these characteristics secondarily take on "roles" and gain selective value – but that this new meaning is additional and comes later.

The new twist in morphology testifies already today that the organism uses a large part of its energies to build structures that serve foremost to manifest its character, to serve self-expression. Nobody will ignore the role of maintaining functions. But the exaggerated attention they get thanks to an interpretation of nature that accentuates usefulness and purpose is being highly limited by the emerging new morphology.

This survey of the field of life research sees a new front of biology emerging. During this account I have left no doubt that the main front of this field of research serves the goals of life science oriented toward chemistry and physics. When I emphasize here so strongly the indications of a new front and individual sections of its combat area, I make this emphasis by choice, by a decision.

In such a survey, an individual researcher can never play the role of an instrument that shows what is happening at the moment. We are not just seismographs registering a new intellectual quake and calculating the location of its origin. Certainly, we should be able to do this if necessary, but we as active people have a much more important task. I see this task in making decisions in the bustle of possibilities, in making choices in the confusion of reality, in deciding which way to go in the future. We must not see the future of our development, of our mind and our life forms as an inevitable, fixed process that research has to recognize in advance. Nobody knows what will happen in the future. That is why we have been given the gift of judgement, the great freedom of choice, and we are forced to decide if we want to be active.

This element of decision and valuation is therefore present also in my account of contemporary biological research.

I have mentioned possibilities that still currently belong to the more hidden tasks but which seem to me crucial for the future of biology. It should be mentioned that biology is more than the mediation of new methods of technical progress, new ways of intervention and exploitation and thus of power over processes in living beings. It is particularly important to emphasize this today, as many aspiring nations see Western science – and thus also Western biology – purely as a tool of power and therefore strive to acquire such means of dominance.

It is therefore a special task of our time to remember that biological research involves subjects other than mastering natural processes, that it is also concerned with capturing the image of nature and seeing more clearly our position in the world. The front of the biological examination of nature whose structure I try to describe here serves these goals, and an image of living beings can only be comprehensive and valid if it contains the knowledge that has been won on this front.

Translated from Neue Fronten der biologischen Arbeit. In *Transparente Welt. Festschrift zum 60. Geburtstag von Jean Gebser*. Edited by G. Schulz. Bern, Huber 1965, s. 23–37.

Chapter 3
The Language of Nature Reread

Marjorie Grene

In introducing Portmann's account of the characters of living things, I started from Galileo's famous passage about the language of nature. Contrasting Portmann's view with Galileo's, I went on to outline a new conception of nature and of man's place in nature. This new conception has been, and is being, developed, I suggested, by a number of scientists and philosophers whose thoughts can be seen to converge on an urgent task of fundamental philosophical reform. Yet the resultant point of view is still very much a heresy among biologists and philosophers of science. Nor can it prevail unless it includes, centrally and emphatically, a new theory of knowledge, and indeed, inseparable from this, a new ontology. For it was from a theory of knowledge, and of the method of acquiring knowledge, that modern naturalism took its rise; only through a renewed philosophy of knowledge can we free ourselves from the disastrous consequences of this naturalism, overcome our alienation from nature and find ourselves once more, not disinherited, but at home in the living world. It is extremely important, I believe, that philosophers should take account of the reflections of scientists like Portmann, Buytendijk or Straus; but the central philosophical job that wants doing none of these thinkers has quite accomplished. The missing philosophical keystone is supplied, however, in my opinion, by Michael Polanyi's theory of tacit knowing, and by the ontology which it entails. I want therefore this evening to round out my sketch of a new view of man in nature by outlining the principal points of this philosophy insofar as it bears on our present theme.

Polanyi, like Portmann, philosophizes against the background of a distinguished scientific career. With Portmann, it was the very subject matter of his science, the very existence of what he has come to describe as "authentic phenomena," that was called in doubt by modern naturalism. Polanyi's former field of research, physical chemistry, belonging to the sacrosanct "exact sciences," was seemingly safe enough

Marjorie Grene (1910–2009)

M. Grene

F. Jaroš, J. Klouda (eds.), *Adolf Portmann*, Biosemiotics 23,
https://doi.org/10.1007/978-3-030-67810-4_3

from the corrosive implications of the orthodox philosophy of science. But what moved him to embark on philosophical reflection was, in the first instance, the problem of the *organization* of science. Just after the war, when there was a vigorous movement in England for planning in science, he was concerned to justify to himself and to others, if they would listen, his belief that centralization is incompatible with the life of science. Although the administrative underpinning of modern science is necessarily vast – depending as it does on expensive equipment, long and costly training of apprentices, allocation of research funds and so on, nevertheless scientific discovery – which is after all the heart of science and of scientific progress – must be *spontaneous*. But it is just this spontaneous core of scientific life – on which all the rest depends for its significance, which must be ignored and even denied by traditional naturalism, with its emphasis on objectivity and explicitness as the be-all and end-all of knowledge. To justify his conviction of the central role of discovery, and hence of spontaneity in science, therefore, Polanyi had to develop a new philosophy which would take due account of the nature of discovery, and more generally of problem solving as the paradigm case of knowledge. As he puts this himself:

> Upon examining the grounds on which science is pursued, I saw that its progress is determined at every stage by indefinable powers of thought. No rules can account for the way a good idea is found for starting an inquiry; and there are no firm rules either for the verification or the refutation of the proposed solution of a problem. Rules widely current may be plausible enough, but scientific enquiry often proceeds and triumphs by contradicting them. Moreover, the explicit content of a theory fails to account for the guidance it affords to future discoveries. To hold a natural law to be true, is to believe that its presence may reveal itself in yet unknown and perhaps yet unthinkable consequences; it is to believe that natural laws are features of a reality which as such will continue to bear consequences inexhaustibly. (Polanyi 1969: 138)

What I hope to do tonight is first to tell you about the theory of knowledge which has grown out of this initial problem, and then to show you how this theory leads us to an ontology which once more allows man as knower to fit smoothly into the world he is concerned to know.

I think I can best take you rapidly but simply to the center of Polanyi's doctrine by going back for my starting point once more to our first Galilean passage. For Polanyi, like Portmann, although in a different direction, is trying to overcome the difficulties and indeed the contradictions inherent in the Galilean program. Let me then once more take Galileo's statement as my text.

> Philosophy is written in this grand book, the universe, which stands continually open to our gaze. But the book cannot be understood unless one first learns to comprehend the language and read the letters in which it is composed. It is written in the language of mathematics, and its characters are triangles, circles and other geometrical figures, without which it is humanly impossible to understand a single word of it; without these, one wanders about in a dark labyrinth. (Galileo 1957: 237–238)[1]

[1] We submit a standardized quotation of Galileo's text, relieved of Grene's rhetorical insertions and minor deviations from the original.

Let me first remind you briefly of the themes we discovered in this passage in connection with Portmann's account of living things and then proceed to show you how the key to Polanyi's reform of epistemology is also to be found there.

First, we must beware, I said, of taking too sweepingly the claim that Galileo's "geometrical" characters constitute the sole language of nature. In the context of my discussion of Portmann, what was important here was the limitation of a mathematicizing approach to organic phenomena, as against the rich reality of color, sound, smell and taste that makes up the sensory plenum of living nature as experiencing and experienced. The language of mathematics, I pointed out, is a second, a foreign language, and our acquisition of it and our understanding of it are always grounded in the primary life world that is our first and natural dwelling-place. Mathematics is an artefact which develops within and must be sustained by the fuller reality of our existence as human beings, and as living beings within a richer, not exclusively mathematical natural world. A similar thesis is prominent in Polanyi's arguments, especially when he deals with what he calls "the limits of formalization." I shall return to this point later.

A second point arises from the first one: the contrast between Galileo's "mathematical language" and our understanding in and through the life world should make it plain to us that there is no one language in which nature's truth is eternally and unambiguously written. Indeed, we need only compare our present text with Galileo's own achievements in dynamics to demonstrate that this is so, that is, that there is no single language, not even a single mathematical language, which is uniquely appropriate to convey the truth about the natural world. Let me explain. (I am following here, by the way, an argument of Polanyi's in the first chapter of his book *Personal Knowledge* published in 1958). In our passage Galileo invokes geometrical figures – triangles and circles – as the characters in which the natural language is written. Now a mathematical physics that elicits natural laws out of the simple, whole-number ratios typical of geometry had indeed been characteristic of the ancient Pythagorean tradition, which was again influential in Galileo's time. This kind of physics, which seeks for basic geometrical harmonies in the universe, was, Polanyi argues in *Personal Knowledge*, in a manner re-introduced into physics in this century with Einstein's theory of relativity. But in the meantime, during the great period of classical dynamics, physics had been dominated not by the theme of simple geometrical shapes and ratios, but by the notion of *measured* quantities – and this *new* mathematical language, which is not geometrical, not Pythagorean, but quantitative in a different sense, is first and classically exemplified in Galileo's own discovery of the constant of acceleration. Galileo himself, in other words, first introduced the language of measurement, which was to be perfected by Newton in the calculus, and which remained the language of modern physics up to relativity. But this language was not one of "triangles and circles," such as Galileo had pronounced the language of nature to be. Even the language of mathematics, then, through which as Galileo claims we read the book of the universe, is not a single language. There is no one language, a mathematical one, through which alone nature is to be understood.

This thesis, which one might perhaps call the principle of the *plurality of truths*, is of course common to many modern philosophies, even though few of those who espouse it have followed through consequentially its most general and radical philosophical implications. To this theme I shall also return shortly in the context of Polanyi's theory of knowledge.

But it is yet another theme suggested by our Galilean text that will take us straight to the heart of Polanyi's theory. Galileo tells us about the "book" in which "philosophy" is written, and the language we must learn in order to read the letters in which it is written. Now this metaphor about reading the book of the universe was a common one in the seventeenth century; it occurs also in one context or another in Bacon and in Descartes – and it is an apt metaphor, for that *is* what scientists are trying to do. Indeed, if we think of contemporary efforts to "crack" the genetic code, that is quite literally what some of them are doing, and in every kind of research it is some code, in a more general sense, that research workers are trying to decipher. They are trying to find significant patterns, meanings, where others have been "wandering in dark labyrinths."

Fair enough. But what Galileo and others in the Galilean tradition failed to think about was the nature of reading. When Galileo tells us that great book the universe is forever open to our gaze, he seems to be suggesting that once we have learned the mathematical language – which to be sure we must painfully acquire – but once we have learned our mathematical alphabet – we shall have all nature's secrets *explicitly* laid bare before our eyes – or better, as mathematicians, before our minds. The new science was to reject all opacities, to bring the whole truth about the natural world squarely and totally to the center of the intellectual stage. The man who gave its classic formulation to this program was Galileo's younger contemporary Descartes; for him, what comprised knowledge was a series of "clear and distinct ideas," each wholly self-contained and transparent, and the process of discovery consisted, first, in eliminating all that was dark and doubtful from our mind's periphery, and then in concentrating, one after another, in the right order, on a series of mental contents, of increasing complexity, to be sure, but no one of which dare outrun the boundaries of our clear, unclouded focal consciousness. Later the emphasis on intellectual clarity was to diminish in favor of a more mechanical explicitness. The fulfilment in principle of the Galilean program, 150 years after Descartes's death, was Laplace's dream of an omniscient observer, who, given the particular location of all the particles in the universe at a given time, could predict their location at any future time, and hence, Laplace believed, could know all there was to know about everything. This is reading the book of the universe with a vengeance! But the Laplacean observer need not be a mind, as Descartes's mathematician was, he is himself, in Laplacean terms, only a collection of particles. Indeed, he could be, and is, in twentieth century versions of the Laplacean ideal, neither a mind nor a man but an artefact, that is, of course, a computer. What is common to all these versions, however, is their demand for total *explicitness*, for the exact specification of everything that is or can be known. For all thinkers in this tradition – our dominant intellectual tradition through the last 300 years – it should be possible, as Polanyi has put it, to write out on an infinite blackboard all the knowledge of all men past, present and to come. Nothing is hidden. *But is it possible at all to write knowledge on a blackboard?*

Look again at Galileo's metaphor of reading. What are you doing when you read a book, say *War and Peace*? Watching in rapid succession a series of marks on a page? In a way; but unless it's very bad print you aren't thinking about those marks; you are not *attending* to them as such, but *relying* on them in order to attend to the story: to Natasha's nursing of Prince Andrey, or Napoleon's retreat from Moscow, or Nicholas Rostov's married life with Princess Mary. Through *relying* on the printed page, in other words, you attend to what it means. Or, in the terminology which Polanyi introduced in *Personal Knowledge*, through your *subsidiary* awareness of the words you are *focally* aware of what they mean. You are following, not the words themselves, but the story *through* the words. Similarly, when you hammer in a nail, you are subsidiarily aware of the pressures in your hands and arms, while paying focal attention to the nail and the wood into which you are driving it. It is Polanyi's central thesis that all knowledge has this dual structure. There is no wholly explicit knowledge, no knowledge which is completely, on one self-explanatory level, present to the mind. In reading, as in tool-using, and whether you are reading a novel or a mathematical formalism, the *Milwaukee Journal* or the book of the universe, your awareness of meanings is always on two levels at once. Polanyi calls this universal dual structure *tacit knowing*. Even in the most articulate knowledge – as in the formalisms of mathematical physics – this structure is present as the unspoken ground of the explicit characters we use. The scientist relies on his formulae and equations to understand through them the natural order he is investigating. This shows us, in the structure of knowledge, why, as we have already noticed, formalization is always limited. Wholly precise, wholly explicit speech would be empty: there is always, in all discourse, however formal, the resonance of what is *meant*.

Much knowing, moreover, is *wholly* tacit. Hence the importance, for example, of laboratory experience in mastering a scientific discipline. The lab instructor can show you what to do, but he cannot *say* it; you have to learn by doing it yourself. In this tacit component, therefore, there is a continuity from the most humble skills to the most precise and sophisticated theoretical knowledge.

The most basic example of tacit knowing and the one from which indeed all the achievements of science develop is the transaction between our bodies and the world through which we *perceive* events and objects. For instance, as I'm sure you all know, the images on our retinas are two-dimensional, upside down and reversed as between right and left; yet of these two backward upside down flat pictures we *make* our perception of properly oriented three-dimensional objects out there, in a uniform 3-dimensional space. To do this, to focus on objects out there before us and around us, we have to rely on all sorts of performances by our bodies: by our eye muscles, our central nervous systems, our semi-circular canals, and so on, performances of which we are not, and for the most part, cannot be focally conscious. Yet they are the clues through which we achieve ordered perception of an objective world. The nature of this achievement has been clearly exhibited through the experiments of the transactionist school of psychologists and also, especially dramatically, by the experiments on vision with inverting spectacles, first performed by Stratton

70 years ago and continued more recently by Ivo Kohler and his colleagues in Innsbruck. There is, for our purposes here, one especially illuminating consequence of this recent work. Ever since Stratton's first experiments it has been clear that if a subject puts on inverting spectacles he is, as you might expect, at first terribly disoriented: he cannot keep his balance or move about properly in his environment, but after a few weeks things come right again. Now it used to be thought that what had happened was that the image which had been inverted had righted itself again, and that the subject had succeeded once more in having a right-side up picture of things. But Kohler's associates have discovered that this is not how it works: a subject who has succeeded in overcoming the initial confusion caused by wearing inverting spectacles, is asked if he now sees, say, that house over there right side up: he answers: I wish you hadn't asked me, now that I think about it, it *is* upside down, but I was managing all right until you called my attention to it. In other words, instead of the *image* in *itself* coming right again, what has happened is that the perceiving person has succeeded in using new and confusing clues to make sense of his environment and to "see," in the transactional sense, a house right side up even though the picture of it that he *strictly* "sees," in a passive sense of seeing, is still wrong. However, confused the clues – up to a point of course – he can still, with learning and resolution, rely on them in order to attend to external objects in an ordered and effective way. Now this is just what the scientific discoverer does too: through a series of puzzling data he perceives a general pattern, a law which makes them fall into place in an orderly way. This is the kind of thing we are always doing in all our contacts with the world, the simplest or the most complex, the most routine or the most original and daring.

From the structure of tacit knowing a host of important consequences follows. It should be plain from what I have just said about perception, how Polanyi has been able to derive from his account of tacit knowing a rational theory of discovery: not, to be sure, a rational theory in the Cartesian or Laplacean sense: where truth follows truth automatically by some wholly impersonal method. In this perspective discovery simply has to be left out of science as an irrational addendum. But a rational theory of discovery in that the work of the creative scientist is seen to be an extension and elaboration of the way in which all men grope their way forward to find ever new patterns, ever new significance in experience. In short, Polanyi's theory of tacit knowing should produce a radical and, I think, a very fruitful reorientation in the philosophy of science.

But what I want to concentrate on here are those consequences of the structure of tacit knowing which bear most directly on our knowledge of living things and of ourselves. Let me mention some of these, first in relation to knowledge itself, and then to the ontology implicit in this theory, i.e. the view it suggests of the realities that there are for us to know.

First, it is characteristic of all subsidiary knowledge that a shift of focal attention to what was subsidiary changes and often destroys the previously focal pattern. Admittedly, in perception, the primordial case of tacit knowing, many of the clues are not only subsidiary but necessarily unconscious. We cannot watch our eye muscles or our optic nerves at work. In many other cases, however, we can shift attention from what is normally focal to the subsidiary. Thus, in our example of reading, we can, if we are interested in printing, look at the page for its own sake instead of for its meaning. In that case we are looking focally at the letters and the meaning fades into the periphery of our awareness. *But we cannot focus on both at once.* Thus, for example, in recognizing someone we know we are subsidiarily aware of his gait or features or dress while focally acknowledging that's Hank or Jane or Peggy. But as we have all discovered at some time, I imagine, the medical specialist who looks focally at eyes or feet or noses tends to forget the presence of the human being as a whole; yet if he were not, in the periphery of his knowledge, aware that there were people – and people who have not only eyes or feet or ears but pay-checks or bank-accounts – he would not trouble to practice his profession at all. Similarly, microbiologists, who are professionally concerned with minute parts of the organism, tend to forget the existence of the whole: indeed, under the influence of a Galilean philosophy of science, they may even, and often do, insist that there is no such thing: really there are only the particles which in given conditions react physically and chemically in given rather elaborate ways. But at some time he has nevertheless acknowledged focally that there are organisms, otherwise he could not have begun to analyze them; and he still acknowledges this subsidiarily – he is still all the time relying on his common sense accrediting of the existence of the organic world in order to pay attention to its minute parts.

This fact, that what was subsidiary may be moved to the center of attention, and what was focal become subsidiary, confirms what I have already emphasized: that no knowledge can be completely precise, exhaustively formulated. As in the ambiguous figures of the psychologists, a shift of attention may always reveal another pattern, and relegate to background what was formally locally perceived. We see here, I think, *why* human existence has the basically *ambivalent* quality I stressed last time. It may be objected that I have taken perception as the paradigm case of tacit knowing, and animals too perceive. Why, it may be asked, should the figure-background relation produce these shifting and equivocal results only for *us*? I spoke last time about the unique character of human perception, and I might just remind you again of the experimenter and the chicken[2]; but let me give you in the

[2] Here Grene evidently refers to Wolfgang Köhler's experiments with chickens' perception of colour shades. She tackles this topic in *Knower and the Known* where she writes: "A well-known experiment has shown, for example, that chickens fed on the lighter of two grey squares will always peck on the lighter of a pair." (Köhler 1929: 167–8, 233–4). The experimenter, however, can not only distinguish the relative light and dark of each pair, but can see that a constant mid-grey is first lighter and then darker than its mate. He can match a single colour in two cases against a standard. Without such matching, without evaluation, he could not say with confidence, 'this is grey', or 'this is the same colour'. Far from having first primitive perceptions of particulars, from which we abstract to form, by convention, 'general concepts', relational in import, we perceive, primarily, relations, and to perceive particulars is already to have transcended the relations in which, as perceived, they imbedded. For to 'see' this particular as red we must know how to isolate the new particular from its context and judge it, instead, by the general standard of redness. This is to say, with Kant, that concepts are rules for organising experience, not indeed, formally stateable rules, but norms to which judgement is implicitly referred, by which it is tacitly governed. And where I say: 'this here is red', I am already applying such a rule – the rule of colour judgement – to

present context a different and perhaps more fundamental answer. It is this. Animals also of course rely on interior clues to perceive things out there; but as we also saw last time, their world is a more limited one, or to put it in our present terms, their points of focal orientation are sharply circumscribed by nature. In contrast, we live much of our lives in and through *cultural* objects – houses and towns, language, laws, science, religion. All these, since we can stand back from them, look at them, make and unmake them, seem to have attained a kind of independent existence, if only as artefacts. But cultural objects are by their very nature ambiguous in the sense of the focal-subsidiary distinction: we can attend to them or use them. I can build a house or live in one, study religion or practice it. And even science, as I have just been suggesting, must be not only inspected by a scientist but practiced by him. He interiorizes the maxims, the fundamental concepts, the experimental techniques of his specialized discipline and relies on them in order to resolve the particular problem with which at the moment, or for years, or for his mature lifetime he happens to be focally concerned. Yet many of these maxims and concepts and techniques, now interiorized and subsidiary were focally attended to, while he was acquiring them. Indeed if, like Portmann or Polanyi, he turns in his later years to reflect *about* his science, he will again try to draw into the focus of consciousness the presuppositions he had once assimilated to himself; and now he will rely, subsidiarily, for this purpose, on the many problems to which, throughout his years of active research, he had been focally attending.

This matter of interiorization brings me to my second point. Through all the interplay of focal and subsidiary knowledge, the subsidiary pole is always an aspect of myself, *from* which I attend *to* something beyond myself, to something *in the world*. From perception all the way to the highest achievements of art or science, the clues through which we feel our way forward to make sense of our experience, are performances of our own bodies, skills, memories, crypto-memories, hunches of our own. Through the instrumentalities of our own bodies, and through tools, which are extensions of our bodies, we reach out into the world around us and at the same time, inseparably from this outward reaching, assimilate the world to ourselves. Polanyi puts this by saying that all knowing is *indwelling*. In existentialist language, it is being-in-a-world, or better, being-*with*-a-world. From birth to death each of us constantly both shapes his world, relying on a host of clues to reach out to the world, and interiorizes into himself what he learns in and through the world. There is an inherent tension radiating from the individual, which forever holds linked together the subsidiary and focal, inner and outer poles of knowledge.

But you may well ask, is not this inner-outer business very much exaggerated? To take our case of reading again: surely the letters are just simply there on the page – as Galileo believed the triangles and circles were just simply there in nature, waiting for us to read them. Why should we say that the clues through which we understand – in this case, the marks on paper, are parts of *ourselves*? The answer is

---

this particular instance. Were I not able to exercise such complex evaluations, I should be bound, like the chicken, to the direct impact of a relational context, and unable to isolate it from the particular I wish to report" (Grene 1974a: 162–3).

that in fact letters and words would be *only* marks on paper unless we had first assimilated to ourselves the language to which they belong. The child learning to speak is not simply learning by rote and repetition a correlation of a given sound with a given object; *before* he can do this he must come to *live in language* – that is to transform himself into a language-user, to accept, as his own, life in the medium of rational discourse. I spoke to you last time about Portmann's concept of the first year of life as a period of social gestation; and it is precisely the process of interiorization of the cultural which is established at this time in its first and basic stage: in the first steps we take to meet our fellows, both literally when we learn to walk, and metaphorically, in the acquisition of our mother tongue. Each further step we take, whether to read the language of nature or the languages of man, or indeed to rewrite them both through creative discovery in science or art or statesmanship – each further step still carries on the fundamental pattern, of which the bodily orientation of the subject-person was the first, and remains the primary, center.[3]

All this means, thirdly, that knowledge is always personal. The impersonal aspect of knowledge arises from and returns to personal participation in the search for and acceptance of the object to be known. Only the explicit, formulable core of knowledge can be transferred, neutrally, from person to person. Its implicit base, in contrast, since it cannot be verbalized and cannot be formulated and so impersonalized, must be the groping, the orientation and reorientation of *someone*. Even the greatest *de*tachment, we saw last time, the objectivity of a judge or a scientist, must be responsibly accepted and interiorized by its practitioner. He must have taken into the very tissue of his life the impersonal standards by which he, personally, admits or rejects evidence for a judgment or a theory. Human centricity, in other words, has implications running far beyond the individual's needs and appetites, but it is still centricity.[4] All knowledge, because acquired through the allegiance to standards accepted as universal through the responsible decision of the knower, *claims* universality. But this claim can never be detached from the personal responsibility of the individual knower himself, in which it has its roots.

Hence also, of course, the principle I have already referred to, of the plurality of truths. There is only one universe, but there are many persons seeking to know it, often through the shared standards of a single discipline, but still each from his own personal perspective and on his own responsibility. Each is doing his best, in his

[3] Grene discusses Portmann's conception of social gestation more widely in her book *The Understanding of Nature*. She concludes there: "In short, the whole biological development of a typical mammal has been rewritten in our case in a new key: the whole structure of the embryo, the whole rhythm of growth, is directed, from first to last, to the emergence of a culture-dwelling animal – an animal not bound within a predetermined ecological niche like the tern or the stag or the dragonfly or even the chimpanzee, but, in its very tissues and organs and aptitudes, born to be open to its world, to be able to accept responsibility, to make its own the traditions of a historical past and to remake them into an unforeseeable future." (Grene 1974b: 288)

[4] The word "centricity" is used by Grene as an English translation of Portmann's notion of *Innerlichkeit*. Portmann acknowledges this translation in the letter from 10 April 1965 (UB Basel, NL 345, Nachlass Adolf Portmann, C II e42).

own situation, to grasp the truth, but each situation is limited. Out of no single historical situation can the *whole* of reality be grasped.

***

So far, I have been talking about the theory of tacit knowing as it bears on our view of knowledge. This is an important, indeed a central, issue in our conception of man, if only because it is as knowers that we have been traditionally said to excel the achievements of other species. But my theme in these lectures is man *in* nature. What philosophy needs to offer us at this juncture of our intellectual history is not *only* a theory of knowledge, but a theory of knowledge that will exhibit once more the rootedness of our knowledge in our natures, and of our natures in living nature as a whole. In particular, it is the relation of ourselves as knowers to the nature of living beings as such that is the crucial question. In the Galilean tradition, as I argued last time, man, unrooted from his place in the finite medieval hierarchy of beings appears either as pure mind divided from purely corporeal nature, or as nothing *but* a collocation of material particles, or as producing over against such a one-level material nature a proliferation of cultural artefacts, of mere conventions. Portmann's description of living nature and of man's place in it seems to me, as I said then, closer to the phenomena of the natural world and of our own experience; but Polanyi's theory of tacit knowing, as I hope to show you now, can provide a *philosophical* accounting for this more adequate description. It brings man and nature together, not through a new account of knowledge and/or a better description of organic phenomena, but by directing us *from* the structure of tacit knowing to the related and even identical structure of the entitles thus known.

And this is indeed a necessary step in the conceptual reform on which we are engaged. In connection with Portmann's concept of centricity, I admitted that if the world were in fact Galilean, our acceptance of centricity as characteristic of life would be but a concession to subjectivity, to untruth. The same holds of the theory of tacit knowing. If the world *were* a Galilean world, such a personalized re-reading of our cognitive situation, far from curing us of our alienation from reality, would only aggravate it the more. It would be a re-echoing of Kierkegaard's "subjectivity is truth," cutting us off irrevocably from the outside, objective world of science and common sense. No; what we need is a pointer *from* the structure of personal knowledge *to* the changed reality which our changing conception of knowledge shows us that we know. It is this new *ontology*, grounded in the theory of tacit knowing, and in particular, this new ontology of *living things*, that I want to tell you about now.

The pointer we need from the knower in our sense of personal knowledge to the reality he seeks to know: this pointer is to be found in the *analogy* between the structure of an act of comprehension and the structure of the kind of "comprehensive entity" – as Polanyi calls it – that is the object of such an act. Polanyi has developed this analogy in a recent lecture on "*The Emergence of Man*."[5] He has started,

[5]This lecture took place at Duke University on the 2nd of March 1964. Polanyi's lectures from Duke were later published in revised form in *The Tacit Dimension* (Polanyi 2009).

as we have seen, from the thesis that all knowing has a structure continuous with that of skills. I took hammering and reading as two such examples. But the analogy shows up best in a complicated performance like playing a game. Consider such a skillful performance as playing football, for example. Your understanding of the playing of the game parallels the structure of playing the game itself. Not that you exert yourself muscularly in the stands in precisely the same way as the players on the field. But you do have to know the rules in the sense of having to some extent interiorized them, as the players have, and you rely on this knowledge, as well as on your awareness of the position and movement of particular players, to attend to how the game is going. Moreover, if you attend too carefully to the particular position of a given player, who happens for instance to be a friend or relation, you lose track of the game. The interplay of focal and subsidiary features involved in following the game has the same structure as the more active performance of playing the game itself. Similarly, Polanyi argues, we can reasonably surmise that, for all cases of tacit knowing, "the structure of comprehension mirrors the structure of that which it comprehends." We can go even further "and expect to find the structure of tacit knowing duplicated in the principles which account for the stability and effectiveness of all real comprehensive entities" (cf. Polanyi 2009: 33–34).[6] In other words, where we confront a complex set of events like a game of football or chess, a piece of scientific research, *or* a living organism - in all such cases, we may expect to find the same congruence, the same conformity of pattern, that we find between *following* a game and *playing* it.

In what would such a parallel consist? In tacit knowing, Polanyi reminds us, *first,* we rely on our awareness of the parts of a whole in order to attend *from* the particulars to the whole. And secondly, as we have also noticed, if we attend focally to the particulars, we lose sight of the whole to which we had been attending. Comprehensive entities, if they mirror out tacit knowing of them, should display an analogue of both these properties: (1) As we rely on subsidiary clues to attend to the objects of our focal attention, so the principles governing comprehensive entities rely for their operation on the laws governing their particular constituents; and (2) just as attention to the particulars in themselves dissipates our apprehension of the whole on which they bear, the particulars forming the necessary conditions for the existence of a comprehensive entity cannot in themselves account for the existence of the entitles whose conditions they are.

Now living things, with which we are here chiefly concerned, do in fact display this double parallel with the activity of knowing. We recognize a species, for example, by relying on its features to attend to what it is. An eminent zoologist has described how he recognizes a new species of worm in the field: at first, he has a rather uncomfortable feeling, then he exclaims, "Why it is rhynchodemus, but it's not bilineatus, it's an entirely new species"[7] (Pantin 1954: 592). The clues in terms

[6] Grene seems to refer to the original text of Polanyi's Duke lectures, which differs slightly from the published version.

[7] Pantin's original sentence differs: "It is a Rhynchodemus all right, but it is not bilineatus–it is an entirely new species!" (Pantin 1954: 592). Grene uses this modified version also in *The Knower and the Known* (1974a: 203), whereas Polanyi (1974: 351) quotes it correctly.

of which he achieves this recognition are not explicit, nor if made so would they add up to the organism in question. They are features of it, in Portmann's language, aspects of its *display*, which have their being *as* display of a living creature in the way in which they express its centered existence. Similarly, in the interpretation of genetic events, research workers rely on the particulars of a series of organic bases in order to understand the message they convey; and the *organism* relies on the same DNA code to grow into the type of organism it is. The cognitive relation, between subsidiary and focal, is mirrored in the organic relation between part and whole. Moreover, the particulars themselves would not constitute a message except in the context of the organism whose growth they subserve any more than the enumeration of a particular string of organic bases would constitute in and of themselves an understanding of genetics. The particulars without the organism could not add up to the organism of which they are the parts. Only the unity of the whole makes the part the parts they are. In general, where we are dealing with living things and their particulars, this parallel obtains. Indeed it obtains precisely because we *are* living things, and in understanding living things we are exemplifying in that achievement the very structure of all organic life, its *reliance* on particulars to achieve an over-all aim, and its transcendence of these particulars in the significant form which it expresses through them, but which they in themselves would be impotent to express.

What Polanyi is saying here, in other words, as applied to living things, is that organisms always exist on at least two levels at once. This character of living things has sometimes been called "double determinateness." That is, the whole depends on the parts as conditions of its existence, but the parts exist as parts only as so constituted by the unifying principle of the whole. And, as Polanyi has argued at length in *Personal Knowledge*, the insistence that organisms are machines only serves to confirm this thesis, since it can easily be shown that machines themselves are essentially two-level existences: they function in accordance with the laws of physics and chemistry as specified, not *by* those laws – which in themselves leave open a wide range of possible blue-prints – but by the operational principles according to which they perform their functions as machines.

Thus, over against the tradition which insists that only one kind of reality is real at all, Polanyi is clearly hoping to reinstate, in some form, a conception of *grades* of reality. But before I go further, I think I should insert a warning here. The acknowledgement of levels of reality is in one form or another a fundamental level for any viable philosophy of biology. Such hierarchical distinctions, I should like to emphasize, can be made on more or less complex, more or less generalized lines. Most philosophers who introduce distinctions of level do mark out, in some way or other, matter, life, and mind as main stages of "ascent," but beyond this very general division there are a variety of perspectives from which such a survey may fruitfully be made. If, indeed, as I have repeatedly insisted, our situation as knowers is ineradicably ambiguous, if we are questioning beings, if the very fact of speech springs from a question mark, we may well find the degrees of being whose reality we risk asserting, to be themselves equivocal. What is most important here, I believe, is not that there are precisely such and such kinds or grades of being in the world, so many and no more, but *that there are beings of more kinds than one*. There are many ways of being, as many, Aristotle said, as there are kinds of unity.

It is worth looking at the world from the point of view of that Aristotelian maxim. Wherever we acknowledge that a set of events or characters form in some – any – stable and interesting way a unity, we acknowledge the existence of some being. But such constitutive unities may be of very different sorts: as in a range of biological examples, for example, there is the unity of a cell, the unity of a multicellular animal, say, a dragon-fly, the unity e.g. of the dragon-fly's mating behavior, the unity of the entomologist's performance in studying the dragon-fly, and so on.

Further, such wholes are sometimes related to one another, in turn, as higher to lower, both in terms of existential dependence and of relative complexity of organization. But sometimes they are *just different one from another*. The tendency of modern naturalism has been to reduce all existence both to one level *and* to one sort; both these reductions must be overcome. For the world is much richer in its ways of being than Boyle or Newton were willing to admit. There are indeed some things that are more highly organized, more "real" than others, and there are also things that are simply different from others, different not only in number but in their way of being. To break the stranglehold of reductivism, we must acknowledge once more *the multiplicity of forms of being*. That liberating acknowledgment achieved, we can then recognize as well the stratification of some of these many forms relatively to one another. True, such multiplicity can be recognized only in terms of double determinateness: it is the two-levelled structure of comprehensive entities that exhibits the inadequacy of a metaphysics for which things just simply *are* or *are not*, and therefore permits us to recognize once more the pluralism of *kinds* of being. Thus, a one-*level* ontology and a one-type ontology, and *a fortiori*, their negations, are logically related. But we should nevertheless keep separate three aspects of the situation: (1) the double determinateness of comprehensive entities, including living things and the achievements of living things; (2) the multiplicity of kinds of such doubly determinate entities; (3) the question of their stratification relatively to one another.

But there is a further problem. For the variety of entities whose existence we thus acknowledge, we need to ask what we *mean* by their being, or their reality. Such a question was out of place in the tradition of modern naturalism, where events simply were or were not, on a single level of existence. But once we allow levels, or grades, of reality, the question arises: what, if not bare existence, do we mean by the concept of "reality," what is it that is present at all these levels? Is there no more to be said than the bare and metaphysical assertion: that there are kinds of things, some more real than others? Indeed, does such an assertion mean anything at all? Or are we perhaps once more smuggling in existence as a predicate as Kant has classically forbidden us to do? One thing can be harder or hotter or older than another but *more real*? In his Duke University lectures of 1964[8] Polanyi has developed further a criterion of reality, or of comparative reality, already implicit in the argument of

[8] See the footnote no. 5 above. Grene originally incorrectly states the year 1963 here. This mistake is fixed later in *The Knower and the Known* (1974a: 219).

Personal Knowledge. One entity is more real than another, he suggests, when *it carries the possibility of a greater range of interesting and unexpected consequences*. He has been showing how the same structure of tacit knowing holds good in a number of different kinds of cases. Thus, we rely on clues in order to attend to entitles as diverse as a scientific problem, another person, or an inanimate object like a cobblestone. He continues:

> The structural kinship between knowing a person and discovering a problem, and the alignment of both with our knowing of a cobblestone, call attention to the greater depth of a person and a problem, as compared with the lesser profundity of a cobblestone. Persons and problems are felt to be more profound, because we expect them yet to reveal themselves in unexpected ways in the future, while cobblestones evoke no such expectation. This capacity of a thing to reveal itself in unexpected ways in the future, I attribute to the fact that the thing observed is an aspect of a reality, possessing a significance that is not exhausted by our conception of any single aspect of it. To trust that a thing we know is real is, in this sense, to feel that it has the independence and power for manifesting itself in yet unthought of ways in the future. I shall say, accordingly, that minds and problems possess a *deeper* reality than cobblestones, although cobblestones are admittedly more real in the sense of being *more tangible*. And since I regard the significance of a thing as more important than its tangibility, I shall say that minds and problems are more real than cobblestones. This is to class our knowledge of reality with the kind of foreknowledge which guides scientists to discovery. (cf. Polanyi 2009: 32–33)[9]

This statement suggests, on the one hand, views so orthodox as to be almost trivial, and, on the other, consequences so revolutionary as to make it seem to many almost absurd. Let me consider its implications in both directions.

On the one hand, the reference to future consequences is reminiscent of reductivist theories like pragmatism or positivism. Thinkers who wished to evade the problem of truth have often insisted that theories are acceptable not for their conformity to some elusive and mysterious reality but simply for their fruitfulness: for the empirically observable consequences which may follow from them. Polanyi attacked this theory in *Personal Knowledge* and it may seem odd that he should himself now use a formulation superficially so similar to it. But let us look at the situation a little more closely. It seems reasonable to suppose that Copernicus believed his theory to be, not simply a device for "saving the phenomena," but a statement of the truth about the heavens. He *could* not himself accept it *because* of its fruitfulness, since he could not himself know, in the explicit sense of "know," how fruitful it was to prove. He accepted it because he believed it to be true, and he would have hoped that it would prove fruitful *because* of being true, because in formulating it he had come into contact with reality. If truth is in some sense a congruence of thoughts and things, it follows that a true theory may have unforeseen consequences because things themselves may unfold themselves in unforeseen ways, while a false theory will display no such fruitfulness because it is not founded on the real. Fruitfulness works ultimately, therefore, as a criterion of the truth of

[9] The original differs slightly: "[…] I shall say, accordingly, that minds and problems possess a deeper reality than cobblestones, although cobblestones are admittedly more real in the sense of being *tangible* […]" (Polanyi 2009: 32–33).

theories because it is a criterion of *reality*. Discoverers assert their theories because they believe them to be true, not because they are (now) fruitful (which they must prove themselves to be hereafter). But if they are true, they *will* be fruitful; if false, empty of consequences. There is the same sort of confusion here in the positivist theory that there is both in positivism and pragmatism about the relation between *meaning* and *truth*. Fruitfulness is a *test* of truth; positivists make it not a test, but the essential nature, not of truth only, but of meaning as well. A theory is supposed to be *nothing but* a device for drawing out further observations. This, as we have just seen, is absurd, since no one can equate a theory stated now with its non-existent future results. What its discoverer judges it by, however, is the *confidence* he feels in its truth, that is, his confidence that he has made contact with a significant aspect of reality, and that therefore out of his explanation many unforeseen consequences may flow.

As against the positivist view, moreover, that a theory is simply a device for prediction, Polanyi's concept of fruitfulness stresses *un*foreseen consequences. What marks off important theories is that they prove so much more fruitful than their discoverers could possibly have foreseen. It is not predictability, but *un*predictability that distinguishes the most powerful and most interesting discoveries, because it is unforeseen consequences that distinguish the real from the unreal.

But, it may be objected, is it not its very predictability that distinguishes the real world, for example, from that of dreams? If I dreamt, I had won the Irish sweepstake, nothing would follow from my dream; if I had really won, I should be rich, and a solid and coherent set of consequences would follow from this fact. Dreams are unreliable, realities reliable – not, as Polanyi's definition seems to suggest, the other way round. And it is true, of course, that reality is also full of *predictable* consequences; but the point is that as against the unsubstantial, the artificial or the false, these real consequences are *inexhaustible* and *therefore* unpredictable. They are indeed systematic and therefore up to a point predictable, but real systems are also richer than our predictions and so capable of *indefinite* consequences as well. Our dreams, on the other hand, being arbitrary figments, are soon exhausted. That is why, again, great discoveries often prove to have more, and more kinds of consequences than their discoverers know or could have known. All this fits in well enough with ordinary concepts of "real" and "unreal."

The really surprising aspect of Polanyi's criterion, however, emerges from its *comparative* use. An entity is the *more* real, the more unforeseen consequences could flow from it. But *ideas*, like the Copernican theory, often contain more unforeseen consequences than do ordinary external things, like the cobblestone. Can anything mind-dependent be *more* real than what is out-there, measurable in space and time? Thus Polanyi's thesis, seems to go perversely against common sense; and in particular it goes against what we believe to be scientific common sense. A psychologist, Dr. John Beloff, himself opposed to behaviorism and so sympathetic with Polanyi's attempt to renew the philosophy of mind, nevertheless professed himself shocked by the case of the problem and the paving stone. What makes the cobblestone real, he said, is just that it is *independent* of any mind's perceiving it, while the problem has not this independence. What about this objection? If to be real is by definition to be independent of mind, then whatever *is* mind is non-real. Psychology is here coming down squarely in favor of the unreality of mind. Or if psychology is not wholly to deny the existence of its own subject-matter, it must add to the external real the ghostly reality of an inner mind and take refuge once more in Cartesian dualism. This is in fact what Beloff argues in his book, *The Existence of Mind* (Beloff 1962). But this refuge of despair is unnecessary if we apply Polanyi's criterion. Human minds turn out to be, not a mysterious addendum to the inventory of more obviously real things, but one of the more real of the many kinds of things there are. And so do the mind's achievements, theories, works of art, inventions, laws. The paradigm of the problem and the paving stone may hold for any skillful performance or for its product.

Surely, however, it will be objected again, the basic distinction to be made here is that between the natural and the artificial: all human achievements are *cultural*, that is, at bottom, the products of convention. Are we not making *fictions* more real than *facts*? In answer, let us take a case literally from fiction and see how Polanyi's criterion might apply to it. Jane Austen is dead, and no further unforeseen consequences can flow from her character. Even if some new discovery were made about her, say, that she was secretly married, that would be an unforeseen consequence for us, not for her. Emma, though a fiction, is alive; she is the kind of live character who continues to develop between the lines of the novel, the more one is acquainted with her. Being, as E. M. Forster remarked of Jane Austen's creatures, a round character, almost the *round* character *par excellence*, she shows just the kind of living richness exhibited by our friends, at least the more interesting among them. But Emma's creator did exist and Emma never did (cf. Forster 1985: 62). Are we to call such a fiction more real than its inventor? Yes and no. We may, and should, distinguish, as I have already suggested, between different ways, different *dimensions* of being. What Polanyi's criterion applies to we might call *intensity*, or, as he calls it, *depth* of being. Existence and non-existence, the sheer *that* of being or non-being, corresponds to what Sartre calls *facticité* or Heidegger *Geworfenheit*. It is an *extensive*, linear dimension. In making a judgment of reality we count this as one factor against others, and also, we count it in a different way. When we say Jane Austen is dead we are referring to the historical non-existence of an individual; when we say Emma is alive we are speaking of the depth of being of a non-historical individual. No direct comparison between them is possible or intended.

Yet, in general, Polanyi's criterion does enable us, I believe, to handle conceptually the difficult problem of the *reality* of works of art. Berenson in a famous essay has described how Giotto's treatment of bodies makes us *feel* more intensely than we otherwise do the three-dimensionality, the sheer solidity of objects around us (Berenson 1902). This is not just a pointer *from* the painting to the observer and his perception of bodies. The solidity is felt through the painting. In the heavy kneeling or slightly turning figure, the solidity of bodies *is* more intensively than in the ordinary objects of the ordinary world, even though, even because, we "know" it is "only" painted, only "represented" on a two-dimensional surface before us. This would be true if the Arena Chapel were to vanish tomorrow, if the human race were to vanish tomorrow. This is a dimension of being as unrelated to contingent fact or non-fact as are the relations expressed in mathematical formulae.

Of course, it will not do, on the other hand, to absorb, as Hegel seems to do, extension into intension, contingency into depth of being. That is why it seems to me also important to insist that it is not one fixed, universal hierarchy which the recognition of degrees and dimensions of being seeks to restore. It is by no means true that all that is real is rational, and that, conversely, all that is irrational is unreal. There is no one whole System of Being which, happily, all the sufferings of countless individuals are intended to realize, of which they are instruments. Such a conception is as false as it is unfeeling. But neither is the nonsignificant, neither is sheer givenness, the only way things are. The achievements of all living things, the achievements of human minds, are more than tiny superscripts on a single monotonous succession of mere facts. They are enrichments of being itself.

Only such an acknowledgement, finally, will enable us to see knowledge itself as a real achievement of real beings. Let me put before you very briefly the series of levels involved here. We ask what a person is doing in a cognitive performance – and we find, first, that the recognition of gestalt or pattern is common to all acts of knowing. For, in every such act, as we have seen, the knower is relying on a variety of clues, grasped only subsidiarily, in order to attend to the focal object of his knowledge. This is everywhere the structure of *comprehension*. But when the knower is engaged in the business of knowing living things, further, the object of his knowledge itself is, in Polanyi's language, a *comprehensive entity*. This is what we found to be the case with skills, and also with living things. Not only the process of knowing but the thing known consists of an organic whole whose parts have their meaning, and their existence as parts, through their bearing on the whole. Now from simpler comprehensive entities, or individuals, like snails or dragon-flies, e.g. we can proceed by the same kind of process, to understand more deeply centered individuals, like dolphins and apes, and, finally, responsible persons, i.e. human beings, including the knower himself, whose activities we have been all along inspecting. Thus we find the task of knowing knowers smoothly assimilated to the task of knowing life. To know life is to comprehend comprehensive entities; to know knowing is to comprehend those particular achievements of living things which consist in their acts of comprehension. Mind is once more a natural reality, and nature once more both the medium and the object of mind's activity.

This is high metaphysical language but don't let it frighten you. What I am asking, in the last analysis, is simply that, as philosophers, we take seriously the routine admission – granted in one breath by biological writers but rescinded in the next – that there *are* levels of organization of real entities which are at the same time levels of achievement – and achievements of the same general nature as our achievement in understanding such entities It must be noted, in conclusion, however, that this is *not* to reinstate either a Cartesian dualism or any other ontological pluralism of this kind. It is precisely the alternative between materialism and mind-body dualism which we are trying to overcome, and which the nature of the phenomena demands that we overcome. What we need is to articulate an *analytical* pluralism, a metaphysic which will allow us to acknowledge the existence of a rich variety of realities, not all of which need exist in identifiable, spatio-temporal separateness. Minds are not separate from bodies, yet persons capable of "minding" are richer and more highly endowed than persons, or individuals, not so capable. And achievements of responsible persons, such as laws, works of art, or forms of worship, may again be richer in reality than those persons themselves. That does not mean that such performances, such products of human skill, somehow exist "in themselves," separately from the existence of those who contrive, support, and also depend upon them. The alternative "separate mind" or "no mind," two reals or one real only, has been too long dominant over Western thought. We need to recognize once more the richness of thought in comprehending what cannot be wholly reduced to so explicit a pair of formulae. And equally, we need to recognize the richness of reality, including the achievements of human persons and human traditions. For this transcends even the profoundest acts of comprehension, harboring for future knowers consequences not yet imagined: harboring them just because the nature of these realities is more complex, involves more strands and levels and ways of being, than we had thought. Not only in extension, but in intensity, in depth of being, the world has more to say to us, in a greater range of types of discourse, than our Galilean imprisonment has allowed us to believe. Let us try to reread it in this more hopeful sense.

Editorial Note

Filip Jaroš and Matěj Pudil

The typescript of the following lecture is preserved in the archive of Adolf Portmann, University Library Basel, NL 345 (Nachlass Adolf Portmann). The typescript has 33 pages, with the heading "Lecture Three," and the subtitle "The Language of Nature Reread." Bibliography and references were added by the editors; all editorial changes are marked with footnotes. This text has not been published previously. For the generous permission to publish it, the editors give thanks to the heirs of M. Grene, prof. Ruth Grene and prof. Nicholas Grene (USA) and to the University Library Basel (Switzerland).

As the heading "Lecture Three" suggests, the lecture was part of a series; details about the location and dates of the series are not listed. From the text itself we are informed that the series was devoted to the topic of "man *in* nature" (p. 32) and the preceding lectures were specifically related to the philosophical biology of Adolf Portmann and the opposition of his approach to the Galilean scientific ontology. "The Language of Nature Reread" continues the topic and focuses on the epistemology of Michael Polanyi, whose theory of tacit knowing is argued to provide a "philosophical accounting" for Portmann's biological program (p. 32).

According to Phil Collins (personal conversation), Portmann was consulted by Grene in 1964 and 1965 on launching the Study Group on Foundations of Cultural Unity (funded by the Ford Foundation). However, Portmann is not listed as a participant of either of the two meetings of the group in August 1965 and 1966 at Bowdoin College, Brunswick, USA. Following the correspondence between both thinkers (Nachlass Adolf Portmann), Portmann received a manuscript from Grene on 31 March 1965, parts of which later became the "Adolf Portmann" chapter of Grene's book *Approaches to a Philosophical Biology* (Grene 1968). Based on the titles of Polanyi's works and lectures cited (Polanyi's lecture on "The Emergence of Man" presented at Duke University on 2 March 1964, cf. p. 32), allusions to the Irish milieu (see p. 37), as well as on the exposition of Portmann's work, it is clear that the lecture was delivered during the late period of Grene's appointment as lecturer in philosophy at Queens University, Belfast (1960–1965).

Grene used some passages of her lecture later in her book *The Knower and the Known* (1974), especially in the chapter "The Multiplicity of Forms" (cf. Grene 1974: 218–221, 222–224), though the context is different. Portmann refers to Polanyi's concept of tacit knowledge as pre-verbal recognition and association, common to animals and humans, particularly children who have yet to master language (Portmann 1970).

References

Beloff, J. (1962). *The existence of mind*. London: MacGibbon & Kee.
Berenson, B. (1902). *The study and criticism of Italian art*. London: G. Bell and Sons.
Forster, E. M. (1985). *Aspects of the novel*. New York: Harvest Books.
Galileo, G. (1957). The assayer. In S. Drake (Ed.), *Discoveries and opinions of Galileo* (pp. 229–280). New York: Doubleday & Company.
Grene, M. (1968). *Approaches to a philosophical biology*. New York: Basic Books.
Grene, M. (1974a). *The knower and the known*. Los Angeles: University of California Press.
Grene, M. (1974b). *The understanding of nature. Essays in the philosophy of biology*. Dordrecht: Springer.
Köhler, W. (1929). *Gestalt psychology*. New York: Liveright.
Pantin, C. F. A. (1954). The recognition of species. *Science Progress, 42*, 587–598.
Polanyi, M. (1958). *Personal knowledge. Towards a post-critical philosophy*. Chicago: University of Chicago Press. [Polanyi, M. (1974). *Personal knowledge. Towards a post-critical philosophy*. Chicago: University of Chicago Press].
Polanyi, M. (1969). The logic of tacit inference. In M. Grene (Ed.), *Being and knowing. Essays by Michael Polanyi* (pp. 138–158). Chicago: University of Chicago Press.
Polanyi, M. (2009). *The tacit dimension*. Chicago: Chicago University Press.
Portmann, A. (1970). Der Weg zum Wort: Stufen lebendiger Kommunikation, In *Eranos-Jahrbuch 39, Mensch und Wort* (pp. 397–424). Leiden: Brill. [Portmann, A. (1990). The path leading to words: Levels of living communication. In *Essays in philosophical zoology by Adolf Portmann: The living form and the seeing eye* (pp. 97–124). Translated by R. Carter. Lewiston NY: The Edwin Mellen Press].

Part II
Biology, Biosemiotics, Anthropology

Chapter 4
Adolf Portmann: The Researcher and the Teacher

Roger A. Stamm

Adolf Portmann was a zoologist, and an expert in morphology, studying embryonal, young animal and human developmental stages. In the era of biology he represented, the main focus was on morphology and physiology. In his time – in the middle of the twentieth century – every student at German-speaking universities had, in one or more semesters, to learn independently about all main animal types – through careful observation and using dissection. In plant science, the requirements were about the same. The organization of the studies was usually determined by a single full professor in the specific field; Portmann was also one of these single full professor cases. If this mentor was mainly focused on physiology, then this discipline was more widely discussed; however, morphology represented the first and fundamental base of the studies.

How far away from this model we are today! Now the studies are usually determined by molecular biology, microbiology (research of single-celled life forms), biochemistry and especially, and more extensively, by the science of hereditary issues, genetics. Floods of new discoveries have been brought by those scientific fields that revolve around cell structure and function. Such a variety of structures, such a richness of functional, chemical and genetic connections in the finest details were discovered using modern methods, firstly and mainly by many thousand times greater magnification of the visual image provided by electron microscopy, which the scientist in the past decades, when Portmann was active, could not have had any idea about.

Does this indicate that Portmann has been overthrown? Has it become redundant to talk and write about him? Should his works not be studied anymore? This book

The chapter was translated by Jana Krtková and Andrew G. Christensen.

R. A. Stamm (✉)
Professor Emeritus, Basel, Switzerland
e-mail: rogeralfredstamm@magnet.ch

F. Jaroš, J. Klouda (eds.), *Adolf Portmann*, Biosemiotics 23,
https://doi.org/10.1007/978-3-030-67810-4_4

will show that Portmann, in many aspects of his thinking, his research and his writing, is still highly relevant.

In this overview, first, we want to briefly introduce Portmann's areas of study and his achievements in his core research on animals and humans. Further, we will try to outline Portmann's more comprehensive understanding of life – more clearly: his philosophy of life. We want to grasp why Portmann is today – during the time of growing awareness of environmental disruption, in which all of us are involved – as relevant as he was during the time he dwelled with us, thinking and performing his active research.

Researcher and Zoology Teacher

We might begin with a brief chronological outline of Portmann's research interests. By 1920: Doctoral dissertation on dragonflies in the area around Basel (Portmann 1921), supervised by his teacher Friedrich Zschokke, an internationally recognized hydrobiologist and zoogeographer. The study on dragonflies falls into a set of about 30 dissertations concerning the fauna of the Basel area, which is an intersecting subset of all European zoogeographical-ecological regions. Portmann was methodically influenced by the model of the Danish entomologist and freshwater researcher Wesenberg-Lund. Portmann's work also illustrates larval development in dragonflies, the first of many embryological studies he would publish.

1922–1927: Residence at the *Station biologique* in Banyuls-sur-Mer (Pyrénées Orientales). Portmann spent a brief period as an assistant in Geneva, where he gained insight into modern techniques in histology, and held a short fellowship in Richard Hertwig's lab in Munich, when he came to know the coasts of Heligoland and Roscoff (Brittany). Following this, Banyuls-sur-Mer became one of the most important sites for Portmann's further work. Here, he conducted various small studies on octopus and sea snail embryology. As a spin-off from his general learning about the animal and plant worlds at these seashores, he also published a report on locomotion in the sessile sea anemone (Portmann 1926). His illustrations from this study are still cited in many textbooks to this day. There was only a narrow gap between his actual career path and another potential, totally different path more physiology oriented. The possibility loomed that Portmann could follow his colleague Anna Bidder to Cambridge and continue his career there as a marine biologist. However, Portmann habilitated and thus qualified to become a university teacher in Basel in 1926. A call to replace Friedrich Zschokke, who was seriously ill, followed in 1927. With this, his return to Basel was set. However, this brought him up to the chair of the department after his teacher definitively departed from this life.

1932–1939: The first task of a department chair at that time was to give students an overview of his subject – so in this case (all of!) zoology – in a lecture course of 5–6 h per week over two semesters. The students involved were not only future zoologists (scientists or teachers at the middle or higher levels of the school curriculum), but also first-year medical students, who also had parallel classes in botany, chemistry and physics – a reminder of the fact that in the nineteenth century, scientific subjects were part of the medical faculty.

For these classes, Portmann developed two lectures, "General Zoology" for one semester (general biology was taught similarly at most German-speaking universities; see the textbook by Alfred Kühn; Kühn 1955) and "Comparative Vertebrate Morphology" for the other semester – the latter he developed especially in regard to medical students. In one of the semesters, the lectures were supplemented with practical labs for 4 h a week. As already mentioned above, for future zoologists, there was a further lab section to learn about all forms of animal life and for independent study (the lab was available all day long and was attended ca. 20 h a week over the two semesters); this activity, however, was organized and supervised by the first assistant. Above all, Portmann made sure he had time for a special lecture, 1 h a week, where he reported about his own research and, finally, for a 4-h laboratory class on special topics.

The second task was to supervise doctoral students, preferentially future researchers in zoology. For these students, the main focus was on morphology and developmental science. Portmann set questions on topics from vertebrate morphology, but also on topics he had learned as a marine biologist. Most of the more than 70 dissertations supervised by Portmann came from this area. Some students could choose among topics from different areas of zoology; in these cases, they were usually co-supervised by professors in different fields (e.g. chemistry). Let us follow and highlight the topics that determined Portmann's research area as well as those of his students: (1) Embryonic development in octopuses and snails. (2) Development of bony fishes, reptiles, birds and mammals: embryonic stages, post-embryonic development, development of the brain and sensory organs, the importance of juvenile stages, feather development in birds. (3) Portmann also supervised some students who studied behavioral topics.[1]

The developmental studies resulted in two important overviews of general importance:

1. An overview of different types of development in general. Here, we find a distinction between primary and secondary larva (see Geigy and Portmann 1941).
2. A correct morphological characterization of both types of development in altricial (nidicolous) and precocial (nidifugous) birds and mammals. An important comment was included: in birds, the primary evolutionary type is precocial, whereas in mammals, the earlier ("primitive") evolutionary phase is altricial, and in terms of evolution, the precocial type needs to be considered the higher trait. Both of these different evolutionary paths can be explained in terms of evolution and ecology (Portmann 1935, 1936, 1939).

   Ultimately, these attempts at large overviews resulted in the inclusion of humans as a research topic.

[1] For a detailed list of individual projects throughout these works see Stamm and Fioroni 1984.

The Decade of Syntheses

In the 10-year period starting ca. 1937, Portmann attempted to provide large overviews of three different areas: (1) in biological anthropology, he published the book *Biological Fragments of An Anthropology* (Biologische Fragmente zu einer Lehre vom Menschen, Portmann 1944), (2) in animal appearance morphology, *Animal Forms and Patterns* (Die Tiergestalt, Portmann 1948), (3) in comparative morphology, the textbook *Introduction to The Comparative Morphology of Vertebrates* (Einführung in die vergleichende Morphologie der Wirbeltiere, Portmann 1948a). These three works were developed as follow-ups to his teaching activities: the anthropological book as a follow-up to some special lectures and several seminars, the book about animal appearance as a follow-up mainly to the practical laboratory exercise, and the morphology textbook as the fruit of his annually repeated major lecture for the students of biology and medicine.

Introduction to the Comparative Morphology of Vertebrates

The textbook starts with a discussion on fundaments for the comparison of different forms (homology/analogy, ontogeny, series of forms, paleontology and classification groups sequence in terms of natural history). The main part is dedicated to the comparison of organ systems (locomotive organs, orientation system, digestive track, reproductive organs, skin structures). The chapter about skin leads to the extraordinarily original, and for Portmann typical, understanding and representation of the appearance of animals. The text is demanding and makes the reader constantly keep comparative aspects in mind. It is not a simple list of facts. The reader should join in thinking and be aware of the evolutionary chain step by step. The quality of the images is striking. Portmann closely guided and carefully mentored his illustrators. It is didactically important that all the illustrations are labeled with numbers – a huge help for learning the names.

Portmann as Anthropologist

1938–1945: The comparison of development in birds and mammals resulted in the question of how it could be in humans. Newborn humans do not yet have the body proportions of an adult; they cannot move independently – so altricial? On the other hand, their eyes are open, which is a trait of precocial species. The data on embryology and youth development in humans had already been available; however, a conclusive integrating model was lacking. Through a comparison of human growth curves with those in closely related mammals, Portmann recognized typical human features. He expressed the type of human development in two terms which shortly after became catchwords in discussions on the exclusive position of man:

"extra-uterine first year" (*extrauterines Frühjahr*) and "secondary altriciality" (*sekundärer Nesthocker*).

Both features have something to do with the specific way of foetal growth in humans: the brain and thus the head grow relatively quickly, the length of the torso and the extremities fall proportionally behind. If we compare body weight gain during the foetal period, it is significantly higher in humans compared to apes. The rapid growth still continues during the first year of life. Subsequently, however, there is a deceleration rather atypical for primates. Until puberty, the body grows slower in humans compared to their primate relatives. There is a clear reason for this: the extraordinary position of a man! The unique features of human development reflect the individuality of a living creature *whose nature involves culture as a fundamental element*.

The first year in human life represents a phase which prepares (1) upright walking, (2) a basis for learning to talk, (3) a reasonable way of acting that far exceeds instinctive impulses. Portmann describes these links individually and highlights them, rather than mentioning them solely as an aside. Portmann's openness to the essentiality of culture in the biological nature of humans results in an impressive description of all events and connections. Zoologists and cultural scientists usually see the status of a man in the realm of living things with totally different eyes. Typically, zoologists grasp a man in a similar way as they do the so-called higher animals: anatomy, physiology, cell biology, genetics, behavior. They see man in a special way, of course, but not fundamentally differently than how they would think about any other animal species. For them, man is a case within a systematic order, displaying similarities with and differences from his biological relatives, and, having evolved from the yet unknown common ancestor, is connected with all other animals on Earth. According to the now generally accepted theory of evolution, this is obvious. Among cultural and social scientists, man is understood in a completely different way. For those who survey the immense richness of human cultural representation (languages, worldviews, religious concepts, history, art and science, technology, sports etc.), man is utterly unique, sometimes to the extent they miss the point that, despite all uniqueness, man is – in body structure, organ function, in many behavioral aspects like aggression, sexuality, maternal behavior and ability to make friends – principally and in fundamental tendencies the same as other mammals, distinguished only by a further developed mental superstructure.

In this area of tension, the bridging accomplished by Portmann is unique. He does not refer to the uniqueness of human development as one case of development among others, but rather as the *sine qua non* of culture and the specifically human coexistence in complex, structured societies.

Animal Forms and Patterns

Adolf Portmann the morphologist was not satisfied with just a simple description of body structure and organ functions and their evolutionary adaptation. While studying the physical build and color patterns, he did not regard them as simple evidence of adaptation or as a mere tool for systematic classification. Rather, he studied them as an *appearance* displayed by the entire set of characteristic features, i.e. as an integrated whole in its representation value (*Darstellungswert*). With this unique point of view, typical for him, he could, firstly, suggest a distinct way for the research of specific appearance, and secondly, pose research questions with respect to the ontogenetic origin of this appearance and, in the end, also the developmental and genetic forces behind it.

Consider the "ritualized" behavior patterns used in discussion or courtship – with regard to appearance, but also behavior, which ultimately depends on the movement of appearances. Portmann reminds us that each form of life – bacteria, fungi, plants, or animals, including humans – always has a relationship to the world "from the inside out."

At this point, there comes immediately the question of the *biological function* of the appearance, and subsequently the question of the evolutionary-historical cause of its emergence. Portmann in no way refuses such questions; on the contrary, he faces their full weight. He repeatedly gives clear examples of such consideration in his works – for example in *Animal Forms and Patterns*, *Animals as Social Beings* (Das Tier als soziales Wesen, Portmann 1953), *Animal Camouflage* (Tarnung im Tierreich, Portmann 1956) etc., and further also in his other works and countless lectures.

Some of Portmann's colleagues were unpleasantly puzzled when he repeatedly warned of premature assignments and generalizations and when he, with cautionary words, always put functional and evolutionary explanations in their place. And he was completely right! Those who have experienced how hard it is to give experimental evidence that the emergent black-colored morphological variety of peppered moth (*Biston betularia*) really provides a selective advantage will quickly accept that many statements determining the advantage of trait maintenance are often very superficially justified; they seem plausible, but they are hardly truly evidence based, as Julian Huxley noted (Huxley 1942: 23). Today it is more often recognized that the traits often (generally) represent a compromise between different functional and selective forces. So, for example, if a clutch of eggs with an effective camouflage coloration is placed in a nest where their camouflage effect is negated, it may be advantageous for another functional reason – like protection against high water levels in the primary hatching biotope. Moreover, justified assignment of functional advantages does not yet explain a specific form of appearance and should in no way result in neglect of careful and conscientious research of the pure morphologically explainable representation value (*Darstellungswert*). The same is valid for its physiological and, ultimately, genetic analysis. Portmann strongly supported the latter.

However, Portmann complicates such a "trivializing" understanding by his use of the terms inwardness (*Innerlichkeit*) and self-presentation (*Selbstdarstellung*). To make inwardness a bit more understandable, he starts expressly from the human inner being and accepts that the presence of conscious and unconscious experience in natural systems, i.e. in animals that are distant from humans, is very hard or even impossible to prove. However, it would be erroneous to assume that while thinking about inwardness or self-presentation he takes into account some other-wordly force elements, such as *élan vital* (Bergson) or *entelechy* (Aristoteles, Driesch). Portmann only insists that inwardness and self-presentation be recognized as real phenomena that point to the mystery of emergence. He strongly warns against the self-righteousness of the scientist who believes he can, at least in principle, explain or comprehend these phenomena.

Portmann and Theoretical Biology

Let us linger a little on the two terms that provide Portmann the large framework for his understanding of life: world-relationship through inwardness and self-presentation.

Relationship to the World Through Inwardness

For Portmann, world-relationship (*Weltbeziehung*) is the superior term through which the essence of life can be understood, superior to the set of specific functions of lifestyle. Organisms are bodies that seem to be separated from the outer world by their surface border structures (skin, cell membrane). But in many ways and under strictly controlled conditions, the border enables constant exchange with the outer world or environment (*Umwelt* – a profound term used for the first time by Jakob von Uexküll more than a century ago – however not in today's ecological sense). The exchange is manifested in metabolism, in sense-perception, in all behavior. However, an organism is not just a "physiological bag" – an old phrase mocking a too narrow view of the organism – on the contrary, it is organized, structured on all levels and from all perspectives in which it can be observed. This organization is powered from two different directions. From one side, by heredity (genetics), from the other side, by the experience that is processed in the respective organism – indeed, in each organism – and can be taken into account during future actions.

If these abstract terms and formulations were substituted by the whole abundance of life events, the obvious richness of these natural principles could be clearly seen. As stated above, the term "world-relationship" should not be misunderstood by any biologist. Portmann further supplements the "world-relationship" in its whole diversity with the word "inwardness," which can be explained as a general term describing "from inside out." Thus, genetic regulation as well as the consequences of "experience" are involved. Then, developmental (ontogenetic) regulation is also included. As a result, there should not be any controversy over this very broad term. Portmann derives "inwardness" unambiguously and always from human experience, and he ascribes a certain form of experience to all animals and plants. It is definitely not only about awareness. Experience is more about going through, interiorizing or taking into account in a broader and more general sense – not only about what is called motivation or action preparedness by an ethologist. All demonstrations of life in animals and plants give evidence of the inner control. In other words, these processes are those described in the language of cybernetics. Today, some researchers are less hesitant to talk about animal cognition in a systemic sense (Range 2009; Brown et al. 2011).

Self-Presentation

Portmann uses the term self-presentation (*Selbstdarstellung*) centrally and primarily for the processes related to the growth and establishment of a species-specific appearance. The appearance, morphological as well as behavioral, is always the result of an inwardness. We should not seek for any mystical events behind the term *inwardness*; crossing the boundaries in this sense was for Portmann completely unfamiliar. Those who are uneasy regarding this expression can very well understand it in the sense of genetic-physiological regulation. With this, Portmann wanted to point out that the self-organization of an organism is not an isolated phenomenon; on the contrary, it is deeply based on the relation to the outer world, on the world-relationship in the broadest sense. The color pattern of the four wings in butterflies that appears as a completed pattern in their natural posture, or the coloring in marine nudibranchs (a group of marine molluscs) that are not even clearly visible under water in low-light conditions appear as if those patterns, in their contrasting colors, were meant for a "looking eye." Does it always have to be a result of a selecting power? Portmann doubts it. It has indeed rarely been proven by precise observation. According to precise observation it is not, indeed. There is no definite evidence that all these forms and patterns are produced by genetic control; it is rather an urgent research task.

Portmann, an Outstanding Teacher

Portmann was a masterful pedagogue. He did not follow any standardized, structured schedule imposing a variety of media and generating excitement with multiple tricks, as in modern didactics. He was rather a teacher deeply rooted in the essence of pedagogy – in the practice of observation and thinking. He led students to nature, he let them observe what it offered, he trained them to write precise, conscious reports of what they had seen. He let the students marvel at the appearing nature and ask questions related to it. Mental imagery originates above all from pondering about the thing presented in words, not after seeing a more or less well understood picture or film. What made Portmann's representation of natural phenomena so convincing was his art of presentation in words and drawings. Speech was never simply formulaic, it was custom-made for nature, no matter what it described – an animal or a research process. Portmann often reached for the latter while describing at length the historical journey of science during the nineteenth century, in the early modern period, or back to the Middle Ages, when the view of the world was still often determined by mythology. He used his drawings as illustrations. While speaking, he sketched with steady hand an object onto the board– a drawing as a record of the perception of appearing nature. He rarely projected photographs, only once or twice during the semester. The grand phrase used by Goethe, "a manifest secret," became the great experience for each of Portmann's mesmerized listeners. A natural object – an animal, a plant or a landscape – became conceivable in words through his presentations, understandable in its reality. Nature revealed itself, threw away its veil. The ancient Greek ἀλήθεια – unhiddenness (*Unverborgenheit*) according to Martin Heidegger – could be experienced here. Ultimately, however, nature always remains a secret to us, no matter how deep and precise our analyzing insight penetrates into it, and something always remains hidden in its very basis. Portmann was not afraid to announce this fact, very often arousing mockery or anger in his narrowly rationalist colleagues. The natural scientist Portmann should be honored for always keeping in mind the limits of natural research.

Portmann's masterful teaching was linked with his deep insight into the essence of educational processes during the development of a young person. He stressed the polarity between the views of the world in respective stages. During childhood years, learning is determined by the *primary view*, in which the key role is played by sensory observation of the surrounding world. Only during the prepubescent stage is it gradually taken over by the *secondary view of the world*, which is dominated by thinking – understanding and reason. He considered an early emphasis on knowledge transfer to be harmful. Portmann leaves no doubt that these facts should be taken into account by educational policy. Today we read with admiration that shortly after World War II Portmann had already warned against the flood of images and their harmful effects on the development of healthy relationships with fellow human beings and to the world as a whole. A "flood of images" practically before the age of color films and television? And today? No further word is needed!

Portmann, Public Teacher

Portmann had a deeply ethical and social mission to be a teacher for the general public. He developed superior skills in reporting what nature is, and what biological research had achieved through precise work, to a world beyond the specialists. He reported on his own studies and the results of other experts in a comprehensible but precise manner that was highly appreciated in German-speaking regions. He was a pioneer in radio lectures. He was discovered early by the director of the Basel station of Swiss Radio (1928). He appeared in hundreds of broadcasts as speaker or discussion partner. There were small, 15-min insights into different aspects of animal life (published in numerous collections), his fabled contemplative appeals at the end of the year, his participation in discussions on political and social topics of the day (for example, the right to vote for women, abortion), his repeated appearances on the broadcast "What Do You Think, Professor?" and much more.

Portmann appeared at many places as a speaker, from the Aquarian Association to colloquia with physicians, philosophical discussion clubs and big anniversary events of scientific societies. Some recordings of these lectures are still available. Therefore, his voice, his way of speaking and thinking can still be analyzed today. Twenty years ago, most of the radio stations in German-speaking countries still had recordings of Portmann's lectures in their archives. Two hundred of them are preserved in the Portmann Archive at the University of Basel. Among the most important documents are his commemorative lecture given at the 600-year anniversary celebration of the University of Basel in 1960 (the other speaker that day was the philosopher Karl Jaspers) and a lecture on the history of biology in the nineteenth century given in Hamburg. Of special significance were his 2-h performances during the Eranos meetings in Ascona (1948–1978). In 1962, he took over the scientific leadership from the founder Olga Fröbe-Kapteyn. In these lectures, all his proficiency was manifested when he put the discoveries of modern biology into the anthropological-cultural-historical context of the conference (Portmann 1956a, 1965). Portmann's social motivation to engage in public was also demonstrated during the decades (1938–1964) when he was a leader of the *Volkshochschule* (adult education courses) in Basel, and in his participation in the governance and animal care in the zoo and the animal park Lange Erlen in Basel. He was a co-founder and for many years the president of Swiss Youth in Science (*Schweizer Jugend forscht*).

Portmann, Artist and Philosopher

In his basic attitude, Portmann was an old-style scholar, not in the sense of the universal scholar of earlier centuries, but in the sense that he was a scientist, artist and philosopher in one. He was an artist not only because of his talent for drawing, but especially because of his artistry in giving speeches. This was recognized by the Sigmund Freud Award for writing in the sciences by the German Academy for Language and Literature in Darmstadt and by the Schiller Award by the Swiss Schiller Foundation. With his texts, he captivated his listeners, amateurs as well as professionals. His German was rich, but clear at the same time, his sentences were logical and deliberate. Portmann also spoke and wrote fluent French. He was a philosopher in that he had the courage to ask questions beyond the borders of scientific research. While a young student of Portmann, my colleagues would strongly remind me of standard paradigms. Science has strict rules concerning questions to be asked and statements to be made. "Morphology, behavioral research, those are only descriptions, not yet science. Analysis in physiology, biochemistry and genetics is the real science!" Or more blatantly: "Questions that cannot be answered should not be asked!" This approach is completely wrong. Questions arising from subject matter *must* be asked! According to the scientific paradigms in natural sciences, these questions should, if possible, be processed via clear, reproducible and objective observation. If necessary, they should be supplemented with statistical evaluation in order to make them more reliable, experimental evidence should be given, and the answers should be declared as natural law that could be always experimentally verified. It is the task of philosophy, another important research method, to think beyond this, to place things into larger contexts (integrated thinking). For fundamental and complete discovery, it is crucial to integrate also into this field. An ability to perform this style of research is not given to everyone, that has to be accepted. Therefore, the scientist should never be sniffy and make fun of this approach. Portmann's philosophical orientation was formed by the intensity of his experience with nature during childhood. The traces are clear: from his own book of animals that he started to draw when he was only 10 years old, then, longer participation in the local youth club around cultural philosopher Rudolf Maria Holzapfel (as revealed in Portmann's sole preserved diary), to the student discussion group on Max Scheler's *The Human Place in the Cosmos* (Die Stellung des Menschen im Kosmos, Scheler 1928), visited in 1929 by Hans Mislin, who was one of Portmann's first students, to his officially announced anthropology seminars in the 1940s and his membership in the Basel Philosophical Society. The effort to reflect on something, to search for the reasons behind the state of things, even without any special philosophical education, shows his passion to ask and to seek. Apparently, longer discussions with Hungarian philosopher Wilhelm Szilasi, working in Freiburg im Breisgau, followed the publication of *Biological Fragments*. Portmann was a helping hand in the reopening of the University in Freiburg. The fact that he was a contemporary of the important followers of philosophical anthropology – Max Scheler, Helmuth Plessner, Arnold Gehlen, Frederik J. J. Buytendijk – was not without consequences, even though there is no biographical evidence for closer personal relationships. Portmann was a natural philosopher because he asked about the *whole*, about the *links* beyond what has already been explained or is explainable by science, because he knew about the deepest *mystery* of life and was able to explain it to his listeners and readers. Moreover, the fact that he supported dialogue among faculties at the University of Basel and building bridges between different fields of study, manifests his deep philosophical mission.

Portmann, Pioneer of Nature Conservation and Environmental Protection

Portmann was a pioneer of modern nature conservation. Someone might ask: "What did he do for this?" From an external, technical or political perspective, perhaps very little – he helped to found the Swiss Ornithological Institute in Sempach. Its first director for many years was his student Alfred Schifferli. Portmann's attitude to nature and his teaching awakened in his listeners the preparedness to see, to listen to, to gain experience with nature. Under his supervision, one was taught the ability to perceive, to respect and to care for life in all its forms, from the tiniest and most unnoticeable forms to human beings. His lectures on the adaptation of the rabies virus or on the life cycle of the slime mold (*Dictyostelium*, Myxomycota) were impressive. It was amazing how we saw the fullness of life while exploring a mountain stream during field trips guided by Portmann: from astonishing planaria, larvae of caddisflies, mayflies, stoneflies, dragonflies – who could have anticipated that in 50 years, all of this complex richness of life will be almost lost as a consequence of thoughtless and inconsiderate human behavior? Portmann had already warned of the destruction of nature before forest degradation, urban sprawl, waste contamination or global warming were hardly mentioned by anyone. "Nature Conservation Will Become Human Conservation" was the title of a small work he published in 1971 (Naturschutz wird Menschenschutz, Portmann 1971). Instead of the "environmental protection" that became more popular later, Portmann spoke of "nature conservation" – an important detail. Because protection of the environment reflects an anthropocentric perspective – a protection of human life conditions. Conservation of nature, however, includes broader horizons. Each life, and nature on the whole, should be protected out of respect for and awe of creation. Nature in its fullness is an inexhaustible source of experience, joy and small discoveries for us: an important basis of our intellectual and spiritual way of life.

Personality, Grandeur and Limits

Portmann was an important person, self-confident in his presentation, but always moderate and very humble. His analytical skills, ability to reflect different points of view during scientific or political discussions, his delicacy in all his statements and expressions quickly ensured him a leading position on boards and panels he belonged to, whether it was the Faculty of Science in Basel, the leadership of the International Ornithological Congress in Basel in 1954, or engagement at the University in Freiburg im Breisgau, in the Association of Swiss College Professors (he was its honorary president), in the International Association of University Professors and Lecturers (I.A.U.P.L.) (he was its leader for 1 year), in one of the UNESCO committees, and others. Therefore, some of the honors he was awarded were not only for his scientific research achievements, but also for his general engagement in the culture of dignity. Nonetheless, honesty commands us to mention also the limits of Portmann's thinking. He cannot be blamed for the fact that he was not as fully oriented in the already immense fields of biology as he was in morphology, developmental biology and extensively also in the behavioral research of the time. Compared to many experts today, he had both an outstanding overview and especially the capability to gain insight also into distant fields of biology. But, there are two areas – and these directly belong to his central concerns – in which he reached his limits: first, in the issue concerning the special status of man, and second, in modern evolutionary theory. Let us try to analyze these two limits. In discussions with his teacher, the author of this chapter also saw these limits and depicts them here in trying to understand Portmann's standpoint.

The Special Status of Man

Portmann was very much aware of the impact he would have with his interpretation of early development in humans. He had sent out copies of his *Biological Fragments* to numerous professors of philosophy and medicine. (The list of recipients could be reconstructed only to a limited extent – depending on whether there are any responses preserved in the Portmann Archive, for example by Heidegger, Szilasi, Jaspers.) The effect was astounding. The work struck those recipients like a bomb. To understand this effect, we have to understand the discussion on the special position of man. For simplicity, let us identify two poles: the "natural-sciences perspective" and the "humanities perspective." For a zoologist, i.e. natural scientist, man is simply another animal species, of course with individual morphological and ecological traits, similarly as every other animal. One often hears that "among animal species, there are much bigger differences than between man and his closest mammal relatives." For a humanist, man is unique, superior to any other living being because of his intellectual and cultural achievements. In 1960, an extreme statement by one of the leading German psychologists struck me: "Portmann is the only biologist to admit that man does not originate from apes." Surely a wrong view!

If we would attempt a reconciliation of both perspectives, first, we have to admit that both are correct. It is necessary to merge them. But how should we formulate this? Portmann was, of course, convinced that humans evolved from animal ancestors, even though genetic and ethological arguments then were insufficient to identify precisely a close human relative. The data he presented were mainly cited from the literature, and he used them for arguments in a way that would be valid for any other mammal. Thus, the specific features mentioned above were revealed. They were supplemented by further morphological data: development up to puberty, walking upright, the specific constitution of speech organs. The way of working and thinking here was that of the zoologist, the first pole. Portmann's analysis of human ontogenesis showed that humans occupy a special position also from the zoological point of view – and that precisely the unique features of human development represent the indispensable cornerstones of human culture. This statement convinced the cultural scholars. But there was one point in the natural-scientific classification of man where Portmann was oddly inaccessible. And that was regarding the fact that in behavior, especially in social behavior, there are features already clearly manifested in higher primates. We need to remind ourselves that before 1960, there was barely any knowledge of the behavior of higher primates, and no knowledge at all of their behavior in the wild. Later, he would have been more open to the discoveries made by Rudolf Schenkel, George Schaller, Jane Goodall and others, because exactly these comparisons became an immense task for behavioral research in primates. The consequences of these discoveries were not disparaging to the human special position. On the contrary, they opened the eyes of cultural scholars to the unique and grand evolution resulting in man. Portmann emphatically calls the human newborn a precocial individual, similar to other primates. He ignores, though, that there is at first an absolute dependence on mother, especially in the great apes. The newborn chimpanzee can hold tight to its mother, but without any assistance, it is not able to change its position alone (Hess 1996). The stage of the "carried" (Tragling: Hassenstein) is common to all primates. However, Portmann was apparently so convinced of the special position of man also in the biological sense that in the 1956 pocket issue of his main anthropology work (Portmann 1956b), a mistake occurred. A human baby (ethological stage infant (1) is shown in a picture together and in contrast with a much further developed baby macaque (ethological stage infant (2) (see Portmann 1956b: Fig. 8, Fig. 14). In the third issue of the main edition (Portmann 1969), these images are not present anymore; however, the text does not reflect at all the details of evolutionary prehistory (see Portmann 1969: 57).

The General Theory of Evolution

Portmann was, of course, convinced of evolution as a process of natural history and that humans evolved according to this process. He also accepted that mutation and selection are two fundamental factors in this process. However, he always had a critical judgement on evolutionary-biological functional explanations. There were two main reasons for this:

1. His observations of animal appearance were discriminating but also integrated. These, for him, and especially his observations of body color patterns, could be explained as evolutionary adaptations only in very specific cases (see Portmann 1956). The possibilities of interpretation as functional adaptation had hardly been verified by precise observation and experiments; rather, it was too often claimed as a plain statement, as Julian Huxley observed (Huxley 1942: 23). The emergence of optical regularity and symmetry in patterns was very underresearched by physiology and genetics. Portmann's artistic sense and his fundamental devotion to the organism in its essence made him strongly reserved. He almost suffered under his colleagues' tendency to pass over the variety presented in appearance and coloration and consider it relevant only as a taxonomic trait. (See the statement quoted above: "Simple description is not yet research!")
2. It was clear that although he was thinking about evolution, he was not at home in this research area that was rapidly developing during his middle research years. This is already indicated by his persistent use of the term "neo-Darwinism." During his early years, also as a consequence of his strongly manifested attention to natural history, Portmann was particularly influenced by the discussions and rapidly changing positions (dogmas) between saltationists (large mutations enable leaps in evolution) and followers of micro-evolution (evolution is driven by small steps that on a large timescale enable big differences as well).

In the meantime, a new model was developed which far exceeded previous extremes and stressed the fundamentally synthetic character of evolutionary theory – *numerous fundamental factors in concert.* There was an essential tipping point towards a focus on populations and their interaction dynamics. Further, impacts of geographical factors affecting spatial and ecological divisions were considered. However, the most important – regarding the aspect of organism in evolutionary theory, i e. the perspective that puts the whole organism and its social relationships into the center of attention – was ecology in all its facets. Only from this perspective are organisms revealed in their entire fullness of life! The key term here is *ecological niche* (in the dynamic meaning of the term: filling a niche). An important term in biogeography – "island biogeography" – includes distribution barriers around an island and among islands, but also the barriers formed by streets and paths that block the distribution and conservation of soil life – an important topic for nature conservation. From the perspective of molecular biology, the most crucial role today is played by genetics. (Do not forget about organ physiology, cell biology, biochemistry and biophysics!) The result of the synthesis of all these parameters and perspectives, achieved by only a few biologists, is the "synthetic evolutionary theory" that could be easily called the "general theory of evolution." The term was coined by prominent English zoologist Julian Huxley in his study *Evolution: The Modern Synthesis* (Huxley 1942). Further, George Gaylord Simpson deeply incorporated the perspective of paleontologists, and finally, Ernst Mayr presented all the evolutionary processes again in an extensive and critical overview in his authoritative work *Animal Species and Evolution* (Mayr 1963).

Being a synthesis of research from all fields of study, the synthetic theory proves useful, since it can be extended and expanded and can also include all results of modern genetics. However, it asks important critical questions of genetics that remain to be answered in the future. The synthesis blossomed especially in the fields that witnessed huge advancement then – ethology, ecology, especially population ecology, and nascent genetics. The fact that by the end of the twentieth century, genetics would become the fundamental science of biology, was hardly anticipated then. Morphology, behavioral science, physiology and microbiology still represent essential, standard fields of study, even if they have been overshadowed by the dominant genetics. Recent genetics addressing mainly gene-protein interactions needs to open itself to a new, more comprehensive outlook. So far, there are no attempts to explain, for example, the forms of appearance, or the "intellectual" informational structures of innate behavior. In the future, the "classical" fields of study will need to find their appropriate value and position. With his view of evolutionary processes, Portmann did not join the modern synthesis, which is even more unfortunate, since he often expressed his opinion regarding evolution.

Though Portmann turned his back to this field, we cannot draw the misleading conclusion that he was not modern. The broadness of his view in the fields of developmental biology, theory of appearance, behavioral science and anthropology was impressive and did not need to fear comparison with those of his more specialized mainstream colleagues. For thinking in terms of general evolutionary theory, it is not sufficient to operate just with the two fundamental terms of mutation and selection. The phenomenon of recombination as a consequence of sexuality already enlarges the field substantially. Further, it is essential to take the three levels – *gene, individual* and *population* – into full and balanced consideration. Above these, the researcher has to get a clear idea about each type of animal or plant species, and about the ecological requirements that reflect the main functions of their relationship to the environment that ensure their survival (fundamental topics of ecology): protection against natural risks, finding food, defense against predators (or against other impacts of organisms, for example root competition, infectious diseases), collaboration with social partners (intra- and interspecies), reproduction (collaboration of sexual partners) and parental investment (protection and care). Importantly, spatial conditions need to be considered as well: along with necessary spatial, geological-geographical and climate conditions enabling finding a niche, spatial distribution (biogeography) needs to be reflected– while always considering competition and co-operation with other species. Even if all these parameters are considered, it cannot be ruled out that many of the established adaptations evolved from *compromises* between often contradictory selection factors. For example, Tinbergen and his colleagues gave evidence that the egg camouflage of the Black-headed gull acts as a way to avoid predators (Tinbergen et al. 1962). However, the eggs are laid in a noticeable nest, which is in contrast with the egg camouflage. Nonetheless, such a nest is necessary to protect the eggs against wet conditions at the primary nesting sites in wetlands and on seashores. Ultimately, there can be no doubt that there may be traits that are neutral regarding all selection factors. Yet, it remains for genetics to elucidate the development of such traits.

The general or synthetic theory of evolution is one of the possible ways to focus, instead of on a single species or an isolated individual, on living communities and the integrity of biological systems in natural history as well as today. Consequently, genetic-molecular research will be able to make further advances and overcome the narrow field of gene-protein interactions where it is often stuck today. Portmann certainly thought in broad horizons and therefore also correctly defended those aspects that were ignored by purely functional views and by superficial observation. Despite this, he did not appreciate the excellent qualities of synthetic evolutionary biology as a comprehensive, integral concept. This theory highlights the functional aspect during the process of development, which, considering his work, was not unfamiliar to Portmann. If researchers of evolution neglect the descriptive-analytical aspect, they should be rightly criticized, as Portmann did. True general biology requires that both functional and descriptive studies gain their full recognition. Description is essential for all aspects and levels of biological research, not just the first stage of functional-rational analysis. Appearance in Adolf Portmann's sense represents not just how the body looks, but also behavior (Stamm 1986) and – if viewed correctly – also genetics.

The Postulate of Organismal Biology

Recently, there has been more encouragement to counteract the dominant molecular biology and genetics with organismal biology, i.e. with a concept that has an integral organism at the center of its attention. Regarding this, there is no either/or; instead, both polar standpoints should be balanced. An organismal or integral perspective starts with and essentially focuses on the research of body structure and behavior – the primary and immediate characteristics we can notice in any organism. In this research field, Portmann is exactly at the right place. As a morphologist, he also had an open mind for the neighboring field of study – ethological research. How often he expressed his opinion on behavioral issues! Organismal biology is neither a closed concept, nor a field of study with set borders. It could be better characterized as open to all directions, especially to both poles of descriptive and analytical research: on one side to the width of ecology and general evolutionary theory, on the other side to the depth of the molecular world of submicroscopic processes. In the decades since Portmann studied and worked as a teacher and researcher, organism-oriented biology has made gains in its scope and content in all directions.

In Portmann's era, cell biology and genetics, but also ecology and developmental science, were still in an early stage of their development. Since then, detailed knowledge of immense range has increased in all fields of study. It is sometimes now said that for a single person, an overview is no longer possible. In this sense, rightly, the old "professor ordinarius of zoology" as Portmann was, is no longer conceivable. But we cannot agree with such a pessimistic view. Let us only consider the task and commitment to educate teachers for all levels, from elementary education through university, and also to provide information for citizens who should adopt a posture on the current climate catastrophe and other practical topics. How should teachers of all levels provide facts about nature and its laws to their students, listeners and readers, if they cannot look beyond a single isolated date? We know well this selective way of journalistic representation from TV news and newspapers. That which Portmann carefully and temperately formulates as inwardness and self-representation is of great importance for *applied biology*. In the future there will again be stronger awareness that the approaches to understanding life occur at two levels: strict natural-scientific analysis is accompanied by the intellectual-spiritual approach to life, something that Portmann indicates, using a modification of common lexical meaning, with the French *savoir par cœur* (in German *Weisheit des Herzens*), the wisdom of the heart that follows from intensive and focused observation and a compassionate attitude. This dual perspective is important for all fields of applied biology: for animal breeding and animal training, agriculture and horticulture, medicine, pedagogy, jurisprudence. The practical impact makes us consider both these perspectives as serious ways of understanding also in research. Researchers can perform their work successfully only if they focus on a specialized topic.

However, people educated in science and philosophy working as teachers or practitioners should pursue the widest possible view. They need to be *generalists*, to keep wide horizons and to seek structural fundamentals and their order in their respective field of study. They should see the structure of the whole in clear logic, be able to mentally sketch in a few deliberately arranged lines, and search for clear and strong images to illustrate all aspects and present them in brisk language. Of course, this task will become more difficult as more phenomena and data are brought by research. The results need to be assessed critically. Therefore, teachers of all levels need to be scholars as well, which means they need to keep looking, observing, classifying and considering. The erudition and education of teachers at universities is an important and necessary task. No crafty didactics or teaching method can hide that there is something missing – an understanding of the subject, the integral organism and the wholeness of the nature.

Perspectives

What Portmann discovered and described as a researcher goes into the comprehensive body of knowledge and will get more and more lost there. Most of this will become a *dormant knowledge* that will be recalled as required and interconnected with new discoveries. It could also disappear in the darkness of ignorance, as often happens to old topics in modern research. However, it should keep us heading towards what Portmann strived for and the research tasks he sketched out. These are not only tasks of specialized clarification of facts – they regard the basic standpoint we adopt towards nature. In the future, people will need to learn to avoid the egoistic, egocentric handling of nature for scientific reasons and to put their own economic interests and safety into the background. In the future, they will also need to recognize the community of all living creatures and consider themselves a part of a magnificent organization. In their own behavior and life experience, people should feel close to the behavior and putative life experience of animals. In this regard, let us follow Portmann as our role model!

Appendix

I. Biographical Data

1897 Born in Basel on May 27. Attends schools in Basel.

1916 A-levels at the higher secondary school. Studies natural sciences at the University of Basel.

1921 Graduates from Basel. Dissertation: "Dragonflies in Basel's Surroundings," supervised by Professor Dr. Friedrich Zschokke. From 1919 (–1925) Subordinate Assistant at the Educational Institute of Zoology (subsequently on leave).

1921–1922 Studies in Geneva, among others under Emile Guyenot (histology). Assistant of E. André (limnology). In 1921 during summer vacation, the first research fellowship on Heligoland.

1922–1923 Studies in Munich (at the Zoological Institute directed by Richard Hertwig; performs his research under, among others, Richard Goldschmidt and Karl von Frisch). During summer vacation in 1922, the second research fellowship on Heligoland.

Summer semester 1923 Studies in Berlin under Oskar Hertwig, Richard Hesse and H. Nachtsheim.

Winter semester 1923/24 Studies in Paris under A. Prenant. In winter 1923/24 the first research fellowship at the Station Biologique in Roscoff (Brittany).
1924–1927 Research fellowship in Laboratoire Arago in Banyuls-sur-Mer.
1926 First Assistant at the Educational Institute of Zoology in Basel. Habilitation in zoology in Basel.
1928 Becomes director of the Educational Institute of Zoology and takes over compulsory classes as deputy from Friedrich Zschokke.
1931 Professor extraordinarius and chair of the Educational Institute of Zoology.
1933 Professor ordinarius in Zoology. Portmann directed the Institute until he became Professor Emeritus on, March 31, 1968.
1938–1964 President of the Public College in Basel.
1942–1972 Member of the board of directors, Basel Zoo.
1947 Rector at the University of Basel.
1949–1957 Member of the committee section Sciences exactes et naturelles of the National Swiss UNESCO-Committee.
1950–1951 President of the International Association of University Professors and Lecturers (I.A.U.P.L.).
1954 General Secretary of the 11th International Ornithological Congress in Basel.
1962–1982 Director (together with Rudolf Ritsema) of the Eranos circle in Ascona as a successor of its founder, Olga Fröbe-Kapteyn.
1962–1969 President of the Conference of Swiss University Rectors.
1964–1982 President of the Hermann-Hesse-Foundation.
1967–1975 Founding president of the Swiss Youth in Science foundation.
1968 Professor Emeritus at the University of Basel.
1974–1977 Lecturer in The Biological Bases of Communication at the University of Bern.
1982 Dies in Binningen, close to Basel, on June 28.

II. Selected Bibliography

Dissertation

1921 Die Odonaten der Umgebung von Basel: Beitrag zur biologischen Systematik der mitteleuropäischen Libellen. Lörrach: [Private print]. Reprint (shortened) in A. Portmann (1967), *Zoologie aus vier Jahrzehnten*, pp. 11–62.

Articles (selected)

1925 Der Einfluß der Nähreier auf die Larvenentwicklung von Buccinum und Purpura. *Zeitschrift für Morphologie und Ökologie der Tiere*, *3*(4), April 1925, 526–541.
1926 Der embryonale Blutkreislauf und die Dotterresorption bei Loligo vulgaris. *Zeitschrift für Morphologie und Ökologie der Tiere*, *5*(3), March 1926, 406–423.

1926 Sur la structure et les affinités de Chondranthus denudatus, actiniaire nouveau de la Méditerranée. *Bulletin de la Société Zoologique de France*, *51*, 1926, 19–30.
1926 Die Kriechbewegung von Aiptasia carnea: Ein Beitrag zur Kenntnis der neuromuskularen Organisation der Aktinien. *Zeitschrift für vergleichende Physiologie*, *4*(5), 1926, 659–667.
1928 Yolk-absorption in Loligo and the function of the embryonic liver and pancreas. *Quarterly Journal of Microscopical Science*, *72*, New Series, 1928/29, No. 286, October 1928, 301–324.
1930 Die Larvennieren von Buccinum undatum L. *Zeitschrift für Zellforschung und mikroskopische Anatomie*, *10*(2), 1929/30, 401–410.
1933 Die Bedeutung der Schale für den Wasserhaushalt der Eier von Enten und Haubentauchern. *Revue Suisse de Zoologie*, *40*(fasc. 1, no.7), January 1933, 179–186.
1935 Die Ontogenese der Vögel als Evolutionsproblem. *Acta Biotheoretica*, Series A, *1*(1–2), 1935, 59–90. Reprint In A. Portmann, *Zoologie aus vier Jahrzehnten*, 63–90.
1936 Die Ontogenese der Vögel als Evolutionsproblem. *Verhandlungen der Schweizerischen Naturforschenden Gesellschaft*, 117. Jahresversammlung vom 28. bis 30. August 1936 in Solothurn, 224–241.
1938 Beiträge zur Kenntnis der postembryonalen Entwicklung der Vögel. I. Vergleichende Untersuchungen über die Ontogenese der Hühner und Sperlingsvögel. *Revue Suisse de Zoologie*, *45* (fasc. 2, no. 6), March 1938, 273–348. Reprint in A. Portmann, *Zoologie aus vier Jahrzehnten*, 90–145.
1938 Die Ontogenese der Säugetiere als Evolutionsproblem. I. Die Ausbildung der Keimblase. *Bio-Morphosis*, *1*(1), 49–66.
1938 Die Ontogenese der Säugetiere als Evolutionsproblem. II. Zahl der Jungen, Tragzeit und Ausbildungsgrad der Jungen bei der Geburt. *Bio-Morphosis*, *1*(2), 109–126.
1939 Nesthocker und Nestflüchter als Entwicklungszustände von verschiedener Wertigkeit bei Vögeln und Säugetieren. *Revue Suisse de Zoologie*, *46* (fasc. 3, no. 12), 385–390.
1941 (with Rudolf Geigy) Versuch einer morphologischen Ordnung der tierischen Entwicklungsgänge. *Die Naturwissenschaften*, *29*(49), 734–743.
1942 Die Biologie und das neue Menschenbild. *Schriftenreihe der Studentenschaft der Universität Bern*, Heft 2. Bern: Lang.
1945 Die Ontogenese des Menschen als Problem der Evolutionsforschung. *Verhandlungen der Schweizerischen Naturforschenden Gesellschaft*, *125*, Jahresversammlung vom 1. bis 3. September 1945 in Freiburg, 44–53.
1946 Études sur la cérébralisation chez les oiseaux. I. *Alauda*, *14*, 1946, 2–20.
1947 Études sur la cérébralisation chez les oiseaux. II. Les indices intra-cérébraux. *Alauda*, *15*, 1947, 1–15.
1947 Études sur la cérébralisation chez les oiseaux. III. Cérébralisation et mode ontogénétique. *Alauda*, *15*, 1947, 161–171.
1948 Von der Idee des Humanen in der gegenwärtigen Biologie. *Basler Universitätsreden*, Heft 22, 3–25.

1950 numerous chapters in Pierre-P. Grassé (Dir.), *Traité de Zoologie*, tome XV: Oiseaux. Paris: Masson.

1950 Der Zahn als Objekt vergleichend-biologischer Forschung. *Verhandlungen der Naturforschenden Gesellschaft in Basel, 61*, 114–127.

1951 Les bases biologiques d'un nouvel humanisme. In Émile Guyénot, Raymond Ruyer, Adolphe Portmann et Jean Baruzi, *Le Problème de la Vie* (pp. 59–79). Neuchâtel: Editions de la Baconnière.

1952 Die allgemeine biologische Bedeutung der Cerebralisations-Studien. *Bulletin der Schweizerischen Akademie der Medizinischen Wissenschaften, 8*(3), 19, 253–262.

1955 Die postembryonale Entwicklung der Vögel als Evolutionsproblem. *Acta XI Congressus Internationalis Ornithologici*, Basel, 29. V. – 5. VI. 1954 (pp. 138–151). Basel, Stuttgart: Birkhäuser.

1955 La métamorphose "abritée" de Fusus (Gast. Prosobranches). *Revue Suisse de Zoologie, 62*(fasc. suppl., no. 14), 236–252.

1955 Um eine basale Anthropologie. In Gottfried Eisermann (Ed.), *Wirtschaft und Kultursystem: Alexander Rüstow zum siebzigsten Geburtstag* (pp. 286–300). Erlenbach-Zürich and Stuttgart: Rentsch.

1957 Unterwegs zu einem neuen Bild vom Organismus. pp. 24–46 In Jean Gebser (Ed.), *Die Welt in neuer Sicht* (pp. 24–46). München-Planegg: O. W. Barth.

1959 Zur Philosophie des Lebendigen. In Fritz Heinemann (Ed.), *Die Philosophie im XX. Jahrhundert: Eine Enzyklopädische Darstellung ihrer Geschichte, Disziplinen und Aufgaben* (pp. 410–440). Stuttgart: Klett.

1960 Généralités sur les Mollusques. In Pierre-P. Grassé (Dir.), *Traité de Zoologie. tome V, fasc. II. Bryozoaires, Brachiopodes, Chétognathes, Pogonophores, Mollusques* (pp. 1625–1654). Paris: Masson.

1960 Naturforschung und Humanismus. *Basler Universitätsreden*, Heft 42/43, 31–56. [Double issue, together with Karl Jaspers, Wahrheit und Wissenschaft, 3–29].

1960 Biologie und Anthropologie. In Golo Mann (Ed.), *Propyläen Weltgeschichte: Eine Universalgeschichte, Bd. IX: Das zwanzigste Jahrhundert* (pp. 559–594). Berlin, Frankfurt/M., Wien: Propyläen-Verlag bei Ullstein.

1963 Die Vogelfeder als morphologisches Problem. *Verhandlungen der Naturforschenden Gesellschaft in Basel, 74*(1), 106–132.

1963 Die Hoffnung auf die Vervollkommnung des Menschen in der Sicht der heutigen Naturforschung. *Universitas, 18*(9), 921–930.

1964 Dimensions et croissance relatives des Octopodidés méditerranéens. *Vie et Milieu, Supplément, 17*(volume jubilaire dédié à Georges Petit), 213–233

1964 Die Stellung des Menschen in der Natur. pp. 437–460 In Ludwig von Bertalanffy (Founder), *Handbuch der Biologie. Bd IX: Der Mensch und seine Stellung im Naturganzen. Zweiter Teil: Der Mensch und seine Umwelt* (pp. 437–460). Konstanz: Akademische Verlagsgesellschaft Athenaion.

1965 (with Esther Sandmeier) Die Entwicklung von Vorderdarm, Macromeren und Enddarm unter dem Einfluss von Nähreiern bei Buccinum, Murex und Nucella

(Gastropoda Prosobranchia). *Revue Suisse de Zoologie*, 72 (fasc. 1, à la mémoire d'Émile Guyénot, no 10), 187–204.

1965 Über die Evolution der Tragzeit bei Säugetieren. *Revue Suisse de Zoologie*, *72*(fasc. 3, no 31), 658–666.

1966 Die Sprache im Schaffen des Naturforschers. *Jahrbuch 1965 der Deutschen Akademie für Sprache und Dichtung Darmstadt*, (pp. 61–75). Heidelberg and Darmstadt: Lambert Schneider.

1968 (with Pio Fioroni) Zur Morphogenese der Verdauungsorgane und der Larvalorgane von Fusus (Gastropoda, Prosobranchia). *Revue Suisse de Zoologie*, *75*(fasc. 4, no 46), 833–882.

1968 Anthropologische Deutung der menschlichen Entwicklungsperiode. In H. Stutte, H. Harbauer, S. Karger (Eds.), *Concilium Paedopsychiatricum: Verhandlungen 3. Europäischer Kongress Pädopsychiatrie Wiesbaden 1967* (pp. 21–32). Basel, New York: Karger.

1978 Vom Geheimnis des Lebendigen: Adolf Portmann im Gespräch mit Udo Reiter. *Scheidewege*, *8*(1), 1978, 38–50.

Books (selected, Portmann published around 58 books and booklets)

1944 *Biologische Fragmente zu einer Lehre vom Menschen*. Basel: Schwabe (3rd, extended edition 1969; in English: *A Zoologist Looks at Humankind*, New York, Oxford: Columbia University Press 1990).

1946 *Natur und Kultur im Sozialleben: Ein Beitrag der Lebensforschung zu aktuellen Fragen*. Basel: Friedrich Reinhardt.

1948 *Einführung in die vergleichende Morphologie der Wirbeltiere*. Basel: Schwabe (4th, revised and extended edition 1969; in 2013 in a new version: David G. Senn, *Einführung in die Morphologie der Wirbeltiere. Begründet von Adolf Portmann*).

1948 *Die Tiergestalt*. Basel: Friedrich Reinhardt (2nd enlarged edition 1960; in English: *Animal Forms and Patterns*, London: Faber & Faber 1952; New York: Shocken Books 1967).

1953 *Das Tier als soziales Wesen*. Zürich: Rhein-Verlag (in English: *Animals as Social Beings*. London: Hutchinson & Co. Ltd. and The Viking Press, New York 1961; reprinted New York: Harper & Row, Publishers, 1964 – this translation is drastically shortened).

1956 *Biologie und Geist*. Zürich: Rhein-Verlag. [The book includes among others Portmann's first lectures at the ERANOS meetings in Ascona].

1956 *Tarnung im Tierreich*. Berlin; Göttingen; Heidelberg: Springer (In English: *Animal Camouflage*. Ann Abor: The University of Michigan Press, 1959.)

1958 *Meerestiere und ihre Geheimnisse*. Basel: Friedrich Reinhardt.

1960 *Neue Wege der Biologie*. Munich: R. Piper. (In English: *New Paths in Biology*. Joanna Cotler Books, 1963; New York: Harper & Row, 1964.)

1960 *Der Pfeil des Humanen*. Freiburg; Munich: Alber. [Portmann's coping with Teilhard de Chardin].

1965 *Aufbruch der Lebensforschung*. Zürich: Rhein-Verlag. [Other Eranos lectures].

1966 *Kleine Einführung in die Vogelkunde*. Munich: R. Piper.

1967 *Zoologie aus vier Jahrzehnten*. Munich: R. Piper.

1971 *Naturschutz wird Menschenschutz*. Zürich: Arche.

1974 An den Grenzen des Wissens. *Vom Beitrag der Biologie zu einem neuen Weltbild*. Vienna; Düsseldorf: Econ. [With 17 partially colored drawings by the author].

Works on Portmann

Boletzky, Sigurd von (2006). Adolf Portmann, 1897–1982. In David G. Senn (Ed.), *Adolf Portmann. Lebensforschung und Tiergestalt. Ausgewählte Texte* (pp. 11–20). Basel: Schwabe. [Dr. von Boletzky was one of Portmann's PhD students who continued in his marine biology works in Banyuls].

Grene, Marjorie (1974). The biological philosophy of Adolf Portmann. In M. Grene, *The Understanding of Nature: Essays in the Philosophy of Biology* (254– 293). Dordrecht: Boston.

Illies, Joachim (1975). *Adolf Portmann, Jean Gebser, Johann Jakob Bachofen: Drei Kulturforscher – Drei Bilder vom Menschen*. Zürich: Interfrom. [The chapter about Portmann on pp. 16–40].

Illies, Joachim (1976). *Das Geheimnis des Lebendigen: Leben und Werk des Biologen Adolf Portmann*. Munich: Kindler. [The only authentic biography composed from numerous interviews with Portmann as well as some of his students].

Jaspers, Karl (1957). *Zum 60. Geburtstag von Professor Adolf Portmann*. Ex Libris: Zürich.

Koepke, Ewald (1964). *Adolf Portmann, Wegbereiter zu einem neuen Weltbild: Eine naturphilosophische Studie*. Hamburg: Hamburger Kulturverlag.

Krajewski, Viktor (1969). *Heidegger und Portmann, Denker in „dürftiger Zeit"*. Wien (Dissertation).

Kugler, Rolf (1967). *Philosophische Aspekte der Biologie Adolf Portmanns*. Zürich: EVZ-Verlag.

Müller, Helmut (1988). *Philosophische Grundlagen der Anthropologie Adolf Portmanns*. Weinheim: VCA, Acta humaniora.

Ritter, Markus (2000). Die Biologie Adolf Portmanns in zeitgeschichtlichem Kontext. *Basler Zeitschrift für Geschichte und Altertumskunde*, 100, 207–254.

UNI NOVA. Wissenschaftsmagazin der Universität Basel. Issue 79/80, May 1997. [Contains exclusive contributions by Portmann's students on the aspects of Portmann's research].

Wahlert, Gerd von (1972). *Adolf Portmann: Versuch einer Würdigung*. Basel: Friedrich Reinhardt.

References

Brown, C., Laland, K., & Krause, J. (Eds.) (2011). *Fish cognition and behavior*. Chichester: Wiley (1st ed.: 2006).

Geigy, R., & Portmann, A. (1941). Versuch einer morphologischen Ordnung der tierischen Entwicklungsgänge. *Die Naturwissenschaften, 29*(49), 734–743.

Hess, J. (1996). *Menschenaffen Mutter und Kind*. Basel: Reinhardt.

Huxley, J. (1942). *Evolution: The modern synthesis*. London: George Allen & Unwin.

Kühn, A. (1955). *Grundriß der allgemeinen Zoologie*. Wiesbaden: Thieme. (25th edition by Rüdiger Wehner and Walter Gehring in 2013; now called *Zoologie*).

Mayr, E. (1963). *Animal species and evolution*. Cambridge, MA: Belknap of Harvard University Press.

Portmann, A. (1921). *Die Odonaten der Umgebung von Basel. Ein Beitrag zur biologischen Systematik der mitteleuropäischen Libellen* (Diss.). Lörrach.

Portmann, A. (1926). Die Kriechbewegung von Aiptasia carnea: Ein Beitrag zur Kenntnis der neuromuskularen Organisation der Aktinien. *Zeitschrift für Vergleichende Physiologie, 4*(5), 659–667.

Portmann, A. (1935). Die Ontogenese der Vögel als Evolutionsproblem. *Acta Biotheoretica, Series A, 1*(1–2), 59–90.

Portmann, A. (1936). Die Ontogenese der Vögel als Evolutionsproblem. *Verhandlungen der Schweizerischen Naturforschenden Gesellschaft*, 117. Jahresversammlung vom 28. bis 30. August 1936 in Solothurn, 224–241.

Portmann, A. (1939). Nesthocker und Nestflüchter als Entwicklungszustände von verschiedener Wertigkeit bei Vögeln und Säugetieren. *Revue Suisse de Zoologie, 46*(fasc. 3, no. 12), 385–390.

Portmann, A. (1944). *Biologische Fragmente zu einer Lehre vom Menschen*. Basel: Schwabe.

Portmann, A. (1948). *Die Tiergestalt*. Basel: Friedrich Reinhardt.

Portmann, A. (1948a). *Einführung in die vergleichende Morphologie der Wirbeltiere*. Basel: Schwabe.

Portmann, A. (1953). *Das Tier als soziales Wesen*. Zürich: Rhein-Verlag.

Portmann, A. (1956). *Tarnung im Tierreich*. Berlin/Göttingen/Heidelberg: Springer.

Portmann, A. (1956a). *Biologie und Geist*. Zürich: Rhein-Verlag.

Portmann, A. (1956b). *Zoologie und das neue Bild des Menschen. Biologische Fragmente zu einer Lehre vom Menschen*. Hamburg: Rowohlt.

Portmann, A. (1965). *Aufbruch der Lebensforschung*. Zürich: Rhein-Verlag.

Portmann, A. (1969). *Biologische Fragmente zu einer Lehre vom Menschen. Dritte, erweiterte Auflage*. Basel/Stuttgart: Schwabe.

Portmann, A. (1971). *Naturschutz wird Menschenschutz*. Zürich: Arche.

Range, F. (2009). *Wie denken Tiere?* Wien: Carl Ueberreuter.

Scheler, M. (1928). *Die Stellung des Menschen im Kosmos*. Darmstadt: Reichl.

Stamm, R. A. (1986). Ethologische Gestaltforschung: Paarungsnachspiele bei Papageien und Möwen (Aves: Psittacidae, Laridae). In G. Tembrock, R. Siegmund, & M. Nichelmann (Eds.), *Verhaltensbiologie* (Internationales symposium 1983) (pp. 125–131). Jena: Fischer.

Stamm, R. A., & Fioroni, P. (1984). Adolf Portmann, Rückblick auf seine Forschungen. *Verhandlungen der Naturforschenden Gesellschaft Basel, 94*, 87–120.

Tinbergen, N., Kruuk, H., Paillette, M., & Stamm, R. A. (1962). How do Black-headed Gulls distinguish between eggs and egg-shells? *British Birds, 55*(3), 120–129.

Roger Alfred Stamm Roger Alfred Stamm studied zoology, botany, philosophy, and psychology at the University of Basel, graduating in 1960. Between 1961 and 1965 he held several fellowships at Oxford, where he was a member of the Nikolaas Tinbergen research group. His research on behavioral ecology there focused on mating in the black-headed gull (*Larus ridibundus*). From 1962 to 1963 he was an assistant for Prof. Bernhard Hassenstein (animal physiology, cybernetics) in Freiburg/Breisgau. He completed his habilitation in zoology at Basel in 1966 and in 1974 became professor ordinarius of biology and didactics of biology at the Pedagogical College in Niedersachsen, Lüneburg (today: Leuphana University of Lüneburg). He also taught comparative psychology at the University of Heidelberg and social-cultural anthropology at Lüneburg. He has been the research administrator of the Adolf Portmann Archive since 1983 and became professor emeritus in 2000.

Books: Stamm, Roger Alfred & Zeier, Hans (Eds.): *Lorenz und die Folgen*. Zürich: Kindler 1978; Stamm, R. A. (Ed.): *Tierpsychologie* (part from *Lorenz und die Folgen*), Weinheim – Basel: Beltz 1984. Articles about Portmann: Stamm, Roger Alfred & Fioroni, Pio, "Adolf Portmann, ein Rückblick auf seine Forschungen." *Verhandlungen der Naturforschenden Gesellschaft Basel* 94 (1984); Stamm, R. A., "L'intériorité, dimension fondamentale de la vie," *Revue européenne des sciences sociales* 37 (1999); Stamm, R. A., "Nachwort", In A. Portmann, *Biologie und Geist*, Göttingen: Burgdorf (edition Nereïde) 2000; Stamm, R. A.: "Adolf Portmann, akademischer Lehrer und Forscher", In Georg Kreis (Ed.): *Zeitbedingtheit, Zeitbeständigkeit: Professoren-Persönlichkeiten der Universität Basel*. Basel: Schwabe 2002.

Adolf Portmann (in the middle), Prof. Friedrich Zschokke (on the right), approx. 1920

Adolf Portmann (the second one standing from the right), Ramsach, October 1929

Chapter 5
Novel Ecological Mismatches in the Light of Jakob von Uexküll's and Adolf Portmann's Works

Riin Magnus

Abstract The twentieth-century biologists Jakob von Uexküll and Adolf Portmann both discussed the role of stable cognitive and perceptual schemes in guiding and grounding the perception and behaviour of animals. Given the speed of current environmental change, it is rather common that inherited ancestral cognitive structures and recognition templates are not adequate for the environment an organism is born into. Hence perceptual structures that in stable environments have allowed for rapid recognition and behavioural responses might lead the animal astray and result in a novel ecological mismatch between the organism and the environment. This paper will demonstrate the role of such mismatches in recent species decline and animal welfare problems and discuss the relevance of Uexküll's and Portmann's theories for the corresponding research.

Introduction

These days, human-induced rapid environmental change causes biodiversity loss mainly through land-use changes, pollution, unsustainable natural resource use, climate change and invasion of alien species (Slingenberg et al. 2009). All the abovementioned factors are bound with some direct external impact on the survival or well-being of organisms and species. However, less attention has been paid to perceptual and cognitive mechanisms that underlie extinction, although research fields such as cognitive ecology have targeted the problem also from this angle. In the following chapter, discrepancies between the cognitive and perceptual structures of the organism and the conditions of novel environments will be highlighted as specific types of novel ecological mismatches and as significant causes of the decline of the well-being and persistence of different organisms.

R. Magnus (✉)
University of Tartu, Tartu, Estonia

University of Hradec Králové, Hradec Králové, Czech Republic
e-mail: riin.magnus@ut.ee

F. Jaroš, J. Klouda (eds.), *Adolf Portmann*, Biosemiotics 23,
https://doi.org/10.1007/978-3-030-67810-4_5

The retention and maintenance of certain perceptual and cognitive structures was a topic that was also of interest for the twentieth century biologists Jakob von Uexküll (1864–1944) and Adolf Portmann (1897–1982). Therefore, their works allow us to highlight the importance of the semiosic matching between the perceptual schemes and the characteristics of the environment as the ultimate precondition for the persistence of organisms and species. Discussing various origins and functions of the perceptual and cognitive schemes of organisms, Uexküll and Portmann observe how such schemes allow for the development of a particular sign-mediated correspondence between the organism and the environment. The abrupt cutting off or transformation of such a link, and the turning of formerly functional perceptual mechanisms into rudimentary ones, can be considered an instance of semiocide, defined as the destruction of signs relevant for someone else due to malevolence or carelessness (Puura 2013; cf. Maran 2013). In the following, we will investigate in more detail grounds on which the works of Uexküll and Portmann can be reconciled; what their contribution to the discussion of perceptual and cognitive schemes as the mediators between the organism and the environment was; and whether and how their reflections can be connected with current discussions revolving around the causes of human-induced rapid environmental change.

Adolf Portmann and Jakob von Uexküll: Theoretical Alignments and Influences

Jakob von Uexküll and Adolf Portmann met in person at least once – during Uexküll's trip to Switzerland (Uexküll 1964: 161).[1] In addition, there are a number of reasons to juxtapose their works and theoretical approaches. Thirty-three years Jakob von Uexküll's junior, Adolf Portmann published his first books when Uexküll was finishing his last publications. By specialisation Uexküll was a physiologist of marine invertebrates, and his animal behaviour studies were first guided by an interest in the neurophysiological mechanisms of animal behaviour (Uexküll 1894, 1905). Portmann, on the other hand, laid his behavioural studies on morphological, ontogenetic, and evolutionary grounds. Besides empirical observations and studies, both scholars exhibited a significant interest in the theoretical aspects of life sciences, Uexküll authoring one of the first monographs on theoretical biology (Uexküll 1920) and Portmann contributing to the development of philosophical zoology and aesthetics of biological forms. In the most general terms, a reluctance to accept reductionist tendencies in life sciences; an interest in the relations between biological forms and biological interactions; focus on the formation and functioning of meaningful structures in the biological world; and the appreciation of the legacy

[1] There is no indication of correspondence between the two in the archive of Jakob von Uexküll at the University of Tartu. Similarly, no letters were found in Nachlass Adolf Portmann at the University of Basel (Filip Jaroš, personal communication).

of German Romanticism as a potential philosophical framework for biology are shared by both scholars. Their theoretical stances and research interests have been previously compared in connection with the visual communication of animals (Kleisner and Maran 2014), the self-experience of animals (Kleisner and Markoš 2009) and semiotic mechanisms of evolution (Maran and Kleisner 2010).

One can find references to Uexküll's works in several publications by Portmann, although the same does not apply the other way round. In his book *On the Borders of Knowledge: About the Contribution of Biology to a New Worldview* (An den Grenzen des Wissens: Von Beitrag der Biologie zu einer neuer Weltbild), Portmann discusses the influence of Uexküll's writings on the development of his own ideas, mentioning particularly the 1921 edition of Uexküll's book *Umwelt and The Inner World of Animals* (Umwelt und Innenwelt der Tiere) (Portmann 1974: 19). Portmann appreciated Uexküll for introducing natural scientific means for analysing the inwardness (*Innerlichkeit*) of animals and also highlighted his contribution to revealing the diversity of non-human sensory worlds. Portmann wrote the foreword to the 1956 edition of Uexküll's book *Strolls Through the Worlds of Animals and Men: A Book of Invisible Worlds; A Theory of Meaning* (Streifzüge durch die Umwelten von Tieren und Menschen: Ein Bilderbuch unsichtbarer Welten; Bedeutungslehre), where he designates Uexküll's biotheoretical research with the notable phrase of "new biology" (Portmann 1956). In his foreword, Portmann highlights Uexküll's contribution to the recognition of the subject's inwardness as a true biological research object (Portmann 1956: 14), an acknowledgment which is strongly tied to the theoretical focus of Portmann's own studies. The way Portmann defines inwardness is very close to how Uexküll defined umwelt as constitutive of living organisms: "The characteristic proper of biological entities is that they are systems marked by 'inwardness' (*Innerlichkeit*), that is, a capacity to interpret meanings in order to reach goals" (Portmann 1990: 457). Umwelt, for Jakob von Uexküll, was the subjective world of the animal, constituted by the effector world (*Wirkwelt*) and perceptual world (*Merkwelt*), each accessed via the use of signs (Uexküll 1992).

Uexküll was reluctant to accept the Darwinian paradigm of evolution, while Portmann directs his criticism more towards certain neo-Darwinian developments of Darwin's ideas, though both of them particularly oppose the view that the maintenance of life takes place via a constant "fight for survival", a term which Portmann considered to be the only idea of Darwin carried over to neo-Darwinism (Portmann 1960: 200–201, 1965: 91–92). As an alternative, Portmann refers to the ideas of Peter Kropotkin about the prevalence of biological relations that are based on mutual aid (Portmann 1987: 116, 1965: 92–93). Uexküll is rather looking to the works of Romanticist authors (Goethe) to make a claim about the harmonious and plan-based existence of life and living beings (Uexküll 1982). Borrowing metaphors from music, Uexküll develops in his later works an idea about the existence of different organisms in respect to one another and the environment as counterpoint-like phenomena (Uexküll 1982: 63–68; see also Brentari 2015: 167–168).

Both Uexküll and Portmann lay much stress on the ontogenetic processes of living beings as the ultimate proof of the autonomy of living beings and as a cause of the variety of life forms, although their specific reasons are somewhat different. For Uexküll, ontogenetics grounds the umwelt change of one and the same organism in its different stages of life. In the developmental pathways of organisms, Uexküll also finds proof for self-organisation and regeneration that is specific to living beings only and which therefore cannot be covered in mechanistic terms (Uexküll 1982: 40–41). Uexküll found further support for his ideas about the plan-like development of organisms from the experiments of his contemporary Hans Driesch on sea urchin blastomers and of Hans Spemann on the interspecific transplantation of tissues (Uexküll 1973: 314–317, 1937). Portmann's interest in ontogenetic changes was not so much tied to an interest in changes in the perceptual world; he rather explored the role of ontogenetic processes in triggering phylogenetic differentiation (hence the precursory ideas of evo-devo; see also Kurismaa, Chap. 6, in this volume).

Although there appear to be no real controversies between the theoretical premises of Portmann and Uexküll, the focus of interest is partly different in their works. Portmann lays more stress on interspecies communication and the development of biological forms, which allow for the establishment of certain relations between organisms, while for Uexküll the focus stays on the perception and action of an individual organism. Via umwelt analysis, Uexküll maps environmental objects that the organism attends to via the use of signs, organising them under four basic functions or categories of meaning: partner, enemy, food and medium (Uexküll 1992). Portmann, on the other hand, considers an additional function – self-presentation *(Darstellungsfunktion)* – as "the essential characteristic of the living" (Portmann 2000a: 261) and as an integral function of any communicative or expressive organ. Such an inward-directed argumentation for the development of organismic traits and a reference to the "inner self" as a ground of reactions are missing in the work of Uexküll, who rather defines the organism through the meaningful relations it has established.

In line with philosophical anthropology, Portmann was interested in the ontogenetic grounds of human culture, the specificity of humans as biological beings and as cultural beings (e.g. Portmann 1957, 1969, 1970). Despite the fact that humans are not excluded from the umwelt analysis of Uexküll, and despite his detailed account of the spatial and temporal characteristics of the human umwelt (Uexküll 1992, 1920), he does not discuss human culture *per se* and does not argue for a qualitative difference between the human umwelt and the umwelten of other species. Yet, Uexküll's ideas have been considered to be of the strongest influence on the German philosophical anthropology of the twentieth century – e.g. that of Max Scheler and Arnold Gehlen, who referred to his ideas to demonstrate the special place (*Sonderstellung*) of human perception of the world in comparison with other animals (Brentari 2015: 176; Magnus and Kull 2012: 655–656). In line with the claims of the above-mentioned authors, Portmann maintains that animals are confined to their immediate environment (*Umweltbindung*) and hence lack the choices, freedom, and the consequent openness to the world proper to humans, who are endowed with language, imagination and memory (Portmann 1974: 102–104; see also Jaroš, Chap. 7 in this volume). In line with such argumentation, he refuses to overextend the umwelt concept to explanations of human-specific ways of life

(Portmann 1956: 11). Such a contrast of human and animal ways of relating to the world has led to a deterministic interpretation of Uexküll's umwelt idea – something that later biosemiotic interpretations of Uexküll's works have tried to invert by laying stress on the semiotic basis of umwelt formation.

Persistence and Change of Cognitive and Perceptual Schemes in the Works of Portmann and Uexküll

In the works of Jakob von Uexküll, the body of the organism provides a stable ground for perception and action, undergoing relatively little change once it has reached its adult dimensions. He named such an adult stage of the organism the "mechanical phase", distinguishing it from the phase of development, the "technical phase", when the organs are as yet non-functional and hence the developing organism has to rely on another organism and its umwelt (Uexküll 1973: 87–90). Seemingly, the processes of cognition and perception provide flexibility to an organism that is not susceptible to fast morphological change in its adult form. However, cognitive and perceptual processes also uphold a certain stability of forms that help to keep the organism functioning. These stable structures of cognition and perception were of interest to both Jakob von Uexküll and Adolf Portmann. In the following sections we will discuss in more detail how the two authors explained their origins and functions.

In his descriptions of the composition of the umwelt and the inner world (*Innenwelt*), Uexküll distinguished between various factors constitutive of perception and action. The first distinction that Uexküll makes concerns the cues that guide recognition and those that guide motoric activities, respectively called "perceptual cues" and "effector cues" (Uexküll 1992: 322–323). He also provides another type of differentiation between signs that mediate the interactions of the organism and the environment: the signs of content (*Inhaltszeichen*) and signs of direction (*Richtungszeichen*). But it is not through individual signs that more complex animals, at least, really orient themselves in the world. Following Immanuel Kant (and implicitly also Gestalt theory), Uexküll maintains that a "scheme" (*Schema*) is responsible for the shaping of things, as well as for the sketching of images in the imagination (Uexküll 1973: 118). Writing about organisms with more complex nervous systems, Uexküll maintains:

> In the counter-world (*Gegenwelt*), the things of the umwelt are mediated by schemes, which in accordance with the organisational plan of the organism, can be very general and can sum up many types of things. But the schemes can be also very exclusive and can focus on very specific things. The schemes are not a product of an umwelt but are particular tools of the brain, given through the organisational plan, and are always ready to get into action given the stimuli of the outer world. (Uexküll 1980: 236, translation R.M.)

Uexküll further distinguishes between time schemes and spatial schemes (Uexküll 1910). Time schemes are bound with the function of things (a projection of "what can be done and in which sequence"); spatial schemes can be more general or specific depending on the number and kinds of objects covered (Uexküll 1910).

Uexküll does not delve deep into the mechanism and conditions of the preservation or change of the schemes or the perceptual and effector cues of a species, but rather gives anecdotal evidence of how the schemes can guide and shape what is being perceived. When writing about the mechanisms of preservation and maintenance of certain signs and schemes of an organism, Uexküll refers to the morphology and experience of organisms as well as to something which in modern scientific parlance would be designated as evolutionary causes. The persistence of the schemes and signs can first of all depend on the individual experiences of the organism – e.g. the previously perceived shape of an item makes the organism look for similarly shaped objects when the same need is raised. Uexküll uses the term "search image" to denote such an attachment of certain vital functions to experience-bound forms. He brings as an example a toad who, after having been hungry for some time, captures an earthworm; the shape of the latter will serve as a perceptual mould in the search for further food items, so that any item with an elongated, narrow shape will be approached as a food item (Uexküll 1992: 375). Along with the notion of "search image", Uexküll introduces the concept of "search tones" (synonymous with "functional tones"), which guide the perception of animals by making perceptible those objects that serve certain functions for the animal (Uexküll 1992: 375). Despite the centrality of the function of objects, their form is specified by the building plan (*Bauplan*) or the morphological build of the organism – hence, for example, a dog would endow with a "sitting tone" some items that humans would not (e.g. shelves or chairs that have been turned upside down), given the differences in their size and other morphological features (Uexküll 1932). Besides the experiential wiring and imprinting of certain perceptual forms for future use, Uexküll also mentions innate mechanisms which he calls "magical" and which also "make" animals attend to certain signs (Uexküll 1992: 381). He does not specify the biological processes which could ground them, and as a variety of phenomena fall under this category (from the migration restlessness of captive birds to the specific choice of food item by some insect larvae), it is also retrospectively hard to pinpoint some specific biological mechanism behind the "magical". In addition, Uexküll introduces a third type of structure that, unlike the experiential and innate mechanisms that preserve certain perceptual forms and signs, do not connect the organism and environment via direct perception. Uexküll calls these kinds of structures "archetypes", bringing as an example a spider web, which is spun by the spider, who does not know about the visual characteristics of flies. Yet, the web is spun in such a way that it remains invisible to the prey. Various archetypes are in turn united by a comprehensive meaning plan that binds different umwelten into one totality (Uexküll 1982: 43).

Uexküll has been criticised for the rigidity and stability of organism and environment relations that his theory presumes (e.g. Tønnessen 2009). Hence, at first glance his works seem to offer little if the focus is on the transformations and change of biological relations and the mismatch of the characteristics of organisms and environmental conditions. Yet, at a closer look, one can discover descriptions of umwelt change in Uexküll's works where the dynamics of an umwelt are exposed (cf.

Linask et al. 2015), although they are all instances of change falling within the limits of "normality". The umwelt change of an organism is first of all related to the development of an organism and the corresponding changes in morphology –the umwelt of a child would differ from that of an adult, just like the umwelt of a holometabolic insect would differ in its larval stage and adult stages. The same objects may also carry different meanings depending on the moods (*Stimmung*) of the animal, which are tied to the most prominent needs at a particular moment (Uexküll 1973: 195–197). For example, the same crab may use a sea anemone as a shelter or as food, depending on whether the need for protection or for food is more prominent (Uexküll 1992: 357). Moods are in turn bound to changes in the physiological conditions of the organism, which determine the rise or fall of the threshold of irritation for certain stimuli.[2] Moreover, the animal can be taught to make use of different "modes" or "registers" of meanings, depending on the role it is currently carrying out (see also Magnus 2016). Such a recognition served as a basis for the training of guide dogs at the Institute of Umwelt Research (*Institut für Umweltforschung*) in Hamburg, of which Uexküll was director. A working guide dog is attached to the umwelt of a human and is supposed to attend to meanings relevant for humans, whereas when not guiding the handler, a guide dog can freely attend to objects relevant to the canine umwelt only (Uexküll and Sarris 1931).

Although Uexküll did not discuss the possibility of "deficient umwelten", one can imagine occasions where umwelt change not only brings about a different set of signs and meanings but results in the mismatch between the "demands" of the environment and the evolutionarily, ontogenetically or experientially shaped capacities of the organism. In order to analyse such situations, one might juxtapose umwelt with semiotic niche – a term proposed by the Danish biosemiotician Jesper Hoffmeyer, who refers by this to "the totality of signs or cues in the surroundings of an organism that it must be able to meaningfully interpret to ensure its survival and welfare" (Hoffmeyer 2008: 185). In cases where the organism is not capable of responding to the cues of the environment adequately, one can detect a mismatch between the umwelt and the semiotic niche. The latter can result in various consequences for the organism, depending on whether it is capable of making use of alternative cues and learning new ones.

If Uexküll's focus was mainly on the ways animals structure and give meaning to the world, then Adolf Portmann's interest lies more in the ways inwardness is exposed and mediated by the expressions and outlook of the organism. The array of forms and features that could relate the organism to the environment is in principle indeterminate, but determinacy is introduced by the specifics of the inwardness, which serves as an ultimate ground for the expressions (both morphological features and behaviour). Looking for parallels in Uexküll's works, one could point to the concept of "ego quality" (*Ich-ton*) (Uexküll 1982: 34–35), which Uexküll developed in line with Johannes Müller's ideas concerning specific energies of cells.

[2] Portmann also wrote about moods as organismic states that determine attendance to the environment and shifting normalities: "A mood is not the subjective disturbance or distortion of a normal, neutral state, but the basis of a changing 'normality'" (Portmann 1961: 125).

Relying on Müller, Uexküll mainly used the term to point to the specific reaction types of cells of a multicellular organism – cells that belong to the same tissue always respond with a specific reaction irrespective of the type of irritation (e.g. retinal cells transmit the same kind of sensation of light, whether the stimulus be pressure or light) (Uexküll 1931). Although the presence of "ego qualities" serves as proof of the autonomous character of organic units, it is rather the concept of the inner world (*Innenwelt*), with reference to the inner world of the organism in its totality, that bears closer resemblance to Portmann's inwardness. Yet, unlike Portmann, Uexküll does not consider this to be primary to the expressions of the organism, as the inner world and umwelt constitute two sides of the same coin of the subject's existence. The inner world is the intra-organismic counterpart of umwelt, constituted by the impressions and perceptual qualia that are projected on the world as properties of objects. In the case of Portmann, inwardness is coupled with, and exposed by, self-expressions, i.e. certain (potentially) perceptible characteristics of the organism itself. Some of those forms of self-expression imply a particular receiver, while others only a potential receiver. Portmann distinguishes between three major types of expressions: spontaneous expression (*spontane Ausdruck*), temporary expression(organs) (*temporäre Ausdrucksorgane*), and constant expressions (*konstante Ausdruck*) (Portmann 2000b: 21–22). These expressions do not serve the purpose of survival value only but contribute to the way of being of this particular animal. Yet, this does not exclude the possibility that in some occasions they can be detrimental to the existence of the organism. If Uexküll's works turn attention to occasions when "wrong" things are perceived and acted upon, then Portmann's works might supplement this with a focus on occasions when the expressions and means of communication do not make sense in a particular environment. Besides the coupling of the inwardness of the organism with certain expressions, Portmann also writes about certain persistent perceptual and cognitive structures that ground the individual's behaviour. First, he discusses archetypal images (*Urbilder*) or archetypes as potential inherited guides of the behaviour of animals. Along Jungian lines, Portmann adopted the definition of archetype as "the copy of 'a hidden psychic structure which is innately ready and without the help of consciousness releases a behaviour that is a psychic necessity'" (Portmann 1961: 157–158). Although Portmann demonstrates a certain interest in the application of the term in the biological domain, he remains sceptical about its real use for biological research (Portmann 2000c: 127–128; see also Vajdová, Chap. 13, in this volume). The two parts of the "archetype": the inherited origin and the attendance to certain constellations of signals – are discussed via other terms – "Gestalten" and "innate releasers" (in line with Konrad Lorenz), which Portmann more willingly uses in the biological context. Portmann describes Gestalten as certain stable and complex perceptual images that "need more than the sum of stimuli jumbled together; the stimuli have to be arranged in a particular way" (Portmann 1961: 138). As for innate releasers, he notes that the "idea of the releaser has indeed been both fruitful and stimulating, but it has also helped to show the limits of its validity and provide a warning against over-simplification of the problem" (Portmann 1961: 152). The warning against "over-simplification" can be understood from two different angles. On the one hand, it is related to the over-simplified assumption about the evolutionary fixation of cognitive expectations and the stimuli that convert those expectations into actions. On the other hand, it is as if offering the "fake hope" that the character and strength of stimuli which induce certain behaviours can be exactly determined and predicted, while in reality they appear to exhibit significant variability, and hence it is rather difficult to determine the exact scope of a particular stimulus. In this connection, Portmann also discusses supra-normal stimuli, reaching the conclusion that: "We meet the supra-normal in other cases too, which warn us not to consider the normal shapes and colours of effective Gestalten as being all that indispensable" (Portmann 1961: 151). This introduces a certain indeterminacy to the perception and behaviour of the animal and complicates attempts to predict behaviours – the researcher is never able to tell the full array of possible stimuli without the animal really facing them. The latter, in turn, especially complicates the prediction of animal behaviour in novel environments. The unpredictability that is characteristic of human-induced rapid environmental change means that animals have to face conditions and stimuli that they have never encountered in their evolutionary or personal history and that have therefore never been registered before.

In conclusion, in the works of both authors, one can distinguish between two primary domains of stability of organisms – the first relates to the preservation of the organism as a whole (inner world and umwelt [*Innenwelt-Umwelt*], coupled by Uexküll, and inwardness-expression [*Innerlichkeit-Darstellung*] by Portmann), whereas the second refers to the stability of certain perceptual schemes (e.g. search images, archetypal images). Usually, the latter contribute to the maintenance of the organism as a whole, but as will be discussed in the next section, there are conditions in which the persistence of certain perceptual schemes may become unfavourable to the existence of the organism as such.

Human-Induced Rapid Environmental Change and Responses to Novel Cues

In the face of current human-induced rapid environmental change (HIREC), ecologists and behavioural scientists turn more and more attention to the discrepancies between evolutionarily or ontogenetically formed recognition schemes and novel environmental cues that have not existed in the habitat of the species before. For conservation biology, the major question is how sustainable certain established matches between environmental cues and cognitive templates are and whether they are flexible enough to meet the expected environmental change. Those questions are further framed by an interest in predicting whether certain species are able to adapt to current changes and by attempts to take existing perceptual and cognitive matches into account when developing conservation measures.

In theoretical terms, modern (cognitive) biology entertains different pattern recognition theories to explain how cognitive and perceptual templates are formed and applied (Enquist and Arak 1998: 41–42; Hayes 2000: 38–49). What unites them, however, is a presumption about the presence of recognition templates that organisms project on the environment. For example, in the frame of behavioural ecology it has been proposed that recognition works through the application of a recognition template and a matching algorithm, with the recognition template being either learned or inherited (Sherman et al. 2008). The template is seen as being formed in connection with a referent that serves as a model for future perception and cognition. The timing of the template learning is either restricted to a brief period in ontogeny or extended throughout the lifetime, depending on the variability of the cues the organism can possibly encounter during its life. However, whether all previous perceptual images serve as templates, whether a prototype is formed as an average of multiple perceptions or whether only certain features of objects serve as a basis of comparison with features stored in memory are still matters of debate (Hayes 2000). Moreover, one should also consider differences between species before opting for one or the other explanation of the development and functioning of recognition schemes.

Luckily, the diversity of theoretical approaches has not obstructed behavioural scientists and cognitive ecologists in conducting empirical research on the effects of environmental change on the perception and cognition of animals. Yet, these studies also face their own challenges. The necessity of taking into account the kind of environmental change as well as the cognitive rules used by the species is thereby stressed (Trimmer et al. 2017; Sih 2013). So far, one of the biggest challenges in studying the presence and extent of mismatches between recognition templates and environmental cues is related to differences in the decision-making behaviour of organisms under *in situ* and *ex situ* conditions. In a natural habitat, decision making is always contextual, and the same cue might induce different assessment and responses depending on the adjacent signs (Owen et al. 2017). For example, a cue which is recognised as referring to a predator may not induce an escape behaviour in a prey animal if the flight distance is still too big, if the condition of the predator is not assessed to be dangerous, etc. In natural conditions, animals rely on multiple cues, some of which might have a stronger effect than others, but so far this has not been sufficiently taken into account (Sih et al. 2016; but see Fawcett and Johnstone 2003). On the other hand, even if the causes of perceptual mismatches were to be found out and their removal undertaken as a part of conservation measures, this does not exclude the possibility that as a result the animal finds itself in an ecological trap, i.e. in a situation where some crucial resources are missing.

Handling perceptual mismatches caused by HIREC is one of the hardest tasks for conservation biology. First of all, it demands very good knowledge of the animal's sensory physiology – about the cues that are used for decision making as well as the response thresholds in different physiological conditions. In addition, it requires an ability to predict whether the animal is capable of modifying the existing recognition templates and adapting to novel objects and the cues that indicate their meaning. Predictions can also be made – by exploring which organisms could potentially cope in certain environments, given the cues available there. From a biosemiotic stance, Timothy Mullet, Almo Farina and Stewart H. Gage have proposed the

Acoustic Habitat Hypothesis, which allows one to predict (via the use of different soundscape indices) whether a certain acoustically oriented species will inhabit an environment or not, given the acoustic characteristics of the environment (Mullet et al. 2017). Their hypothesis also allows one to predict which species will avoid certain environments because of the acoustic signals indicating unfavourable conditions.

The mismatch between the expectations of the organism and novel cues may concern interspecies interactions as well as an organism's relations with the non-living environment. There are multiple examples of environmental cues that a species has made use of in its evolutionary history but that have become deceitful in the conditions of HIREC. The transformation of habitats entails an introduction of objects and cues that have never existed in the umwelt of the species before – this might entail the introduction of novel cues or novel objects that are mediated by the old cues, the transformation and inaccessibility of biologically necessary cues, or the lack of historically present cues. The case of buprestid beetles (*Julodimorpha bakewelli*) and "mimicry of materials" in Western Australia serves as an example of familiar cues tied to a novel object. It was observed that male buprestid beetles commonly mate with beer bottles, as the latter have a shiny, brown colour and a texture similar to the elytra of the beetles (Gwynne and Rentz 1983). Similarly, asphalt roads that reflect polarised light allure mayflies (*Ephemeroptera*) to lay eggs en masse, as they mimic polarised water surfaces (Kriska et al. 1998). When looking at the anthropogenic mortality of birds, most causes of death are in fact related to an inability to respond adequately to some novel cues in the environment (from wind turbines to power lines and car collisions, cf. Loss et al. 2015). It is well known that light pollution is another source of disorientation for various groups of animals, from insects to birds to reptiles. The hatchlings of turtles are affected by artificial light as they use phototaxis to move towards oceans, and recently it has been confirmed that they are affected by artificial light also when reaching the water (Cruz et al. 2018). Artificial light is known to obstruct the orientation of birds, but it also affects the timing of the singing of songbirds (Da Silva et al. 2014). In addition to the presence of novel cues or the deceptive character of old cues, an animal's survival and well-being may also be affected by the lack of certain cues that in the evolutionary or ontogenetic history of the organism have indicated the presence of certain resources. Examples of the lack of significant cues can be found in the frequent collision of birds with windows, which they fail to recognise as barriers.[3] The categories of birds most vulnerable to collisions include transitory groups and migrants (Arnold and Zink 2011) as well as birds with insectivorous feeding habits (with vertical location of foraging) (Wittig et al. 2017).

Efficiency of coping with novel cues has been associated with various factors, while previous interactions between the organism and the environment (including other species) appear to be of central importance. The first factor for the ease of

[3] Reflective glass may also induce aggressive territorial behavior among birds who take their own reflection in the window for a rival (Klem 1989).

recognising a novel cue is the time a certain cue has not been present in the environment of the species (millions of years vs. a few generations, see Sih 2013). The similarity of a novel object/subject to previously encountered ones is another factor that influences the efficiency of recognition. Whether the species is a specialist or a generalist in its food and habitat selection also influences how prone the species is to accept a novel object. In addition, the specificity of cue templates (e.g. whether the animal uses species-specific odour cues of a predator or cues specific to all carnivores) affects the ease of recognising the novel object and its meaning. Several of the previously mentioned characteristics, in turn, amount to the response threshold of the animal, i.e. how strong a signal has to be in order to induce a response. Characteristics of the species such as phenotypic plasticity and neophobia, which influence the responsiveness and the adequacy of the response in novel situations (van Baaren and Candolin 2018; Sih 2013), are themselves an outcome of the previous interactions with the environment.

Ecological naïveté – the inability to recognise and/or respond adequately to other species with which the animal has never shared the habitat before – is a separate research topic within novel ecological mismatch studies (Verhoeven et al. 2009). Ecological naïveté has been, above all, studied in alien predator and local prey interactions, although in principle this may concern any (antagonistic) interactions (cf. Carthey and Banks 2014: 934). As the biologists Banks and Dickman have pointed out, ecological naïveté is not tied to perceptual mismatch only, although this is often most relevant in the very early phases of the introduction of an alien predator. In developing a multi-level framework of naïveté, they have added to the failure to recognize the novel predator the failure to respond and failure to respond effectively (Banks and Dickman 2007). When writing about naïveté of prey towards novel predators, Cox and Lima have introduced the notion of "predator archetypes," defining the archetype as "a set of predator species that use similar morphological and behavioral adaptations in obtaining prey" (Cox and Lima 2006: 675). Lack of contact with the archetype in the evolutionary history of the prey species results in its ecological naïveté or, in other words, ineffective antipredator responses (Cox and Lima 2006).

In conclusion, the empirical research on novel ecological mismatches has so far mapped the characteristics that determine the adaptive response of animals to HIREC, sought for the characteristics of human-created environments that mimic natural elements and focused on the novel interspecific relations that arise especially when species (re)introduction takes place. Although the stress has mostly been on the negative consequences of novel ecological mismatches, one should not forget that at the same time they might serve as grounds for innovation and might even initiate sympatric speciation events (cf. Thomas 2017).

Conclusions

As was discussed above, the theoretical part of novel ecological mismatch studies mainly focuses on how recognition and decision-making itself works. On this background, the works of Jakob von Uexküll and Adolf Portmann might give an impetus for the development of a broader theoretical stance that would address the role of recognition schemes in the self-maintenance of organisms and would allow for the investigation of the interactions of recognition schemes of different origins (evolutionary, ontogenetic, experiential). The relative rigidity of the models of Portmann and Uexküll might on the one hand make them unfit to explain the processes that current environmental change has induced. However, when ecological naïveté and the mismatches of recognition templates and environmental cues are put into focus, the models may actually provide a good ground to discuss the fixation of certain perceptual and behavioural relations and their functioning in the context of current HIREC. In such a case, not an absolute fixation of the existing perceptual patterns of the organisms would be presumed, but rather a certain stasis evidenced by incompatibility with the current speed of change.

First of all, one might look for reasons for the relative stability of recognition schemes. The persistence and stability of acquired semiotic relations is definitely beneficial for the organism in a stable environment. It saves time in identifying and categorising encountered items and speeds up behavioural reaction. In their semiotic character, such relations are indexical if-then relations and become fully functional once they have passed from an active semiotic phase (with learning involved) to a passive and habitual one, where the recognition and reaction is almost automatised. They help to filter negative effects on the organism by inducing certain behaviours and reactions that help to eliminate the threat or reach the desired object. In environments that undergo constant change, persistent recognition schemes may also serve a purpose, as they might induce fast acceptance of novel meaningful objects. This is possible if the properties of the novel item that are vital or detrimental to the organism are attached to, and indicated by, the habitual cues. However, semiotic indeterminacy – where different objects may be indicated by the same cues and the same objects by different cues – brings along the possibility of erring – the habitual cue might be attached to an object with novel properties or to an object with beneficial characteristics that are attached to detrimental ones.

On the other hand, at least from a certain level of complexity, cognitive and recognition schemes can change much faster than several other characteristics of the organism. Morphological and physiological change needs much more time and more fundamental change, the opportunity for which, however, may be opened up by the changes induced by recognition. Search images, as discussed by Uexküll, offer a stable ground of recognition for the future, yet they are acquired through the experiences of the organism and hence are much more plastic than its building plan (*Bauplan*). Unaddressed appearances, as elaborated by Portmann, may find a receiver at a certain point, and hence the organism may enter into novel ecological relations without having to change its morphological features. For example, it has been demonstrated that the colouration of a predator is not meaningful for the prey animal if they do not share the same habitat (Stankowich and Coss 2007)[4], yet such a meaning may be acquired if co-habitation ensues. Several display behaviours are formed by the partial exposure of former non-addressed behaviours (e.g. the threatening behaviour of wolves, with the exposure of teeth, not followed by an attack, or the incorporation of flight-initiation movements in the courtship displays of birds). Developing these ideas further, one can reach the discussions of epigenetics, with a shift in recognition and the transmission of the shift via learning serving as a prerequisite for possible genetic change. If novel ecological conditions are not paralleled by umwelt change, then it is also highly unlikely that the organism develops some morphological changes that would help it to cope. However, this is not to say that any problem of adaptation can be addressed with an adequate umwelt change, as multiple environmental changes demand physiological and behavioural changes that the sensory transformation cannot compensate.

The partly different focus of Uexküll's and Portmann's works allows us to address different phenomena in the ecological mismatch framework. The works of Uexküll allow us to address phenomena that concern the mismatch between umwelt and semiotic niche or perceptual schemes and the demands of the environment. Portmann, with his focus on the inwardness and expressions of the organism, on the other hand, allows us to target change in the meaning of expressive features in novel conditions. An integration of the two might at least offer an impetus to develop a more consistent platform for the assessment and prediction of HIREC's impact on the semiotic links between organisms and environment. Such an integration would allow for an investigation of the dynamics of meaning loss (if the umwelt and semiotic niche lose correspondence) and meaning gain (if an organism's expressions gain novel significance in the presence of novel receivers) in the conditions of rapid environmental change. These processes underlie the development of novel ecological relations, while the viability of those relations in the face of the continuing pace of transformations should be further studied.

Acknowledgements I thank Filip Jaroš for helpful comments and suggestions on an earlier version of the paper and the participants of the workshop "Adolf Portmann: a pioneer of the eidetic and semiotic approach in the life sciences" (6–7 June, 2019) in *Hradec Královė* for creating an inspiring research atmosphere. This research was co-funded by the Grant Agency of the Czech Republic (GAČR 19-11571S) and Estonian Research Grants PUT1363 *Semiotics of multispecies environments: agencies, meaning making and communication conflicts* and PRG314 *Semiotic fitting as a mechanism of biocultural diversity: instability and sustainability in novel environments.*

[4] I am grateful to Filip Jaroš for turning my attention to this study.

References

Arnold, T. W., & Zink, R. M. (2011). Collision mortality has no discernible effect on population trends of North American birds. *PLoS One, 6*(9).

Banks, P. B., & Dickman, C. R. (2007). Alien predation and the effects of multiple levels of prey *naiveté*. *Trends in Ecology & Evolution, 22*(5), 229–230.

Brentari, C. (2015). *Jakob von Uexküll. The discovery of the Umwelt between biosemiotics and theoretical biology*. Dordrecht: Springer.

Carthey, A. J. R., & Banks, P. B. (2014). Naïveté in novel ecological interactions: Lessons from theory and experimental evidence. *Biological Reviews of the Cambridge Philosophical Society, 89*(4), 932–949.

Cox, J. G., & Lima, S. L. (2006). Naiveté and an aquatic-terrestrial dichotomy in the effects of introduced predators. *Trends in Ecology and Evolution, 21*(12), 674–680.

Cruz, L. M., Shillinger, G. L., Robinsonae, N. J., Tomilloa, P. S., & Paladino, F. V. (2018). Effect of light intensity and wavelength on the in-water orientation of olive ridley turtle hatchlings. *Journal of Experimental Marine Biology and Ecology, 505*, 52–56.

Da Silva, A., Samplonius, J. M., Schlicht, E., Valcu, M., & Kempenaers, B. (2014). Artificial night lighting rather than traffic noise affects the daily timing of dawn and dusk singing in common European songbirds. *Behavioral Ecology, 25*(5), 1037–1047.

Enquist, M., & Arak, A. (1998). Neural representation and the evolution of signal form. In R. Dukas (Ed.), *Cognitive ecology: The evolutionary ecology of information processing and decision making* (pp. 21–87). Chicago: Chicago University Press.

Fawcett, T. W., & Johnstone, R. A. (2003). Optimal assessment of multiple cues. *Proceedings of the Royal Society of London B, 270*(1524), 1637–1643.

Gwynne, D. T., & Rentz, D. C. F. (1983). Beetles on the bottle: Male Buprestids mistake stubbies for females (Coleoptera). *Journal of the Australian Entomological Society, 22*, 79–80.

Hayes, N. (2000). *Foundations of psychology*. London: Thomson Learning.

Hoffmeyer, J. (2008). *Biosemiotics: An examination into the signs of life and the life of signs*. Scranton/London: University of Scranton Press.

Kleisner, K., & Maran, T. (2014). Visual communication in animals: Applying a Portmannian and Uexküllian biosemiotic approach. In D. Machin (Ed.), *Visual communication* (Handbooks of communication science) (pp. 559–676). Berlin: De Gruyter Mouton.

Kleisner, K., & Markoš, A. (2009). Mutual understanding and misunderstanding in biological systems mediated by self-representational meaning of organisms. *Sign Systems Studies, 37*(1/2), 299–310.

Klem, D. (1989). Bird-window collisions. *The Wilson Bulletin, 101*(4), 606–620.

Kriska, G., Horvath, G., & Andrikovics, S. (1998). Why do mayflies lay their eggs en masse on dry asphalt roads? Water-imitating polarized light reflected from asphalt attracts Ephemeroptera. *Journal of Experimental Biology, 201*, 2273–2286.

Linask, L., Magnus, R., & Kull, K. (2015). Applying Jakob von Uexküll's concept of umwelt to human experience and development. In G. Mey & H. Günther (Eds.), *The life space of the urban child: Perspectives on Martha Muchow's classic study* (pp. 177–194). New York: Routledge.

Loss, S., Will, T., & Marra, P. (2015). Direct mortality of birds from anthropogenic causes. *Annual Review of Ecology, Evolution, and Systematics, 46*(1), 99–120.

Magnus, R. (2016). The semiotic challenges of guide dog teams: The experiences of German, Estonian and Swedish guide dog users. *Biosemiotics, 9*(2), 267–285.

Magnus, R., & Kull, K. (2012). Roots of culture in the umwelt. In J. Valsiner (Ed.), *The Oxford handbook of culture and psychology* (pp. 649–661). Oxford: Oxford University Press.

Maran, T. (2013). Enchantment of the past and semiocide. Remembering Ivar Puura. *Sign Systems Studies, 41*(1), 146–149.

Maran, T., & Kleisner, K. (2010). Towards an evolutionary biosemiotics: Semiotic selection and semiotic co-option. *Biosemiotics, 3*(2), 189––200.

Mullet, T., Farina, A., & Gage, S. (2017). The acoustic habitat hypothesis: An ecoacoustics perspective on species habitat selection. *Biosemiotics, 10*(3), 319–336.
Owen, M., Swaisgood, R., & Blumstein, D. (2017). Contextual influences on animal decision-making: Significance for behavior-based wildlife conservation and management. *Integrative Zoology, 12*(1), 32–48.
Portmann, A. (1956). Vorwort. Ein Wegbereiter der neuen Biologie. In J. von Uexküll & G. Kriszat (Eds.), *Streifzüge durch die Umwelten von Tieren und Menschen: Ein Bilderbuch unsichtbarer Welten. Bedeutungslehre* (pp. 7–17). Hamburg: Rowohlt.
Portmann, A. (1957). Die Sonderstellung des Menschen im Reich des Lebendigen. *Universitas, 12*, 337–344.
Portmann, A. (1960). *Neue Wege der Biologie*. München: Piper.
Portmann, A. (1961). *Animals as social beings*. New York: Viking Press.
Portmann, A. (1965). *Aufbruch der Lebensforschung*. Zürich: Rhein.
Portmann, A. (1969). *Biologische Fragmente zu einer Lehre vom Menschen*. Basel: Schwabe Co.
Portmann, A. (1970). *Entlässt die Natur den Menschen? Gesammelte Aufsätze zur Biologie und Anthropologie*. München: Piper.
Portmann, A. (1974). *An den Grenzen des Wissens. Vom Beitrag der Biologie zu einem neuen Weltbild*. Wien/Düsseldorf: Econ-Verlag.
Portmann, A. (1987). Haben Tiere eine Seele? In G. M. Teutsch (Ed.), *Da Tiere eine Seele haben. Stimmen aus zwei Jahrtausenden* (pp. 108–119). Sttutgart: Kreuz Verlag.
Portmann, A. (1990). On the uniqueness of biological research. *The Journal of Medicine and Philosophy: A Forum for Bioethics and Philosophy of Medicine, 15*, 457–472.
Portmann, A. (2000a). Um eine basale Anthropologie. In A. Portmann (Ed.), *Biologie und Geist* (pp. 259–269). Göttingen: Burgdorf.
Portmann, A. (2000b). Die Biologie und das Phänomen des Geistigen. In A. Portmann (Ed.), *Biologie und Geist* (pp. 12–47). Göttingen: Burgdorf.
Portmann, A. (2000c). Das Problem der Urbilder in biologischer Sicht. In A. Portmann (Ed.), *Biologie und Geist* (pp. 119–132). Göttingen: Burgdorf.
Puura, I. (2013). Nature in our memory. *Sign Systems Studies, 41*(1), 150–153.
Sherman, P. W., Reeve, H. K., & Pfennig, D. W. (2008). Recognition systems. In J. R. Krebs & N. Davies (Eds.), *Behavioural ecology: An evolutionary approach* (pp. 69–96). Oxford: Blackwell Publishing.
Sih, A. (2013). Understanding variation in behavioural responses to human-induced rapid environmental change: A conceptual overview. *Animal Behaviour, 85*(5), 1077–1088.
Sih, A., Trimmer, P. C., & Ehlman, S. M. (2016). A conceptual framework for understanding behavioral responses to HIREC. *Current Opinion in Behavioral Sciences, 12*, 109–114.
Slingenberg, A., Slingenberg, L. B., & van der Windt, H., et al. (2009). *Study on understanding the causes of biodiversity loss and the policy assessment framework*. Final report. http://ec.europa.eu/environment/enveco/biodiversity/pdf/causes_biodiv_loss.pdf
Stankowich, T., & Coss, R. G. (2007). The re-emergence of felid camouflage with the decay of predator recognition in deer under relaxed selection. *Proceedings of the Royal Society B, 274*(1607), 175–182.
Thomas, C. D. (2017). *Inheritors of the Earth. How nature is thriving in an age of extinction*. London: Allen Lane.
Tønnessen, M. (2009). Umwelt transitions: Uexküll and environmental change. *Biosemiotics, 2*(1), 47–64.
Trimmer, P. C., Ehlman, S. M., & Sih, A. (2017). Predicting behavioural responses to novel organisms: State-dependent detection theory. *Proceedings of Royal Society B, 284*, 20162108. https://doi.org/10.1098/rspb.2016.2108.
van Baaren, J., & Candolin, U. (2018). Plasticity in a changing world: Behavioural responses to human perturbations. *Current Opinion in Insect Science, 27*, 21–25.
Verhoeven, K. J., Biere, A., Harvey, J., & van der Putten, W. (2009). Plant invaders and their novel natural enemies: Who is naive? *Ecology Letters, 12*(2), 107–117.

von Uexküll, J. (1894). Zur Methodik der mechanischen Nervenreizung. *Zeitschrift für Biologie, 31*, 148–167.
von Uexküll, J. (1905). *Leitfaden in das Studium der experimentellen Biologie der Wassertiere*. Wiesbaden: J.F.Bergmann.
von Uexküll, J. (1910). Über das Unsichtbare in der Natur. *Österreichische Rundschau, 25*, 124–130.
von Uexküll, J. (1920). *Theoretische Biologie*. Berlin: Verlag von Gebrüder Paetel.
von Uexküll, J. (1931). Die Rolle des Subjekts in der Biologie. *Die Naturwissenschaften, 19*(19), 385–391.
von Uexküll, J. (1932). Die Umwelt des Hundes. *Zeitschrift für Hundeforschung, 2*(5–6), 157–170.
von Uexküll, J. (1937). Die neue Umweltlehre: Ein Bindeglied zwischen Natur- und Kulturwissenschaften. *Die Erziehung, 13*(5), 185–199.
von Uexküll, G. (1964). *Jakob von Uexküll: seine Welt und seine Umwelt*. Hamburg: Christian Werner Verlag.
von Uexküll, J. (1973). *Theoretische Biologie*. Frankfurt a. M: Suhrkamp Taschenbuch Wissenschaft.
von Uexküll, J. (1980). *Kompositionslehre der Natur. Biologie als undogmatische Naturwissenschaft. Ausgewählte Schriften. Herausgegeben und eingeleitet von Thure von Uexküll*. Frankfurt am Main/Berlin/Wien: Verlag Ullstein GmbH.
von Uexküll, J. (1982). The theory of meaning. *Semiotica, 42*(1), 25–82.
von Uexküll, J. (1992). A stroll through the worlds of animals and men: A picture book of invisible worlds. *Semiotica, 89*(4), 319–391.
von Uexküll, J., & Sarris, E. (1931). Der Führhund der Blinden. *Die Umschau, 35*(51), 1014–1016.
Wittig, T. W., Cagle, N. L., Ocampo-Peñuela, N., et al. (2017). Species traits and local abundance affect bird-window collision frequency. *Avian Conservation and Ecology, 12*(1), 17.

Riin Magnus is a research fellow at the Department of Semiotics, University of Tartu. Her research interests include human-animal communication, interspecies interactions in urban environments, the biosemiotics of Jakob von Uexküll, and ecosemiotics. She has co-edited a special issue of the journal Sign System Studies on "Framing nature and culture" and a special issue of the journal Biosemiotics on "The Semiotics of Hybrid Natures". She has published articles on topics such as the sign use of guide dogs and their visually impaired handlers, the ecosemiotics of urban trees, the reception and meaning of alien species in the urban context and the implications of Jakob von Uexküll's research for the study of culture.

Chapter 6
Revisiting Basal Anthropology: A Developmental Approach to Human Evolution and Sociality

Andres Kurismaa

Abstract Although the legacy of the Swiss zoologist Adolf Portmann (1897–1982) has left profound traces in the formation of modern evolutionary anthropology and comparative neuroscience, few if any studies have considered the wider context of his comparative research and its potential modern significance. At the same time, current findings and directions seem to offer new tools for this purpose, and doing so may open exciting perspectives in the interpretation of human evolution and neurobiology from developmental viewpoints, as sought by Portmann. Specifically, his bio-anthropological research program, or basal anthropology, challenges established fault-lines between structural (morphological) and functional (physiological) research traditions, and seeks new contacts between them through typological generalizations that draw on the distinct developmental patterns of altricial and precocial species – and humans, as a secondarily altricial species. The concept of secondarily evolved altriciality (in a precocial species, such as humans) may significantly clarify contested issues of neurological, comparative biological, and anthropological research, by recognizing the wide zoological background upon which the human type of ontogeny can be seen as a distinct class among all vertebrates, including higher mammals and primates. By these means, the more abstract concepts of a socially open and symbolically constituted life mode, as discussed in early anthropology, can be integrated with modern frameworks, where a close dialogue is reemerging between developmental and evolutionary, as well as structural and functional research traditions.

Introduction

The Swiss scholar Adolf Portmann is renowned for his fundamental works in anthropology and zoology, as well as for founding new approaches in theoretical biology (Komárek 2009; Kleisner 2008, 2015). In all these areas he made lasting

A. Kurismaa (✉)
University of Hradec Králové, Hradec Králové, Czech Republic
e-mail: andres.kurismaa@gmail.com

F. Jaroš, J. Klouda (eds.), *Adolf Portmann*, Biosemiotics 23,
https://doi.org/10.1007/978-3-030-67810-4_6

contributions, often by integrating into life sciences concepts from other disciplines, including humanities, and forging original, cross-disciplinary ties between them. While its importance for theoretical biology is well known and long appreciated (Komárek 2009; Kleisner 2008, 2015), much less has been done to address Portmann's anthropological legacy, also referred to as *basal anthropology* here, in the perspective of subsequent developments. At the same time, with the recent emergence of evolutionary developmental biology (or "evo-devo") as a major integrative framework, and its increasing influence on anthropology and neuroscience, such reassessment is becoming increasingly feasible and may add important dimensions to current empirical vindications of Portmann's hypotheses, discussed below (Trevathan and Rosenberg 2016a). In particular, this concerns his sociogenic view of human brain evolution, according to which its classical morphological interpretation in terms of maternal obstetrical constraints (the "obstetrical dilemma", see Trevathan and Rosenberg 2016a, b) may need to be significantly expanded, to further incorporate physiological and socio-historical factors that are presumably no less fundamental in shaping human ontogeny and its apparent "preadaptation" to our unique postnatal niche (Dunsworth et al. 2012; Dunsworth 2016).

In this context, reappraising basal anthropology could help to approach human ontogeny in a wider comparative framework, while taking into account both structural and functional dimensions of developmental and brain evolution (Boughner and Rolian 2016; Trevathan and Rosenberg 2016a; Sherwood and Gómez-Robles 2017). Several current challenges and theoretical issues should be noted here.

Firstly, although the complex and diverse involvement of developmental mechanisms in evolutionary dynamics is increasingly investigated, the respective field of evo-devo is yet to address the types of developmental evolutionary divergence detailed by Portmann and his group (Portmann 1944, 1967, 1990; Müller 1969, 1972a, b, c; cf. Starck and Ricklefs 1998; Werneburg and Spiekman 2018). This concerns the wide disparities in offspring maturation by time of birth, which may display qualitative differences even among closely related species of eutherian (placental) mammals. Thus, while the groups of predominantly altricial eutherians are characterized by more numerous but less mature neonates, as commonly seen in "lower" mammals of evolutionarily earlier groups (e.g. many rodents, small carnivores), in contrast, in later evolved eutherians there are typically fewer descendants who are born at a more mature and capable stage (e.g. ungulates, cetaceans, nonhuman primates). Although fundamental and long investigated, such relationships of birth timing to the maturational development of the brain and body pose problems that are still poorly understood in all mammals, including human infants (Clancy et al. 2001, 2007; Workman et al. 2013; Sherwood and Gómez-Robles 2017). The issue is further complicated by the fact that human offspring, although belonging to the precocial group, shows altricial-like immaturity to an extent that distinguishes us from all other higher mammals. In this context, Portmann introduced the term of a secondarily acquired, or "secondary altriciality," and saw this as unique to the genus *Homo*, as shown below (Portmann 1941, 1990).

By reflecting diverse developmental mechanisms, such as neurogenesis timing, gene expression regulation, and cell cycle rates during pre- and postnatal periods, variations in ontogenetic growth modes may indeed play important roles in developmental evolution (Charvet and Striedter 2011; Workman et al. 2013). As a result of such mechanisms, the evolutionary factors driving encephalization could differ not only quantitatively, but also qualitatively across altricials and precocials, and humans as secondary altricials (Charvet and Striedter 2011; Workman et al. 2013). In this context, Portmann's early hypothesis that phylogenetic changes in general somatic and ontogenetic organization may precede major divergences in encephalization remains of particular interest (Portmann 1951; Mangold-Wirz 1966; Müller 1972b; cf. Charvet and Striedter 2011). If supported by further evidence, these studies could cast important light on the evolutionary mechanisms channeling brain growth and maturation across taxa, and moreover, indicate important sources of ethodiversity even as the basic plans of mammalian brain structure and formation remain highly conserved, including in non-human primates and humans (Finlay and Darlington 1995; Finlay et al. 2001; Aboitiz and Montiel 2012).

Interestingly, the question of whether humans qualify as a predominantly altricial or precocial species has long been a source of controversy and confusion in the literature. This is probably due to the unique combination in human neonates of traits that are either at markedly early formative stages (e.g. motoric, neurocognitive, many morphological ones), or on the contrary, more typically precocial at birth (such as open sense organs and homeothermy). Moreover, by the time of birth, the human brain has attained ca. 25–30% of its adult size, a proportion that lies close to the mid-range between altricial and precocial group norms, thus further complicating respective assessment. From a phylogenetic perspective, however, Portmann's arguments for placing humans firmly in the precocial group (Portmann 1967, 1990) and seeing their altricial features as secondarily derived remain valid (Martin 2007, 2013; Trevathan and Rosenberg 2016a), even as it goes against long traditions of characterizing humans as an altricial species (Fox 1964; Wood et al. 2003; Watson et al. 2006; cf. Clancy et al. 2001). Besides the import of basal anthropology for respective theoretical questions and controversies, it can be noted that until presently, progress in such key applied and translational areas as interspecies age-matching and comparison of developmental timetables across taxa has largely proceeded on an *ad hoc* and single case basis, with limited input from basic biological and evolutionary theory (Rivers and Ashton 2013). In this context, the developmental distinctions noted above could present significant further interest for evolutionary developmental approaches in biology, anthropology, and neuroscience.

While several aspects of Portmann's anthropology have by now found wider currency, such as co-evolutionary views on human biology and culture, there are numerous other issues that remain less investigated or await reconsideration. Besides the status of modal or "typological" patterns in evo-devo, this includes the problem of ongoing and future human evolution in biological and cultural anthropology, and their prospects as a more closely integrated discipline. In this context, Portmann's principled rejection of the more mechanistic accounts of human birth state and timing (by purely obstetrical constraints, i.e. the "obstetrical dilemma" hypothesis) is of particular current interest, as well as his alternative, "sociogenic" view of anthropogeny (cf. Dunsworth et al. 2012; Dunsworth 2016; Trevathan and Rosenberg 2016b). In light of the recent findings discussed below, this socio-historical perspective is indeed garnering significant new interest (Dunsworth 2016; Trevathan and Rosenberg 2016b) and underscores some of the open challenges posed by basal anthropology for current thinking.

Below, a brief outline will be first given of Portmann's legacy in relation to the historical and philosophical contexts framing his approach, before its implications for developmental evolution are analyzed. Along with the above-mentioned aspects, particular attention will be paid to the convergence between functional (physiological) and structural (morphological, evo-devo) research now underway, as this may help to recapture Portmann's own integrative stance on developmental mode differences among mammals. In this context, supplementing the classical morphological focus of evo-devo by more synthetic, organismic and physiological perspectives would be one of the key challenges posed by Portmann's approach to current research. Indeed, reconsidering comparative growth modes may present wide interest for modern evo-devo and its methodological outlooks, as seen below (cf. Jenner 2008).[1]

Theoretical and Historical Background

Adolf Portmann was one of the great scholars of his time, pioneering concepts and studies now classical in anthropology, zoology, and theoretical biology. Although known widely in his lifetime for his fundamental work in comparative vertebrate zoology (Portmann 1948, 1967, 1976), it is interestingly his evolutionary anthropological research that remains the best known and most influential facet of his legacy today (particularly beyond the German-speaking world) (Martin 2007, 2013; Trevathan and Rosenberg 2016a). On the other hand, the wider and timely reception of these works, and the appreciation of their theoretical outlooks, appears to have been significantly hindered by belated and limited translations (Gould 1977, 349), as well as deeper changes in the intellectual climate of post-WWII Europe and America (cf. Jaroš, Chap. 7, in this volume) – shifting from more holistic and "organicist" traditions in life sciences towards more simplified and reductive, mechanistic paradigms (Nicholson and Gawne 2015). This may also explain the scarcity

[1] Historically, the relative exclusion of such modal or "typological" aspects (and their established opposition to "population thinking") may reflect an unfortunate artifact of confusing them with essentialism (e.g. static natural kinds, "species fixism") or other constructs peripheral to typological generalizations (Levit and Meister 2006; cf. Mayr 1982). This conceptual problem has been analyzed by Levit and Meister (2006) regarding the traditions of German structural morphology relevant also for Portmann's approach (cf. Jaroš, Chap. 7, in this volume). In contrast to the former, however, Portmann moved from a descriptive, non-phylogenetic use of typological generalizations to explicit evolutionary developmental hypotheses, as seen below.

(and apparent absence) of works attempting to reassess his anthropological and comparative approach at large.

At the time of his formative years and early career, however, European science and culture were relatively rich in aspirations to forge closer ties between the natural and human sciences (Grene 1968; Danzer 2011). Within biology, this was apparent not only in the diverse early traditions of organicism (Nicholson and Gawne 2015), but also in other bio-theoretical schools, such as early epigenetics and ethology, or the complex approaches of Russian/Soviet life sciences (Kurismaa 2018, 2019; Nadin 2015). The same period witnessed an emergence of organism-centered and phenomenological frameworks also in neurology, psychiatry, and psychosomatic medicine, particularly in Continental and German-speaking Europe (Grene 1968; Danzer 2011). Most relevantly for the present context, related transdisciplinary pursuits included also the search for new, more "anthropologically" oriented lines of research in physiology and medicine (Buytendijk 1974; Blasius 1976; Fagot-Largeault 2009; Wiggins and Schwartz 2013) – forged with the explicit goal to complement their traditionally reductive focus by more biosocially oriented programs. Portmann's search for a new anthropological approach belongs to this theoretical context and *Zeitgeist*. Closely related investigations, concerned with the emergence of socio-cultural and phenomenological aspects of human life from its evolutionary bases, originated in the then contemporary traditions of philosophical anthropology, pioneered by such notable thinkers as H. Plessner, M. Scheler, A. Gehlen, and others (cf. Jaroš, Chap. 7, in this volume; Brentari 2018).

As mentioned, a uniting aim of these diverse traditions was to rethink and overcome the deepening disciplinary divisions of the time – separating not only the humanities from the life sciences, but also different subfields of research from one another even in closely related areas, such as biological and cultural anthropology (Smocovitis 2012). Indeed, for their emergence and evolution in the genus *Homo*, human types of development and ontogeny must have necessarily relied on the growing influence of socio-cultural practices, incorporating not only tool use, but also new forms of behavior and communication increasingly unique to the human species (Portmann 1967, 1990). Although this co-evolutionary dynamics is now more widely appreciated (Deacon 1997, 2012; Schilhab et al. 2012; Foley et al. 2016), it presents problems not only for biological anthropology and neuroscience, but also methodology of science. In theoretical biology, the question of new sources and mechanisms of evolutionary change, emerging in the course of life's history and major transitions, is a critical and still contested issue with regard to language origins, as well as much more basic forms of biological communication, memory and inheritance (Szathmáry 2015; Jablonka and Lamb 2014; Barbieri 2020; Favareau 2010). Some related concerns were in fact expressed early on, particularly in the context of human biology, where broader social, anthropological, and phenomenological analyses were seen as clearly required to extend purely zoological notions of inheritance, evolution, and social behavior. In this regard, the descriptor of "anthropological physiology," introduced by Portmann's colleague F. J. Buytendijk (1974) appears still apt, and encapsulates some of the emergentist and holistic ideas they attempted to introduce in this context (Fagot-Largeault

2009; Wiggins and Schwartz 2013). A related term, used by Portmann himself to qualify his approach to anthropology is *basal anthropology*. As the name implies, it aims to complement the biocultural and phenomenological analyses of human biological makeup (Buytendijk 1974; Wiggins and Schwartz 2013) by wider phylogenetic and developmental criteria, reflecting basal relations among mammalian and primate groups (Portmann 1990). Although the term itself, basal anthropology, remained somewhat tentative and was explicated only in a few writings (Portmann 1956, 1962, 1970), it is retained here to distinguish Portmann's approach from other traditions of evolutionary anthropology (including modern biological anthropology). The specifics of this approach for comparative studies find closer attention in the following sections.

However, the general background outlined above is important to observe, as to a large extent, the scientific concepts introduced in this chapter can be interpreted from various theoretical and methodological perspectives, not necessarily related to Portmann's outlooks. For example, this is most evident regarding the neo-Darwinian interpretations of altricial and precocial patterns as distinct strategies for fitness maximization and adaptation in particular lineages and environments (Pianka 1970; Charnov and Berrigan 1993; Reznick et al. 2002). Such adaptive interpretations have indeed enjoyed considerable success in many cases and demonstrated how variations in population density, age-specific mortality, or other ecological fluctuations can significantly influence physiological and life history evolution (Reznick et al. 2002; Flatt and Heyland 2011; Stearns 2015).

On the other hand, besides adaptive accounts of phenotypic plasticity and variability, their developmental and internal mechanistic aspects are also increasingly incorporated into evolutionary theory, and open up new perspectives not only in morphological, but also physiological, behavioral, and ecological research (Bertossa 2011; Rittschof and Robinson 2016; Dubansky 2018). These advances may help to recast Portmann's legacy in modern light, appreciative of its integrative and organismic (morpho-physiological) stance. Indeed, a bridging of the classical, morphological focus of evo-devo on structural development (Carroll 2005, 2008; Moczek et al. 2015) with more functional agendas, such as physiological and life history traditions, is widely called for and seen as instrumental for future theoretical progress across disciplines (Kalinka 2015; Breuker et al. 2006; Flatt and Heyland 2011; Stearns 2015). Reassessing basal anthropology may contribute several important questions and critical outlooks towards this aim, particularly when organisms belonging to developmentally diverse groups, such as predominantly altricial and precocial mammals, are compered and theoretically investigated. As seen below, closer attention to their differences may highlight important instances of morpho-functional integration and modularity and furthermore, may represent important models for bringing interspecies age-matching and comparisons into closer contact with basic evolutionary and biological theory (Rivers and Ashton 2013; Workman et al. 2013).

Altricial and Precocial Mammals

Supported by comprehensive evidence from the comparative morphology and development of reptiles, birds and mammals, Portmann's early concepts of evolutionary changes in vertebrate ontogeny have become classic both in zoology and anthropology (Werneburg and Spiekman 2018; Trevathan and Rosenberg 2016a; Martin 2007, 2013; Starck and Ricklefs 1998). In particular, his distinction between the modes of development characterizing predominantly altricial and precocial species, with relatively less or further developed neonates, has become widely established and continues to be investigated from various perspectives, including morphological, physiological, social, and cognitive points of view (Workman et al. 2013; Scheiber et al. 2017; Augustine et al. 2019).

Portmann's model of comparative development argued for a relatively constant relationship between the way a group of mammals develops and its adult organization, both in terms of morphological and neurological differentiation (Portmann 1967, 1990). As he observed with colleagues (Müller 1972a, b, c; Mangold-Wirz 1966), in altricial species, which form the majority of extant eutherian mammals, the mothers tend to produce large litters of relatively poorly developed offspring after a shorter gestation period. Their young emerge helpless with sense organs sealed, incomplete homeothermy (e.g. little or no body hair), and poor locomotor ability (Portmann 1967, 1990; Müller 1972a; Derrickson 1992; Werneburg and Spiekman 2018). The brain and body weight of altricial neonates is relatively low, and their adult encephalization (brain-to-body mass ratio) remains below the precocial norms (Müller 1972b; Wirz 1950, Mangold-Wirz 1966; Portmann 1952, 1967, 1990; Martin 1983; Shea 2007). Many species of insectivores, rodents, and small carnivores belong to this group (Table 6.1; Fig. 6.1).

In contrast, the predominantly precocial eutherian groups exhibit notably different developmental features, marked by more differentiated body structures and complex brains. As first detailed by Portmann and colleagues (Portmann 1939, 1944, 1948, 1967, 1990; Müller 1969, 1972a, b, c), and subsequently elaborated in a large number of studies (cf. Derrickson 1992; Shea 2007; Werneburg and Spiekman 2018), precocial species typically produce smaller litter sizes of larger neonates,

Table 6.1 ***Ontogenetic relationships in mammals.*** Characteristic features of altricial and precocial mammals, as representing earlier and later evolved forms of ontogenetic organization

	Early level of organization	Later level of organization
Gestation period	Very short (e.g. 20–30 days)	Long (more than 50 days)
Number of young Per litter	Large (e.g. 5–22)	Usually 1–2 (rarely 4)
State of young at birth	Altricial	Precocial
Examples	Many insectivores, rodents, marten-like carnivores	Ungulates, seals and whales, apes and prosimians

Adapted from Portmann 1990, 22

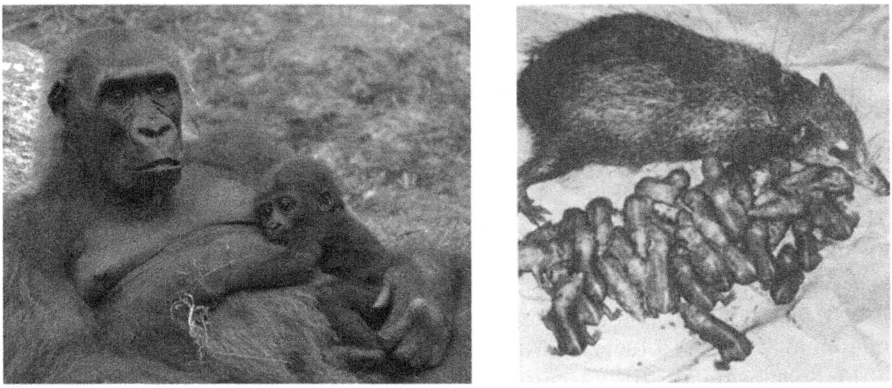

Fig. 6.1 An illustration of contrasting developmental modes in two mammalian taxa. Left: The precocial gorilla bearing a single, well-developed offspring (photo reproduced with the kind persmission of its author Boris Diakovsky ©). Right: The highly altricial tailless tenrec, with a large litter size of immature, highly dependent offspring (Louwman 1973: photo reproduced with the permission of Blackwell publishing)

who after a relatively long period of gestation appear much like adults already at birth, both in form and behavior. This pattern is particularly characteristic of ungulates, seals, cetaceans, and primates (Portmann 1967, 1990), whose offspring are slower to develop, but show higher degrees of encephalization and precocial independence at birth, and as adults retain higher brain-to-body ratios than altricials (Martin 1983, 2007; Jerison 1973; Shea 2007). Apart from relative delays in motor development, primates as an order are generally considered highly precocial (Fig. 6.1) (Martin 1983; Martin and MacLarnon 1990; Shea 2007). While such distinctions between groups are useful for recognizing correlated patterns (e.g. integrated changes in organ systems or functions), they should of course not overshadow or divert from the diverse manifestations of modally intermediate, mixed, or transitional forms in many species and taxa (including humans), as seen below.

Indeed, changes in developmental modes represent a major source of convergent evolution across mammalian taxa, as transitions to precociality have occurred independently in multiple eutherian lineages and distinct ecological settings (Portmann 1967, 1990; Martin 2007). Likewise contradicting any notion of exclusive or categorical "types," the individual factors contributing to developmental independence at birth may show a more or less continuous variation depending on how and which parameters of precociality/altriciality are measured (e.g. thermoregulatory, sensory, motoric, nutritional, or other, see Derrickson 1992), as clearly recognized by Portmann and his later followers (Martin 1976, 2013). The potentially high dissociation of many developmental factors may also explain why, in several mammalian orders, even closely related species can belong to contrasting developmental groups (notably in rodents and carnivores, the most "typologically diverse" or "mixed" taxa, see again Derrickson 1992). Nonetheless, from the structural/mechanistic perspectives of modern evo-devo, it appears particularly interesting that once established, the differentiation between altricials and precocials is a highly conserved evolutionary feature and sufficiently pronounced to usually mark a familial, and often an ordinal trait among living eutherians (Portmann 1967; Martin 1976, 1983, 2013; Martin and MacLarnon 1985, 1990) (although this rule has a small number of exceptions).[2] Given this relative conservatism and recurrence of ontogenetic modes in larger taxonomic units, approaching them in terms of phylogenetic and macroevolutionary constraints rather than ecological context retains interest and value, as pursued in Portmann's traditions, and complements the leading focus on adaptation in most related studies that followed (Gould 1977; Reznick et al. 2002; Stearns 2015).

Regarding the theoretical significance of their results, Stephen J. Gould commented early on that these lines of developmental studies initiated by Portmann deserve to be far better known in the English-speaking world (Gould 1977: 349), and regardless of their occasional departures from the accepted neo-Darwinian and selectionist paradigm, they appear to present materials and merits that can be assessed independently of such methodological questions (cf. also Starck and Ricklefs 1998). Indeed, this seems to have been the case both in zoology and anthropology. In both areas, the differences between altricial and precocial life history strategies have been subsequently interpreted on the basis of various selectionist models, such as the influential *r/K* selection theory, modeling the effects of population density and ecological unpredictability on life history evolution (Pianka 1970; Reznick et al. 2002), as well as the spectrum of fast-versus-slow developmental strategies derived from the *r/K* model (Charnov and Berrigan 1993; Shea 2007). Related works have elaborated upon key problems of Portmann's comparative analysis, albeit from adaptive and ecological viewpoints that differ considerably from his more organism-centered and developmental emphases (i.e. stressing internal integration of developmental processes over selective pressures). On the other hand, with the recent emergence of evo-devo as a major area of integrative biology (Moczek et al. 2015; Love 2015), and its increasing influence on anthropology and neuroscience, this structural emphasis of Portmann's works has become relatively uncontroversial in recent decades. In particular, the modern notions of developmental modularity, morphological integration, and co-regulation of traits all stress the internal complexity of genotype-phenotype maps and show how random genetic variation can be converted into non-random and functionally integrated phenotypes in particular developmental systems (Moczek et al. 2015; Minelli and Fusco 2008; Hlusko 2004). In this case, developmental conditions can both facilitate and constrain the operation of natural selection in specific directions (Moczek et al. 2015). Systematically varying patterns of morphogenetic integration may even arise on the background of relatively minor genetic variations, provided these affect key regulatory networks and transcription factors, responsible for the orchestrated expression of numerous downstream effector genes in specific developmental pathways and time periods (Carroll 2008; Wilkins 2014).

[2]According to subsequent large-scale studies, it appears that only the Rodentia and Carnivora orders contain more than 5 genera exhibiting 2 (or more) developmental categories (Derrickson 1992), and in terms of the neonatal development of their offspring, few orders appear to exhibit such developmental variation in "modal" features (ibid). As Portmann recognized, this low variance testifies to their probably early evolutionary differentiation.

While mainly studied with regard to their morphological effects (Wilkins 2014; Moczek et al. 2015), these developmental factors may cast light also on life history evolution, including the relatively marked bimodal distribution of specific developmental traits in altricials and precocials. Specifically, it appears that a strictly functional and adaptive explanation is still lacking for the markedly discontinuous patterning of a trait as fundamental as eutherian gestation length (Martin 2013) – a highly conserved trait in placental mammals that is significantly correlated with offspring number, body size, and encephalization quotient (Martin and MacLarnon 1985, 1990; Portmann 1951, 1967, 1990; Mangold-Wirz 1966; Müller 1972b). In this regard, it has been shown by Martin and MacLarnon that placental mammals tend to give birth either to one (or few) large offspring with a higher degree of encephalization, following a longer gestation period, or alternatively, to more littermates following a shorter gestation period, with remarkably little overlap between these two reproductive-developmental types (Fig. 6.2) (Martin 1983, 2013; Martin

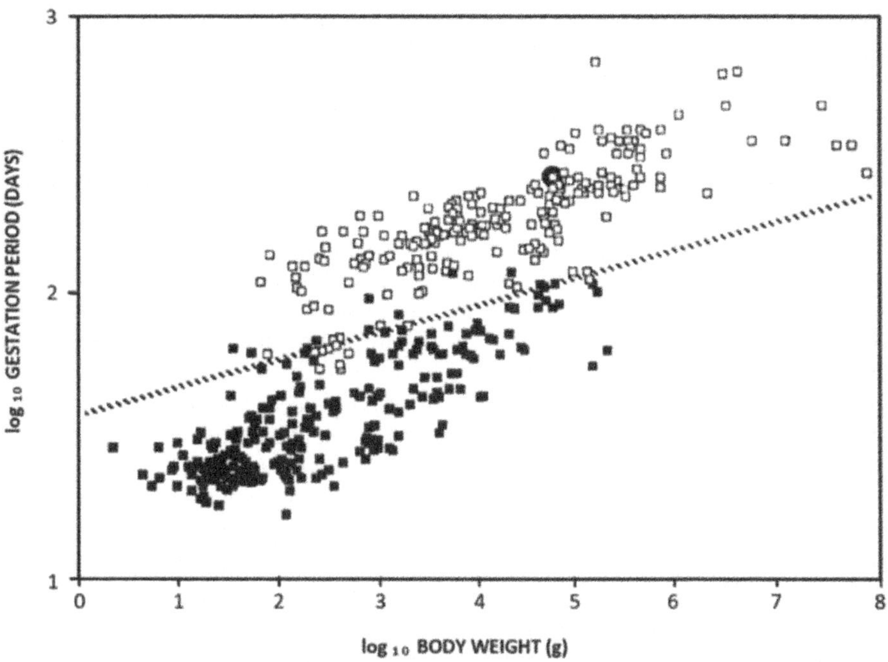

Fig. 6.2 Logarithmic plot of gestation period (days) against maternal body weight (g) for 429 placental mammals, with separate indication for species belonging to predominantly altricial groups (black squares; n = 227) and predominantly precocial groups (white squares; n = 202) (adapted with permission from Martin 1989; Martin and MacLarnon 1990). The predominantly altricial sample contains insectivores, tree shrews, carnivores, lagomorphs, and various rodents (myomorphs and sciuromorphs). The predominantly precocial group contains ungulates, cetaceans (whales and dolphins), pinnipeds (seals and sea lions), elephants, manatees, primates, and rodents (hystricomorphs). The point for *Homo sapiens* is indicated by a black circle and falls within typical precocial norms (Martin and MacLarnon 1990)

and MacLarnon 1985, 1990). However, so far, no obvious ecological explanation seems to account for this bimodality (Martin 2013), which most likely involves a complex interplay of environmental and morphogenetic factors awaiting further study.

In this regard, it is increasingly appreciated that in addition to classical trade-offs affecting mammalian life history evolution, such as the competing energetic demands of body size versus generation time (or size versus offspring number per generation), other and possibly non-selected aspects may be involved in many cases of life history change (Flatt and Heyland 2011). These include potential signaling conflicts and biases in developmental systems, often due to the re-use (exaptation) of structures for new functions (Gould and Lewontin 1979), the multiple and interlinked effects of diffuse signaling substances such as hormones (Ricklefs and Wikelski 2002), and the above-mentioned morphological and genetic integration of traits (including epistatic and pleiotropic genetic effects). By incorporating such biases constraining physiological evolution, new problems and approaches in life history theory may indeed be shifting towards a closer dialogue with evo-devo (Kalinka 2015; Flatt and Heyland 2011; Breuker et al. 2006) and by implication, towards a methodology increasingly more compatible with Portmann's own organismic thinking on the altricial-precocial spectrum and its ontogenetic bases.

A related confusion and criticism regarding altricial and precocial development, still persisting in the current literature, pertains to the fuzzy boundaries between these categories and their perceived vagueness for analyzing many specific cases, including primate and human ontogeny. For example, primates are frequently assigned to the altricial category based on their relative neonatal immobility and dependence on parental care (cf. Scheiber et al. 2017), although by other parameters they clearly qualify as precocials, as noted above (Shea 2007; Martin and MacLarnon 1990). This matter is further complicated in humans, whose offspring are exceptionally helpless among all primates and have therefore been described simply as altricial by numerous authors, particularly in neurological research (Fox 1964; Gottlieb 1971; Wood et al. 2003; Watson et al. 2006; Versace and Vallortigara 2015; cf. Clancy et al. 2001). Nonetheless, according to Portmann's original phylogenetic criteria, both humans and non-human primates should with full justification be assigned to the precocial group, as their offspring number per birth is radically reduced (to a single one in the vast majority of cases), their gestation period is long, and their sense organs are open at the time of birth (or very near it). In this case, the presumed intermediate instances in primates and exceptions to the fundamental altricial/precocial division have been identified on the basis of secondary features (e.g. mobility, degree of parental dependence), reflecting derived variations, rather than in terms of the phylogenetic criteria recognized as basal by Portmann and later evolutionists (Martin and MacLarnon 1990, 52).

It should be noted here that the long rejection and misunderstanding of the developmental distinctions proposed by him may have other causes besides methodological ones. As later indicated, Portmann (1941, 1965, 1967) and his group never conducted scaling analyses to control for the effects of body size on mammalian developmental and reproductive patterns and therefore could not conclusively demonstrate that their differences in the altricial-precocial spectrum survive allometric control (Martin and MacLarnon 1990). This is critical, as many aspects of biological timing and functions are known to scale in a relatively regular fashion with body weight and size (including gestation lengths in altricial and precocial eutherians). In this context, the studies by Martin and MacLarnon again remain pertinent, having shown that once body size effects are controlled for, precocial mammals still have gestation periods nearly four times as long as altricials (Martin and MacLarnon 1985, 1990) and account for their distinct growth modes.

This more qualitative notion of growth types has found considerable support also from subsequent paleontological and comparative evidence, suggesting that modal differences in ontogenetic patterns were laid down early in evolution and thus now appear at or near the ordinal level (i.e. in larger taxonomic units) (Martin 1976, 2007, 2013; Martin and MacLarnon 1990). Moreover, according to recent studies confirming Portmann's original hypothesis (Portmann 1951; Mangold-Wirz 1966, Müller 1972b), the observed modes may phylogenetically precede and constrain changes in encephalization through diverse mechanisms, such as alterations in progenitor cell cycle rates, neurogenesis timing, and gene expression regulation (Charvet and Striedter 2011; Workman et al. 2013). Gaining further clarity on such mechanisms may therefore hold specific interest for comparative developmental modeling, especially when species belonging to phylo- or ontogenetically diverse groups are compered. Given that until recently, basic evolutionary and biological theory have received minor attention within most traditions of comparative modeling in clinical and translational areas (Rivers and Ashton 2013), Portmann's phylogenetic criteria of ontogenetic organization appear still pertinent and may deserve wider attention.

Translating and extrapolating the relative timing of developmental events across species presents considerable theoretical challenges particularly for research on humans, given our highly derived life history and developmental patterns (Bogin 2003, 2012; Schwartz 2012; Hrvoj-Mihic et al. 2013; Sherwood and Gómez-Robles 2017). In this regard, the concept of human "secondarily altricial" slow maturation remains widely employed and influential (Trevathan and Rosenberg 2016a; Sherwood and Gómez-Robles 2017; Werneburg and Spiekman 2018). Perhaps most importantly, by emphasizing the broader context of this concept, it could help to orient comparative analysis in species whose developmental profiles differ in quantitative as well as qualitative parameters (Rivers and Ashton 2013; Clancy et al. 2000, 2007; Workman et al. 2013). For example, it is remarkable that although humans belong to the precocial group in phylogenetic terms, above 95% of common laboratory animals in comparative and clinical research are altricial species, chosen specifically for their rapid development (e.g. zebrafish, mouse, rat). The implications of this choice (or coincidence) for modeling human brain development remain unclear and little studied to this day (Workman et al. 2013; Clancy et al. 2007; Rivers and Ashton 2013). Even as the basic plans and sequence of mammalian neurodevelopmental events remain highly conserved (Finlay et al. 2001; Aboitiz and Montiel 2012), there is much larger variability in their relation to birth timing (Fig. 6.3), and its modifications may reflect an important source of diversity and

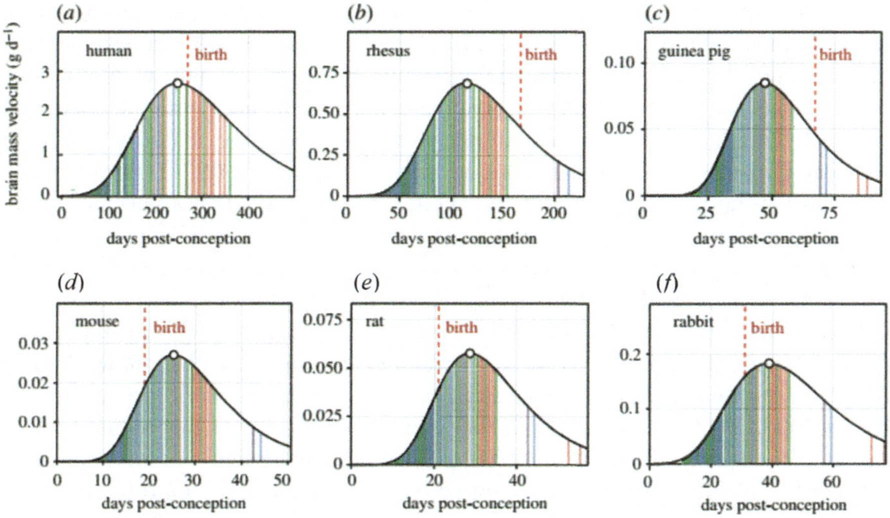

Fig. 6.3 Birth timing variations relative to growth, as reflected in velocity models of fetal and perinatal brain growth in 3 precocial (**a–c**) and 3 altricial species (**d–f**) (image adapted with permission from Halley 2017). Color bars indicate model estimates for the timing of neurogenesis events (green), tract formation (blue), and myelination (red), according to www.translatingtime.net (Workman et al. 2013). Open circles indicate the moment of eye opening, which corresponds well to peak brain growth velocity, showing that altricials are born earlier along their growth trajectories and correlated patterns of neurodevelopment than precocials (Halley 2017)

specialization in eutherians (Portmann 1990; Scheiber et al. 2017). In this regard, closer attention to altricial/precocial patterns and their modular integration may help to guide basic comparative thinking, particularly by underscoring the complexity of morphofunctional patterning in developmental systems (Rivers and Ashton 2013).[3]

[3] These aspects highlight that with differential changes in the rate and/or sequence of developmental events also in *subsystems* of compared organisms, the very concept of a "developmental stage" can break down even in closely related species, and instead, age-matching needs to focus more narrowly on the state of particular systems or functions (Rivers and Ashton 2013), including their maturation sequence (Spicer et al. 2011; Dubansky 2018). As an example of morphofunctional modularity, the case of developmental differences in altricial and precocial avian species is cited, where precocials hatch with functioning legs, in contrast to altricials (Rivers and Ashton 2013). Remarkable instances of morphofunctional modularity were observed by Portmann in intraspecies heterochronies, such as the precocial formation of forelimbs in otherwise highly altricial marsupials (which allows the neonate to crawl from the mother's birth canal to her pouch immediately after birth, cf. Portmann 1990; Keyte and Smith 2010). Such organismic expressions of modularity are indeed widespread and illustrate developmental modes as both ecologically and morphologically integrated systems. Regarding criticisms of linear and non-linear scaling of traits, operating on the simpler ideas of conserved "developmental stages," see Rivers and Ashton 2013; Leigh 2001; Bogin 2003, 2012.

The relevance of these considerations extends beyond comparative neuroscience. In evolutionary anthropology in particular, the (near-)linear scaling of developmental events has been a highly influential model in comparing humans with other primates. In its support, several conservative criteria have been offered, including gestation length, dental eruption timing, age at reproductive maturity, and total lifespan across non-human primates and humans (Schultz 1969; cf. Bogin 2003). This has naturally offered an attractive model, by proposing that all primates have virtually the same life stages (which simply increase in length from prosimians to humans). However, in more recent years, and partly by elaborating on Portmann's traditions (Martin 1983), some inaccuracies of this model for human-primate comparisons have been well documented (Leigh 2001; Bogin 2003, 2012), along with more precise criteria for assessing temporal and structural novelties in the human life cycle by anatomical, physiological, and behavioral markers (see below, cf. Bogin 2003, 2012). These works further confirm that in future theorizing, besides the simpler concepts of scaling conserved "developmental stages" or morphological landmarks (by linear or non-linear models), the principles of morphofunctional integration and modularity will likely predominate, as well illustrated by altricial and precocial traits (Rivers and Ashton 2013).

What are the implications of these comparative problems for biological theory and evo-devo? Addressing this requires a broader perspective on the scope and methods of biology, as a science comprising both individual and unique, as well as widely generalizable expressions of life (Elsasser 1987; Markoš 2002). From a methodological perspective, it is important that both of these aspects can be included in typological classifications (e.g. of developmental evolutionary variations), and this inclusiveness may offer a logical middle ground in studying biological regularities. As suggested by Jenner (2008), the question of developmental modes may have particularly wide ramifications for evo-devo, as a science classically framed in relatively nomothetic (universalistic) terms, perhaps to its disadvantage. In this view, rather than following the probably elusive goal of discovering new, widely generalizable rules or regularities of evolution per se, for evo-devo to form a more distinct and cohesive discipline, focusing on the natural diversity of developmental modes may be a particularly productive strategy and methodological outlook (Jenner 2008).[4]

This reference to methodology of science, while to some extent departing from the main topics discussed above, may be critical for understanding the deeper conceptual and theoretical issues involved in them (and their frequent neglect or misinterpretation, cf. also Levit and Meister 2006). In the following, the reasons for

[4] In this context, the widely integrative nature of evo-devo research across distant explanatory levels can also be noted, ranging from genes and cells to organisms and ecology; as well as its vast disciplinary scope – encompassing developmental genetics, paleontology, phylogenetics, physiology, and other areas (cf. Hall 2012; Moczek et al. 2015). Indeed, such diversity and scope may naturally raise critical questions, especially when the aims of this approach are seen as not only supplementing, but possibly also revising or revolutionizing the classical neo-Darwinian paradigm (Hoekstra and Coyne 2007; Laland et al. 2014, 2015) (as desirable as this may not be in societal aspects, too, cf. Diogo 2020). In this regard, the problem of altriciality/precociality may well supplement the other modes of development highlighted in this context by Jenner (2008).

qualifying human ontogeny as a new class, representing secondary altricial development, will be addressed in more detail, along with the question of its socially conditioned, or sociogenic evolution. This will further highlight the close interrelations between structural and functional aspects of developmental evolution, as anticipated in basal anthropology.

Humans – A Secondarily Altricial Species

The patterns of development evolving in the genus Homo (and possibly earlier hominids) stand out as highly exceptional not only among precocial and altricial mammals, but also the great apes (Thompson et al. 2003; Schwartz 2012; Portmann 1990). Although gestation length has expectably changed little in hominid evolution (Martin 1983) and remains closely comparable across humans and extant apes (Portmann 1941, 1990), the human neonate is enormously large in comparison to the latter group, reaching by birth nearly twice the size of great ape newborns, both in total body size and brain weight (Sacher and Staffeldt 1974; Portmann 1990). Together with the reduced number of offspring per birth (one, with rare exceptions), these features clearly belong to the precocial group (Portmann 1990; Martin and MacLarnon 1990; Trevathan and Rosenberg 2016a). The presence of homeothermy and open sense organs by birth further reflect the precociality of this state, upon which "secondary altricial" features are overlaid as highly derived patterns in more recent evolution (Shea 2007).[5]

The extensive dependence of human offspring on prolonged (allo)parental care is the most general expression of this derived altricial-like condition. It is marked by the unique helplessness of human neonates, whose morphofunctional state resembles a physiological preterm birth by primate standards (Portmann 1941, 1990). Nonetheless, in Portmann's analysis, simple reductions in ancestral developmental rate and timing are probably insufficient to explain the origins of human ontogeny from its preceding forms, regardless of many attempts in this direction (i.e. the neotenic hypotheses of human ontogeny, widely pursued at the time and later: Bolk 1926; Gould 1977; cf. Menke 2015). Specifically, in the basal anthropological perspective, even though human adults do share many features with the juveniles of extant apes rather than their adults (including such conspicuous features as non-protruding jaws, globular skull, large head and brain, etc.), the human form of slow

[5]While the term itself, "secondary altriciality," has been criticized as potentially misleading by appearing to suggest a closer resemblance of human ontogeny to altricial rather than precocial modes of development (Dunsworth 2016), this misinterpretation was excluded in principle by Portmann's original criteria, as seen above. In fact, these sought to highlight that human ontogeny has a number of apomorphic (uniquely derived) features in relation to both preceding mammalian groups. While introducing an entirely new term could have been helpful in this regard (and could still be, cf. Trevathan and Rosenberg 2016b), in its absence the term "secondary altriciality" has been retained here.

maturation is likely too complex and unique to be seen as a simple result of ancestral fetalization, and recent evidence strongly agrees with this critical, more qualified position (stressing modular development) (cf. Somel et al. 2012; Menke 2015). For example, even though its postnatal growth rates are exceptionally rapid and remain so in the first year of life, the human neonate matures extremely slowly (Gunz 2016); and while retaining altricial-like immaturity long after birth in most of its morphological, motor, and neuro-cognitive development (Portmann 1990; Sherwood and Gomez-Robles 2017), it is born with homeothermy and open sense organs, among other precocial features (e.g. the grasping reflex, Werneburg and Spiekman 2018).

In Portmann's estimation, the human gestation period should extend to ca. 21 months if humans were to reach by birth a stage of neuro-motor maturity comparable to great ape neonates. In other words, to attain the typical precocial stage of primates, the period of human intrauterine development should exceed its current duration by more than two-fold (given its present slow pace in comparison to precocials). This pattern is also expressed in the exceptional structure of human brain growth and maturation which, in comparison to precocials, show disproportionately long periods of postnatal expansion and remodeling. Thus, with its nearly 3.3-fold postnatal expansion (Martin 1983; Sakai et al. 2012), the human brain has a multiplier factor (postnatal growth index) lying in between altricial and precocial norms (Fig. 6.4). This shift indicates that in comparison to other primates, major portions of human brain development have been relocated to the extra-uterine environment, with exposure to its complex inputs at very early and sensitive stages of development (by precocial standards) (Fig. 6.4).[6]

The precociality of human development finds further support from embryological observations. Similarly, to primate embryos, the human fetus passes through a phase during its early maturation where its eyes and ears are temporarily sealed by protective membranes (between 3–5 months), and this seems to correspond to the time of altricials' parturition (Fig. 6.4, cf. Portmann 1990; Martin 2007). This remarkable "recapitulation" of ancient developmental programs, in extant primates and humans, served a key role in Portmann's hypothesis on vertebrate phylogeny, stipulating a reversed sequence of ontogenetic diversification in mammals as compared to birds (Portmann 1939, 1951). According to this hypothesis, in contrast to birds (whose earliest representatives were most probably precocials, cf. Charvet and Striedter 2011), in mammalian phylogeny the precocial taxa emerged later, indeed by incorporating the ancestral stage of altriciality into their longer periods of intrauterine development (whereas no such recapitulation of altriciality occurs in

[6] The postnatal multiplier factor of 3.3, obtained in recent studies using ultrasound imaging and tomographic scanning methods (Sakai et al. 2012), is less than the multiplier factor of 4.2 originally proposed by Portmann (1990), or a similar figure suggested later by Sacher and Staffeldt (1974); however, this does not change the general estimates of human brain development as an outlier among precocials in its degree of postnatal expansion, nor its intermediate position between altricials and precocials; the indexes for the latter groups have been similarly reevaluated in various studies (cf. Martin 1983; Halley 2017).

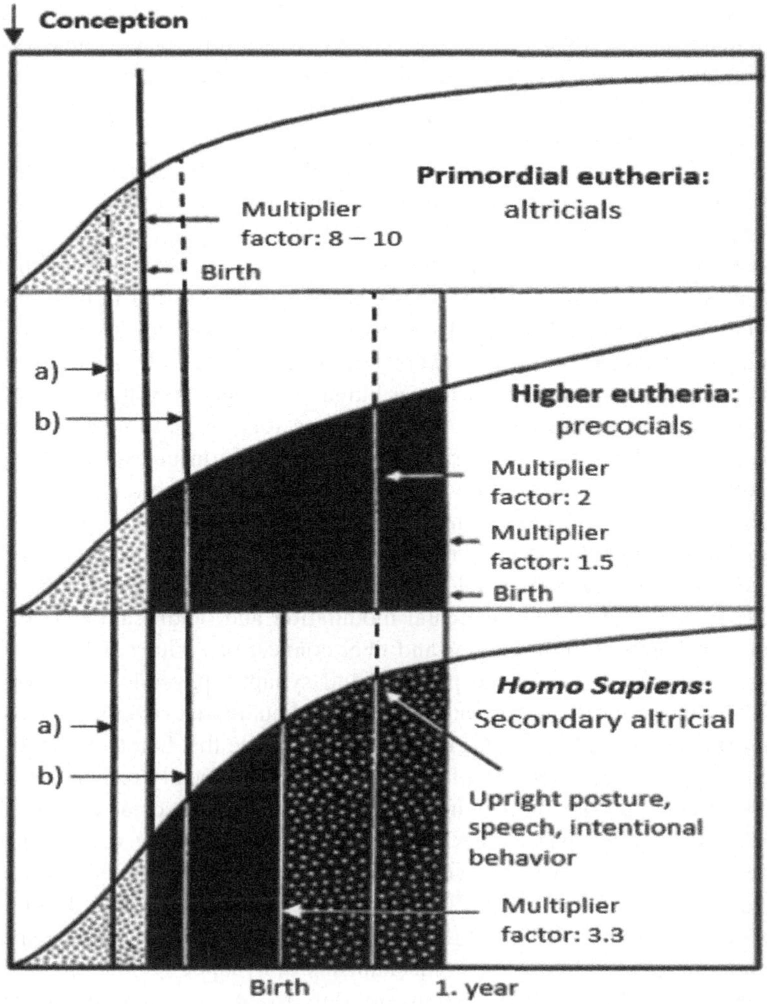

Fig. 6.4 A schema of eutherian brain development (redrawn and adapted from Portmann 1967, pp. 291)
Above: **the archaic altricial type of early birth with closed sense organs; average postnatal brain growth multiplier factor (MF): 8–10.** *Middle:* **The higher eutherian precocial type passes through the above phase in utero; average postnatal MF-s between 1.5 and 2.** *Below:* **a hypothetical higher mammal with human brain weight and body mass, developing according to the middle precocial schema. Relative to this model, our birth is displaced to an earlier point in time, so that postnatal MF attains a value of ca. 3.3**
Symbols: **(a) Temporary closure of sense organs (b) Reopening of sense organs**

precocial bird embryogenesis, see Portmann 1939). While this early hypothesis of mammalian developmental evolution remains valid and fundamental (Werneburg and Spiekman 2018), no clear consensus still exists regarding the precise factors, either ecological or internal developmental, that may have conditioned such transitions to precociality, all the more as this has occurred in diverse lineages and environments independently (Fig. 6.4).[7] Thus, while this relative fetal isolation understandably enables the protection of the offspring from the external milieu during its most vulnerable early periods of development (Portmann 1939, 1967), it still poses significant questions for future research.

Recent findings confirm that the relative timing of birth in relation to brain development can be highly variable both within and between altricial and precocial groups (Clancy et al. 2001; Halley 2017), while brain growth rates and neurodevelopmental sequences in utero remain remarkably conserved across mammalian species (Halley 2017; Finlay et al. 2001). As humans appear to fall in between these two patterns of prenatally attained relative brain development, it follows that in hominid phylogeny, a remarkable reversal has occurred from the ancestral state of primate precociality. As a result, whereas human brain size at birth represents up to 30% of its adult mass (Fig. 6.4) (i.e. multiplier factor ≥ 3.3), the same measure in chimpanzees is about 40%, in gorillas above 50%, and in macaques nearly 70% (Zollikofer and de León 2010; Trevathan and Rosenberg 2016b).

Such shifts in neurodevelopmental and birth timing are thought to have significant functional and structural repercussions. It is known that during its early periods, the formation of functional and anatomical connectivity in the brain is particularly sensitive to environmental modulation and modifications both at the level of broad architectural features and fiber connections (Gunz 2016), as well as the fine regulation of neural morphology and synaptic physiology (Hrvoj-Mihic et al. 2013). Considering its previously-mentioned qualitative divergence from non-human primates (Figs. 6.4 and 6.5), it appears probable that human brain anatomy and function have specifically evolved to be more ecologically responsive at early stages – thus enhancing the extent to which learning and social experience can shape neural circuitry, particularly during the initial periods of maximum neural plasticity and "synaptic pruning" (i.e. as the majority of the initially overproduced synapses are eliminated in the developing brain, see Somel et al. 2012; Halley 2017; Sherwood and Gómez-Robles 2017). Further, such early ecological exposures, apparently anticipated by the basic mechanisms of synaptic regulation and wiring (Greenough et al. 1987), allow humans not only to integrate and internalize complex environmental inputs, but to also take active part in the socio-cultural sphere and receive critical feedback to their own actions (Sherwood and Gómez-Robles 2017). Interestingly, this period of early social integration occurs on the background

[7] In more functional interpretations, this transition has been thought of as providing a possible adaptive advantage, at least in more stable and predictable environments (cf. Pianka 1970; Reznick et al. 2002), or in cases where offspring mortality is low, the offspring is few in number, and maternal investment can be higher (Charnov and Berrigan 1993). Regardless of its causes or interpretation, that a transition to precociality may have occurred in as many as 10 mammalian lineages independently indicates a remarkable case of convergent evolution, moreover in diverse ecological settings, (Martin 2007, 2013). In contrast, there is little evidence of significant reversals to the altricial condition in already precocial species (thereby further challenging ecological hypotheses, see Portmann 1990). The hominid lineage and modern humans seem highly apomorphic in this regard.

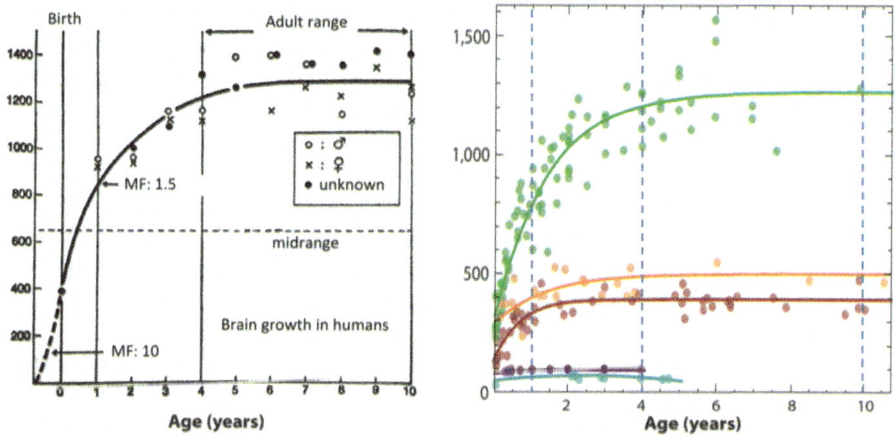

Fig. 6.5 *Left*: Developmental variations in total brain volume (cm^3) in humans, as depicted by Portmann (1967). The table contains figures obtained by his associates (K. Mangold-Wirz) and other researchers, documenting the exceptionally rapid rate of human brain growth in the first postnatal year. The typical multiplier factor (MF) of higher eutherians (1.5) is reached in humans around 1 year after birth, and the adult range of brain size in the 4th year of life. *Right*: Variations in total brain volume (cm^3) during postnatal development in humans (green), mountain gorillas (orange), chimpanzees (red), rhesus macaques (purple), and tufted capuchins (blue), as redrawn from a summary figure by Sherwood and Gómez-Robles (2017) and readapted for comparison with the left-hand diagram

of significantly accelerated brain growth, which, as pointed out, continues in humans with near-fetal rates well into the first year of life, in remarkable contrast to other primates (whose growth rates decelerate rapidly near the time of birth) (Fig. 6.5).

These differences stress that during the first exposures to the social niche, human brain development appears to go through phases that are still embryogenic in comparison to other primates. Indeed, Portmann compared this early sociocultural scaffolding to a "social uterus" (Portmann 1967, 1990), in which the social group and its complex behavioral dynamics subsume functions that had formerly, in our primate ancestors, been mainly provided intra-corporeally (nutrition, thermoregulation, protection, etc.). This ontogenetic shift may therefore serve as a central impetus in the evolution of new types of cognitive and neuromuscular organization, necessary for upright posture, persistent tool use, and ultimately for the speech apparatus to form. In a similar context, the anthropologist Montagu (1961) later coined the term "extra-uterine gestational period" ("postnatal exterogestation"), reflecting the same early stage of social adaptation and helplessness which Portmann had called the "extra-uterine year" (*extrauterine Frühjahr*) (see Portmann 1944, 1990). This early period anticipates complex physical and symbolic contacts with the world that are in obvious contrast to the more stable and monotonous conditions in which other primates spend the comparable periods of their early development – relatively sealed off from the diverse and fluctuating conditions of the outside environment.

Consistent with this, and drawing on works directly elaborating on Portmann's concepts (Martin 1983, 2007), current studies suggest that to understand the novel features of human ontogeny, it is necessary to focus on the evolutionary differentiation of qualitatively new life history stages within the early hominid life cycle (Bogin 2003, 2012), beyond isolated somatic or neurological changes. In this case, although its genetic bases remain open (Bogin 2003, 2012), it is the *de novo* differentiation of new developmental stages, such as the extra-uterine year and childhood (and respective life strategies in *Homo* adults), that mark the gradual divergence of humans from their closest primate relatives and ancestors (Schwartz 2012). In this regard, the question of how the differentiation of new life cycle phases may differ, in terms of its developmental regulation, from the much better understood cases of body-plan differentiation and segmentation is a highly interesting, though not yet explicitly addressed problem (to our knowledge).

While considering these questions of life history evolution would go beyond the scope of this chapter, it is important to note here some recent findings and approaches confirming basal anthropological views. In particular, Portmann had stipulated the probably early evolutionary origins of the secondary altricial pattern, as well as its possible coincidence with the emergence of the first representatives of *Homo* (Portmann 1976). In line with this, it has been subsequently proposed that upon reaching the cranial capacity of the early *Homo erectus* (850 cm^3, ca. 1.8 million years ago), the constraints imposed by maternal pelvic morphology on the maximum dimensions of the neonatal cranium may have become a limiting factor on its evolutionary expansion. To overcome this maternal constraint, subsequent increases in offspring brain size would have required prolonged phases of postnatal growth (Martin 1983; Walker and Shipman 1997; cf. also Gould 1977). Although this interpretation remains limited by the scarce fossil materials on *Homo erectus* development, the earliest patterns of secondary altriciality may indeed have emerged with the earliest members of the genus *Homo,* according to later consensus (Martin 1983; Trevathan and Rosenberg 2016a). Thus, rather than being an acquisition of our later, "behaviorally modern" stages of evolution, the secondary altricial pattern has probably much earlier origins, significantly preceding the periods of cumulative social changes and "cultural ratcheting" (e.g. by manufactured tools and technological culture, mainly in the last 80,000 years) (Cieri et al. 2014).

It is interesting to recapitulate here the distinction drawn by Portmann between two closely interrelated, but nonetheless different tasks of basal anthropological research. These would respectively seek to 1) identify the organic variations in heredity (i.e. genetic and developmental factors) that have made the transition to a new, cultural-historical type of ontogeny possible; and seek to 2) establish at which stage this new mode of development (characterized by linguistic thought and speech) "began to decisively accelerate human evolution" (Portmann 1962: 14). In line with the accepted mosaic and modular basis of anthropogeny (Foley et al. 2016; Bogin 2003, 2012), the latter sort of evolutionary changes originated probably long after the formation of habitual bipedality (by ca. four million years ago in australopithecines), and much later than the first known instances of tool use (currently dated to 3.3 million years before present, i.e. long antedating the earliest known

Homo fossils from 2.8 million years ago) (Lewis and Harmand 2016). In Portmann's terms, this pertains to the question of at which points new factors may have emerged in the principal determinants of human behavior and its "openness" to the social and symbolic worlds (Portmann 1962, 1990). Also, here, both continuous changes and more sudden grade shifts are likely to be seen (Foley et al. 2016).

In the present context, perhaps most provocative and significant in Portmann's anthropogenic views remains their cultural focus, emphasizing the social nature of human developmental evolution. This suggests that above and beyond morphological constraints (e.g. as imposed by the birth canal on neonatal brain evolution, cf. above), the human type of "physiological early birth" should be seen first of all as a natural correlate and preadaptation to our culturally organized life, with its enhanced demands on learning and lifelong development, in particular. Thus, the notion that "something as mechanical as difficulty in parturition" could have been a leading factor of human birth timing invited strong criticisms from Portmann (cf. also Gould 1977: 369).

Indeed, according to recent findings (Dunsworth et al. 2012; Dunsworth 2016), the roles of anatomical and obstetric factors are not necessarily decisive for secondary altriciality to evolve. These findings emphasize the key role of maternal energetic constraints on gestation length and neonate size (Dunsworth et al. 2012), and further, suggest there is a need for postnatal continuation of human brain growth less due to morphological reasons, and primarily due to the energetic demands of maternal metabolism (the basal rate of which nearly doubles by parturition, cf. Dunsworth et al. 2012). This new hypothesis is particularly interesting, as increases in the rate of maternal metabolism during gestation are specifically related to the requirements of the fetal brain – in neonates, supporting brain growth accounts for up to 73% of their resting metabolic rate (e.g. in comparison to the above 20% of adults, cf. Leonard et al. 2012). Meeting such increased energetic demands of a larger fetus and neonate, with a larger brain, most probably requires collective provision and alloparenting to evolve in the group, and hence changes in social structure and cognition (Fig. 6.6). By these means, in the new hypothesis, changes in birth timing may be intimately related not only to internal energetic, mainly maternal physiological and fetal neurodevelopmental requirements, but also to social factors and organization (Fig. 6.6) (although as pointed out, the "obstetrical dilemma" may remain a complementary, rather than a conflicting interpretation, cf. Trevathan and Rosenberg 2016a).

With good reason, the authors conclude that a major implication of their model of maternal energy investment is the possible vindication of Portmann's views on human evolution (elsewhere also termed "psychogenic," cf. Gould 1977; Jaroš, Chap. 7, in this volume), according to which the decisive factors shaping human early birth may be socio-cultural and neurodevelopmental in nature (Dunsworth 2016; cf. also Trevathan and Rosenberg 2016b). This could mark an important conceptual shift and allow broader hypotheses on human evolution to be accommodated. In this regard, it is also debatable whether the term "psychogenic" is the most apt descriptor of Portmann's outlooks, as clearly, it is only as part of extended evolutionary processes (Laland et al. 2015; Fuentes 2016) that a possible

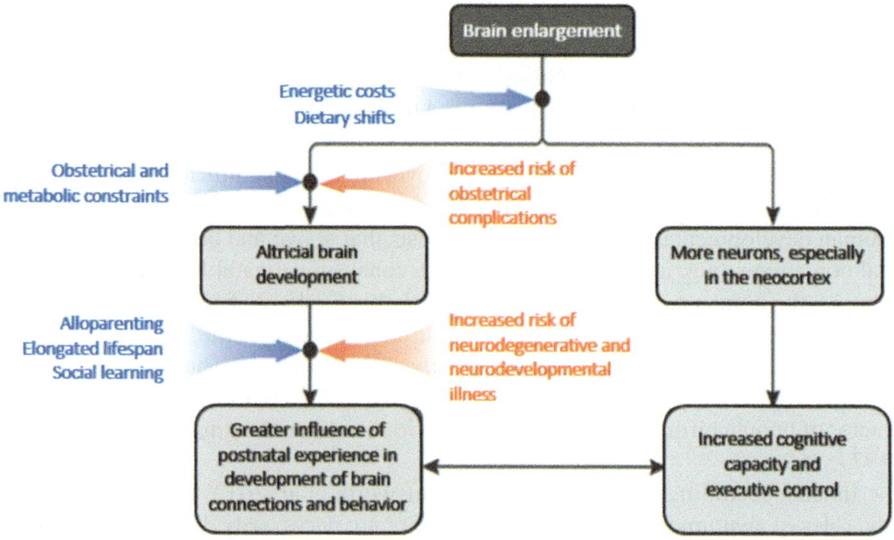

Fig. 6.6 A model of the evolution of human secondary altricial brain development (redrawn and modified from Sherwood and Gómez-Robles 2017). The figure indicates that the evolution of a large and plastic brain in humans is directly integrated with functional life history factors and extended social relations (e.g. alloparenting, social learning and cognition)

co-evolutionary scenario of human development attains its biological meaning (Portmann 1967, 1990; Deacon 1997; Donald 2001). In this view, albeit vastly complex and evolvable, mental and neural processes form but particular units of more complex evolutionary phenomena that emerge only in society, and cannot be understood apart from the latter. It seems that rather than "psychogenic," the term "sociogenic" evolution may be better applied here.[8]

Below, some of the key factors associated with secondary altriciality are summarized (Fig. 6.6). The schema illustrates some advantages and risk factors of this pattern of brain development (cf. Crespi and Leach 2016), as well as its expected social and life history correlates. In particular, attention is drawn to the fact that secondary altriciality most probably co-evolved with other life history changes, such as post-reproductive life-span of care-givers and social cooperation, in particular for collective provision (e.g. high energy food procurement, alloparenting). While these notions have been interpreted mostly from adaptive perspectives, stressing ecological and social selection pressures, it should be noted that such stipulated links between early development and aging are increasingly supported also by molecular biological and developmental findings (Somel et al. 2012). Accordingly,

[8] So far, the term "sociogenic" has been used almost exlusively in medical and epidemiological research, to indicate that besides biological causes, social triggers and factors also participate in disease etiology. However, nothing seems to rule out its wider employment in general biological and evolutionary contexts, particularly in relation to human evolution (but by no means exclusively to it).

ontogenetic shifts in (pre-) humans towards secondary altriciality most probably involved not only substantial delays in the timing of sexual and tissue maturation, but also molecular changes in the aging phenotype and life cycle as a whole, including gene expression patterns throughout life (Somel et al. 2012).

Such emerging connections, drawing together developmental, evolutionary, and human life history changes with complex social and historical factors (Boughner and Rolian 2016; Sherwood and Gómez-Robles 2017) appear to reflect well the integrative potentials of new evo-devo approaches in neuroscience and anthropology. This can provide some of the foundations upon which the prospects of basal anthropology can be further assessed.

In summary, it may be more appropriate to view basal anthropology (and Portmann's research on the whole) as partly a meta-scientific and epistemological enterprise that nonetheless retains a strong scientific foothold. Indeed, Portmann underscored the importance of this objective scientific basis as the *sine qua non* of his approach if it is to inform, from a sufficiently universal perspective, the questions of our socially and historically conditioned life mode (cf. also Jaroš and Maran 2019). Rather than attempting to provide ready-made answers (a "terminal anthropology," see Portmann 1962), a leading motive of basal anthropology is to keep in view the inherent indeterminacy and freedom of this life mode, i.e. its underconstrained hereditary bases and societal forms. Of course, this would freely recognize that many other species, particularly primates, also have significant forms of social inheritance, such as local traditions and animal cultures (Jablonka and Lamb 2014), and may demonstrate cognitive capacities often superior to humans in many respects (De Waal 2016). However, from a basal anthropological perspective, the ways these capacities and traditions are developmentally integrated into their life cycle and neurobiological development would, in comparison to humans, remain divergent and distinct in basic aspects, as seen in this chapter.

Conclusion

This chapter revisited A. Portmann's approach to developmental evolution and his fundamental, pioneering concept of human "secondary altriciality" in the perspective of time. This has served the dual aim of relating his research program of basal anthropology to its historical roots and methodological context, as well as highlighting its interest for current challenges, particularly in the areas of evolutionary (neuro)developmental research and anthropology. In this regard, Portmann's recognition of possible qualitative distinctions between altricial and precocial types of development in eutherian mammals remains of interest and calls for further research on its evolutionary developmental bases (beyond ecological and adaptive ones). Furthermore, assigning humans to the precocial class of mammals, but with secondarily derived altricial traits may help to clarify contested issues of neurological, comparative biological, and anthropological research, where different terminology and criteria have been used, but rarely with an explicit recognition of the wide zoological background upon which the human type of ontogeny can be seen as a distinct class among all vertebrates, including higher mammals and primates. Further, highlighting the apparent socio-historical "preadaptation" of the human phenotype to its postnatal niche led Portmann to a view of anthropogeny that seems increasingly supported by modern evidence, indicating that besides maternal morphological constraints and factors (i.e. the "obstetrical dilemma"), complex physiological, life historical, and social factors were probably no less instrumental in the emergence of the modern human form of development and neurobiology. In these aspects, the basal anthropological approach speaks directly to current requirements, particularly for closer ties between developmental and evolutionary, as well as structural and functional research traditions. It thus seems to retain numerous implications for current theoretical and applied research.

References

Aboitiz, F., & Montiel, J. F. (2012). From tetrapods to primates: Conserved developmental mechanisms in diverging ecological adaptations. In M. A. Hofman & D. Falk (Eds.), *Progress in brain research* (Vol. 195, pp. 3–24). Amsterdam: Elsevier.

Augustine, S., Lika, K., & Kooijman, S. A. (2019). Altricial-precocial spectra in animal kingdom. *Journal of Sea Research, 143*, 27–34.

Barbieri, M. (2020). The semantic theory of language. *Biosystems, 190*, 104100.

Bertossa, R. C. (2011). Morphology and behaviour: Functional links in development and evolution. *Philosophical Transactions of the Royal Society of London. Series B, Biological Sciences, 366*(1574), 2056.

Blasius, W. (1976). *Problems of life research: Physiological analyses and phenomenological interpretations.* Berlin, Heidelberg, New York: Springer.

Bogin, B. (2003). The human pattern of growth and development. In J. L. Thompson, G. E. Krovitz, & A. J. Nelson (Eds.), *Patterns of growth and development in the genus Homo* (pp. 14–44). Cambridge: Cambridge University Press.

Bogin, B. (2012). The evolution of human growth. In N. Cameron & B. Bogin (Eds.), *Human growth and development* (2nd ed., pp. 287–324). Amsterdam: Academic.

Bolk, L. (1926). On the problem of anthropogenesis. *Proceedings of the Section of Sciences, Koninklijke Akademie van Wetenschappen te Amsterdam, 29*, 465–475.

Boughner, J. C., & Rolian, C. (Eds.). (2016). *Developmental approaches to human evolution.* Hoboken: Wiley.

Brentari, C. (2018). From the hiatus model to the diffuse discontinuities: A turning point in human-animal studies. *Biosemiotics, 11*(3), 331–345.

Breuker, C. J., Debat, V., & Klingenberg, C. P. (2006). Functional evo-devo. *Trends in Ecology & Evolution, 21*(9), 488–492.

Buytendijk, F. J. (1974). *Prolegomena to an anthropological physiology.* Pittsburgh: Duquesne University Press.

Carroll, S. B. (2005). Evolution at two levels: On genes and form. *PLoS Biology, 3*(7), e245.

Carroll, S. B. (2008). Evo-devo and an expanding evolutionary synthesis: A genetic theory of morphological evolution. *Cell, 134*(1), 25–36.

Charnov, E. L., & Berrigan, D. (1993). Why do female primates have such long lifespans and so few babies? Or life in the slow lane. *Evolutionary Anthropology, 2*, 191–194.

Charvet, C. J., & Striedter, G. F. (2011). Developmental modes and developmental mechanisms can channel brain evolution. *Frontiers in Neuroanatomy, 5*(4), 1–5.

Cieri, R. L., Churchill, S. E., Franciscus, R. G., Tan, J., Hare, B., et al. (2014). Craniofacial feminization, social tolerance, and the origins of behavioral modernity. *Current Anthropology, 55*(4), 419–443.

Clancy, B., Darlington, R. B., & Finlay, B. L. (2000). The course of human events: Predicting the timing of primate neural development. *Developmental Science, 3*(1), 57–66.

Clancy, B., Darlington, R. B., & Finlay, B. L. (2001). Translating developmental time across mammalian species. *Neuroscience, 105*(1), 7–17.

Clancy, B., Finlay, B. L., Darlington, R. B., & Anand, K. J. S. (2007). Extrapolating brain development from experimental species to humans. *Neurotoxicology, 28*(5), 931–937.

Crespi, B., & Leach, E. (2016). The evolutionary biology of human neurodevelopment: Evo-neuro-devo comes of age. In J. Boughner & C. Rolian (Eds.), *Developmental approaches to human evolution* (pp. 205–229). Hoboken: Wiley.

Danzer, G. (2011). *Wer sind wir? Auf der Suche nach der Formel des Menschen. Anthropologie für das 21. Jahrhundert – Mediziner, Philosophen und ihre Theorien, Ideen und Konzepte*. Berlin, Heidelberg, New York: Springer.

De Waal, F. (2016). *Are we smart enough to know how smart animals are?* New York, London: WW Norton & Company.

Deacon, T. W. (1997). *The symbolic species: The co-evolution of language and the brain*. New York, London: WW Norton & Company.

Deacon, T. W. (2012). Beyond the symbolic species. In T. Schilhab, F. Stjernfeld, & T. Deacon (Eds.), *The symbolic species evolved* (pp. 9–38). Dordrecht: Springer.

Derrickson, E. M. (1992). Comparative reproductive strategies of altricial and precocial eutherian mammals. *Functional Ecology, 1*, 57–65.

Diogo, R. (2020). Quasi-religious belief in Darwin and Darwinism: "Straw-men" scientist believers everywhere. *Current Molecular Biology Reports, 6*, 16. https://doi.org/10.1007/s40610-020-00127-y.

Donald, M. (2001). *A mind so rare: The evolution of human consciousness*. New York: WW Norton & Company.

Dubansky, B. (2018). The interaction of environment and chronological and developmental time. In W. Burggren & B. Dubansky (Eds.), *Development and environment* (pp. 9–39). Cham: Springer.

Dunsworth, H. M. (2016). The 'obstetric dilemma' unraveled. In W. R. Trevathan & K. R. Rosenberg (Eds.), *Costly and cute: Helpless infants and human evolution* (pp. 29–50). Santa Fe, Albuquerque: School for Advanced Research Press.

Dunsworth, H. M., Warrener, A. G., Deacon, T., Ellison, P. T., & Pontzer, H. (2012). Metabolic hypothesis for human altriciality. *Proceedings of the National Academy of Sciences, 109*(38), 15212–15216.

Elsasser, W. M. (1987). *Reflections on a theory of organisms: Holism in biology*. Baltimore: Johns Hopkins University Press.

Fagot-Largeault, A. (2009). Anthropological physiology: von Uexküll, Portmann, Buytendijk. In A. Berthoz & Y. Christen (Eds.), *Neurobiology of "umwelt"* (pp. 1–5). Berlin, Heidelberg: Springer.

Favareau, D. (2010). *Essential readings in biosemiotics: Anthology and commentary*. Dordrecht: Springer Science & Business Media.

Finlay, B. L., & Darlington, R. B. (1995). Linked regularities in the development and evolution of mammalian brains. *Science, 268*(5217), 1578–1584.

Finlay, B. L., Darlington, R. B., & Nicastro, N. (2001). Developmental structure in brain evolution. *Behavioral and Brain Sciences, 24*(2), 263–278.

Flatt, T., & Heyland, A. (Eds.). (2011). *Mechanisms of life history evolution: The genetics and physiology of life history traits and trade-offs*. Oxford: Oxford University Press.

Foley, R. A., Martin, L., Mirazón Lahr, M., & Stringer, C. (2016). Major transitions in human evolution. *Philosophical Transactions of the Royal Society, B: Biological Sciences, 371*(1698), 20150229.

Fox, M. W. (1964). A phylogenetic analysis of behavioral neuro-ontogeny in precocial and nonprecocial mammals. *Canadian Journal of Comparative Medicine and Veterinary Science, 28*(8), 197–202.

Fuentes, A. (2016). The extended evolutionary synthesis, ethnography, and the human niche: Toward an integrated anthropology. *Current Anthropology, 57*(13), 13–26.

Gottlieb, G. (1971). Ontogenesis of sensory function in birds and mammals. In E. Tobach, L. Aronson, & E. Shaw (Eds.), *The biopsychology of development* (pp. 67–128). New York: Academic.

Gould, S. J. (1977). *Ontogeny and phylogeny*. Cambridge, Mass: Harvard University Press.

Gould, S. J., & Lewontin, R. C. (1979). The spandrels of San Marco and the Panglossian paradigm: A critique of the adaptationist programme. *Proceedings of the Royal Society of London. Series B: Biological Sciences, 205*(1161), 581–598.

Greenough, W. T., Black, J. E., & Wallace, C. S. (1987). Experience and brain development. *Child Development, 58*(3), 539–559.

Grene, M. (1968). *Approaches to a philosophical biology*. New York, London: Basic Books.

Gunz, P. (2016). Growing up fast, maturing slowly: The evolution of a uniquely modern human pattern of brain development. In J. C. Boughner & C. Rolian (Eds.), *Developmental approaches to human evolution* (pp. 261–283). Hoboken: Wiley.

Hall, B. K. (2012). Evolutionary developmental biology (Evo-Devo): Past, present, and future. *Evolution: Education and Outreach, 5*(2), 184.

Halley, A. C. (2017). Minimal variation in eutherian brain growth rates during fetal neurogenesis. *Proceedings of the Royal Society B: Biological Sciences, 284*(1854), 20170219.

Hlusko, L. J. (2004). Integrating the genotype and phenotype in hominid paleontology. *Proceedings of the National Academy of Sciences, 101*(9), 2653–2657.

Hoekstra, H. E., & Coyne, J. A. (2007). The locus of evolution: evo devo and the genetics of adaptation. *Evolution: International Journal of Organic Evolution, 61*(5), 995–1016.

Hrvoj-Mihic, B., Bienvenu, T., Stefanacci, L., Muotri, A. R., & Semendeferi, K. (2013). Evolution, development, and plasticity of the human brain: From molecules to bones. *Frontiers in Human Neuroscience, 7*, 707.

Jablonka, E., & Lamb, M. J. (2014). *Evolution in four dimensions: Genetic, epigenetic, behavioral, and symbolic variation in the history of life*. Cambridge, MA: MIT Press.

Jaroš, F., & Maran, T. (2019). Humans on top, humans among the other animals: Narratives of anthropological difference. *Biosemiotics, 12*(3), 381–403.

Jenner, R. A. (2008). Evo-devo's identity: From model organisms to developmental types. In A. Minelli & G. Fusco (Eds.), *Evolving pathways. Keynotes in evolutionary developmental biology* (pp. 100–119). Cambridge: Cambridge University Press.

Jerison, H. (1973). *Evolution of the brain and intelligence*. New York/London: Academic Press.

Kalinka, A. T. (2015). Towards an ecological understanding of morphological evolution. *Journal of Experimental Zoology (Molecular and Developmental Evolution), 324*(4), 383–392.

Keyte, A. L., & Smith, K. K. (2010). Developmental origins of precocial forelimbs in marsupial neonates. *Development, 137*(24), 4283–4294.

Kleisner, K. (2008). The semantic morphology of Adolf Portmann: A starting point for the biosemiotics of organic form? *Biosemiotics, 1*(2), 207–219.

Kleisner, K. (2015). Semantic organs: The concept and its theoretical ramifications. *Biosemiotics, 8*(3), 367–379.

Komárek, S. (2009). *Nature and culture: The world of phenomena and the world of interpretation*. München: Lincom Europa.

Kurismaa, A. (2018). Assimiliating an associative trait: From eco-physiology to epigenetics. *Biosemiotics, 11*(2), 199–229.

Kurismaa, A. (2019). *Towards an ecology of the brain: Reassessing the dominant as a paradigm of organismic and anthropological physiology*. PhD Thesis. Charles University in Prague, Department of History and Philosophy of Science.

Laland, K., Uller, T., Feldman, M., Sterelny, K., Müller, G., Moczek, A., et al. (2014). Does evolutionary theory need a rethink? *Nature News, 514*(7521), 161.

Laland, K., Uller, T., Feldman, M., Sterelny, K., Müller, G., et al. (2015). The extended evolutionary synthesis: Its structure, assumptions and predictions. *Proceedings of the Royal Society B: Biological Sciences, 282*(1813), 20151019.

Leigh, S. R. (2001). Evolution of human growth. *Evolutionary Anthropology: Issues, News, and Reviews, 10*(6), 223–236.

Leonard, W., Snodgrass, J. J., & Robertson, M. L. (2012). Comparative and evolutionary perspectives on human brain growth. In N. Cameron & B. Bogin (Eds.), *Human growth and development* (pp. 397–413). San Diego: Elsevier Inc.

Levit, G. S., & Meister, K. (2006). The history of essentialism vs. Ernst Mayr's "essentialism story": A case study of German idealistic morphology. *Theory in Biosciences, 124*(3–4), 281–307.

Lewis, J. E., & Harmand, S. (2016). An earlier origin for stone tool making: Implications for cognitive evolution and the transition to *Homo*. *Philosophical Transactions of the Royal Society, B: Biological Sciences, 371*(1698), 20150233.

Louwman, J. W. W. (1973). Breeding the tailless tenrec Tenrec ecaudatus at Wassenaar zoo. *International Zoo Yearbook, 13*(1), 125–126.

Love, A. C. (Ed.). (2015). *Conceptual change in biology. Scientific and philosophical perspectives on evolution and development*. New York: Springer.

Mangold-Wirz, K. (1966). Cerebralisation und Ontogenesemodus bei Eutherien. *Acta Anatomica, 63*(4), 449–508.

Markoš, A. (2002). *Readers of the book of life: Contextualizing developmental evolutionary biology*. Oxford: Oxford University Press.

Martin, R. D. (1976). The bearing of reproductive behavior and ontogeny on strepsirhine phylogeny. In W. P. Luckett & F. Szalay (Eds.), *Phylogeny of the primates* (pp. 265–297). Boston: Springer.

Martin, R. D. (1983). *Human brain evolution in an ecological context (James Arthur lecture on the evolution of the human brain, no. 52, 1982)*. New York: American Museum of Natural History.

Martin, R. D. (1989). Size, shape and evolution. In M. Keynes & G. A. Harrison (Eds.), *Evolutionary studies – A centenary celebration of the life of Julian Huxley* (pp. 96–141). London: Palvgrave Macmillan.

Martin, R. D. (2007). The evolution of human reproduction: A primatological perspective. *American Journal of Physical Anthropology, 134*(S45), 59–84.

Martin, R. (2013). *How we do it: The evolution and future of human reproduction*. New York: Basic Books.

Martin, R. D., & MacLarnon, A. M. (1985). Gestation period, neonatal size and maternal investment in placental mammals. *Nature, 313*(5999), 220.

Martin, R. D., & MacLarnon, A. M. (1990). Reproductive patterns in primates and other mammals: The dichotomy between altricial and precocial offspring. In C. Jean-De Rousseau (Ed.), *Primate life history and evolution* (pp. 47–79). New York: Wiley-Liss.

Mayr, E. (1982). *The growth of biological thought: Diversity, evolution, and inheritance*. Cambridge, Mass: Harvard University Press.

Menke, P. R. (2015). The ontogeny-phylogeny Nexus in a nutshell: Implications for primatology and paleoanthropology. In W. Henk & I. Tattersall (Eds.), *Handbook of paleoanthropology* (pp. 177–211). Berlin, Heidelberg: Springer.

Minelli, A., & Fusco, G. (2008). *Evolving pathways: Key themes in evolutionary developmental biology*. Cambridge: Cambridge University Press.

Moczek, A. P., Sears, K. E., Stollewerk, A., Wittkopp, P. J., Diggle, P., Dworkin, I., et al. (2015). The significance and scope of evolutionary developmental biology: A vision for the 21st century. *Evolution & Development, 17*(3), 198–219.

Montagu, A. (1961). Neonatal and infant immaturity in man. *Journal of the American Medical Association, 178*(1), 56–57.

Müller, F. (1969). Verhältnis von Körperentwicklung und Cerebralisation in der Ontogenese und Phylogenese der Säuger. *Verhandlungen. Naturforschende Gesellschaft (Basel), 80*, 1–31.
Müller, F. (1972a). Zur stammesgeschichtlichen Veränderung der Eutheria-Ontogenesen. Teil I. Zur Evolution der Geburtsgestalt: Gestaltstadien der Eutheria. *Revue Suisse de Zoologie, 79*, 22–97.
Müller, F. (1972b). Zur stammesgeschichtlichen Veränderung der Eutheria-Ontogenesen. Teil II. Ontogenesetypus und Cerebralisation. *Revue Suisse de Zoologie, 79*, 501–566.
Müller, F. (1972c). Zur stammesgeschichtlichen Veränderung der Eutheria-Ontogenesen. Teil III. Zeitliche Aspekte in der Evolution der Ontogenesetypen. *Revue Suisse de Zoologie, 79*, 567–612.
Nadin, M. (Ed.). (2015). *Anticipation: Learning from the past. The Russian/Soviet contributions to the science of anticipation*. Cham: Springer.
Nicholson, D. J., & Gawne, R. (2015). Neither logical empiricism nor vitalism, but organicism: What the philosophy of biology was. *History and Philosophy of the Life Sciences, 37*(4), 345–381.
Pianka, E. R. (1970). On r-and K-selection. *The American Naturalist, 104*(940), 592–597.
Portmann, A. (1939). Nesthocker und Nestflüchter als Entwicklungszustände von verschiedener Wertigkeit bei Vögeln und Säugern. *Revue Suisse de Zoologie, 46*, 385–390.
Portmann, A. (1941). Die Tragzeiten der Primaten und die Dauer der Schwangerschaft beim Menschen: ein Problem der vergleichenden Biologie. *Revue Suisse de Zoologie, 48*(11), 511–518.
Portmann, A. (1944). *Biologische Fragmente zu einer Lehre vom Menschen*. Basel: Schwabe.
Portmann, A. (1948). *Einführung in die vergleichende Morphologie der Wirbeltiere*. Basel: Benno Schwabe und CO..
Portmann, A. (1951). Ontogenesetypus und Cerebralisation in der Evolution der Vögel und Säuger. *Revue Suisse de Zoologie, 58*, 427–434.
Portmann, A. (1952). Die allgemeine biologische Bedeutung der Cerebralisations-Studien. *Bulletin der Schweizerischen Akademie der Medizinischen Wissenschaften, 8*(5), 253–262.
Portmann, A. (1956). Um eine basale Anthropologie. In A. Portmann (Author), *Biologie und Geist* (pp. 293–308). Zürich: Rhein-Verlag AG.
Portmann, A. (1962). Preface to a science of man. Translated by H. Kaal. *Diogenes, 10*(40), 1–26.
Portmann, A. (1965). Über die Evolution der Tragzeit bei Säugetieren. *Revue Suisse de Zoologie, 72*, 658–666.
Portmann, A. (1967). *Zoologie aus Vier Jahrzehnten: Gesammelte Abhandlungen*. München: Piper.
Portmann, A. (1970). *Entläßt die Natur den Menschen? Gesammelte Aufsätze zur Biologie und Anthropologie*. München: Piper.
Portmann, A. (1976). *An den Grenzen des Wissens. Vom Beitrag der Biologie zu einem neuen Weltbild*. Frankfurt am Main: Fischer Verlag.
Portmann, A. (1990). *A zoologist looks at humankind*. Cambridge: Cambridge University Press.
Reznick, D., Bryant, M. J., & Bashey, F. (2002). R-and K-selection revisited: The role of population regulation in life-history evolution. *Ecology, 83*(6), 1509–1520.
Ricklefs, R. E., & Wikelski, M. (2002). The physiology/life-history nexus. *Trends in Ecology & Evolution, 17*(10), 462–468.
Rittschof, C. C., & Robinson, G. E. (2016). Behavioral genetic toolkits: Toward the evolutionary origins of complex phenotypes. In V. Orgogozo (Ed.), *Current topics in developmental biology* (Vol. 119, pp. 157–204). Cambridge, San Diego, Oxford, London: Academic.
Rivers, J. R., & Ashton, J. C. (2013). Age matching animal models to humans – Theoretical considerations. *Current Drug Discovery Technologies, 10*(3), 177–181.
Sacher, G. A., & Staffeldt, E. F. (1974). Relation of gestation time to brain weight for placental mammals: Implications for the theory of vertebrate growth. *The American Naturalist, 108*(963), 593–615.
Sakai, T., Hirata, S., Fuwa, K., Sugama, K., Kusunoki, K., et al. (2012). Fetal brain development in chimpanzees versus humans. *Current Biology, 22*(18), R791–R792.

Scheiber, I. B., Weiß, B. M., Kingma, S. A., & Komdeur, J. (2017). The importance of the altricial–precocial spectrum for social complexity in mammals and birds–a review. *Frontiers in Zoology, 14*(1), 3.
Schilhab, T., Stjernfelt, F., & Deacon, T. (Eds.). (2012). *The symbolic species evolved*. Dordrecht, Heidelberg, London, New York: Springer.
Schultz, A. H. (1969). *The life of primates*. New York: Universe Books.
Schwartz, G. T. (2012). Growth, development, and life history throughout the evolution of *Homo*. *Current Anthropology, 53*(S6), S395–S408.
Shea, B. T. (2007). Start small and live slow: Encephalization, body size, and life history strategies in primate origins and evolution. In M. J. Ravosa & M. Dagosto (Eds.), *Primate origins: Adaptations and evolution* (pp. 583–623). Boston: Springer.
Sherwood, C. C., & Gómez-Robles, A. (2017). Brain plasticity and human evolution. *Annual Review of Anthropology, 46*, 399–419.
Smocovitis, V. B. (2012). Humanizing evolution: Anthropology, the evolutionary synthesis, and the prehistory of biological anthropology, 1927–1962. *Current Anthropology, 53*(S5), S108–S125.
Somel, M., Tang, L., & Khaitovich, P. (2012). The role of neoteny in human evolution: From genes to the phenotype. In H. Hirai, H. Imai, & Y. Go (Eds.), *Post-genome biology of primates* (pp. 23–41). Tokyo: Springer.
Spicer, J. I., Rundle, S. D., & Tills, O. (2011). Studying the altered timing of physiological events during development: It's about time… or is it? *Respiratory Physiology & Neurobiology, 178*(1), 3–12.
Starck, J. M., & Ricklefs, R. E. (1998). Patterns of development: The altricial-precocial spectrum. In J. M. Starck & R. E. Ricklefs (Eds.), *Avian growth and development* (pp. 3–30). Oxford: Oxford University Press.
Stearns, S. C. (2015). The concept of phenotypic plasticity and the evolution of phenotypic plasticity in life history traits. In A. C. Love (Ed.), *Conceptual change in biology* (pp. 131–146). Dordrecht: Springer.
Szathmáry, E. (2015). Toward major evolutionary transitions theory 2.0. *Proceedings of the National Academy of Sciences, 112*(33), 10104–10111.
Thompson, J. L., Krovitz, G. E., & Nelson, A. J. (Eds.). (2003). *Patterns of growth and development in the genus Homo*. Cambridge: Cambridge University Press.
Trevathan, W. R., & Rosenberg, K. R. (Eds.). (2016a). *Costly and cute: Helpless infants and human evolution*. Santa Fe, Albuquerque: School for Advanced Research Press.
Trevathan, W. R., & Rosenberg, K. R. (2016b). Human evolution and the helpless infant. In W. R. Trevathan & K. R. Rosenberg (Eds.), *Costly and cute: Helpless infants and human evolution* (pp. 1–28). Santa Fe, Albuquerque: School for Advanced Research Press.
Versace, E., & Vallortigara, G. (2015). Origins of knowledge: Insights from precocial species. *Frontiers in Behavioral Neuroscience, 9*, 338.
Walker, A., & Shipman, P. (1997). *The wisdom of the bones: In search of human origins*. New York: Vintage.
Watson, R. E., DeSesso, J. M., Hurtt, M. E., & Cappon, G. D. (2006). Postnatal growth and morphological development of the brain: A species comparison. *Birth Defects Research Part B: Developmental and Reproductive Toxicology, 77*(5), 471–484.
Werneburg, I., & Spiekman. (2018). Mammalian embryology and organogenesis. In F. E. Zachos & S. N. Asher (Eds.), *Handbook of zoology: Mammalia. Mammalian evolution, diversity, and systematics* (pp. 59–115). Berlin, Boston: De Gruyter.
Wiggins, O. P., & Schwartz, M. A. (2013). Toward a philosophy of life to underpin personhood in medicine. In K. B. Koh (Ed.), *Somatization and psychosomatic symptoms* (pp. 19–30). New York: Springer.
Wilkins, A. S. (2014). "The genetic tool-kit": The life-history of an important metaphor. In J. T. Streelman (Ed.), *Advances in evolutionary developmental biology* (pp. 1–14). Hoboken, NJ: Wiley.

Wirz, K. (1950). Studien über die Cerebralisation: Zur quantitativen Bestimmung der Rangordnung bei Säugetieren. *Acta Anatomica, 9*(1–2), 134–196.

Wood, S. L., Beyer, B. K., & Cappon, G. D. (2003). Species comparison of postnatal CNS development: Functional measures. *Birth Defects Research Part B: Developmental and Reproductive Toxicology, 68*(5), 391–407.

Workman, A. D., Charvet, C. J., Clancy, B., Darlington, R. B., & Finlay, B. L. (2013). Modeling transformations of neurodevelopmental sequences across mammalian species. *Journal of Neuroscience, 33*(17), 7368–7383.

Zollikofer, C. P., & de León, M. S. P. (2010). The evolution of hominin ontogenies. *Seminars in Cell & Developmental Biology, 21*(4), 441–452.

Andres Kurismaa, PhD, is a graduate of theoretical and evolutionary biology at Charles University in Prague and obtained his MSc degree in cognitive science at the University of Osnabrück (Germany). In his work he analyzes current questions in biological and cognitive research from theoretical and historical perspectives, with particular interest in holism as an empirical research paradigm. He has published several works on the significance of Russian/Soviet research programs in this context (Kurismaa 2019).

Chapter 7
Portmann's View on Anthropological Difference

Filip Jaroš

Abstract The most influential of Portmann's concepts up to the present is his characterization of early human ontogeny as secondarily altricial. This finding is coupled with the thesis of a social womb: human children are born prematurely in comparison with other primates, and they find a second womb in a social environment nurturing their healthy development. In Portmann's view, the interconnection of these phenomena forms a basis for the specific position of humans with regards to other forms of life. It is during the first year of extra-uterine life when a specifically human form is acquired in basic charts: it is a triad of an upright posture, linguistic utterances, and rational thought that together constitutes the core of anthropological difference. Although Portmann's position has strong ties to the tradition of philosophical anthropology (H. Plessner, A. Gehlen; humans as beings "open to the world"; critical stance towards Darwinian reductionism), his ambition was to put together the basis for an even more basal and comprehensive account of human beings, which would unite biological, social, cultural, and philosophical aspects. Even though his programme of basal anthropology did not succeed in its original scope, Portmann's concept of a social womb remains a source for valuable insights into the specificity of processes in human history and a point of departure for any comparisons between the social lives of humans and animal species.

Humans as a Specific Form of Life

Portmann's approach to the study of the life form of humans bears two seemingly opposing features. On the one hand, he seeks to bring a morphological picture of the human which could be compared to the biological features of other living beings, namely apes. On the other hand, his writings are distinctive for a motive rather alien to biological investigations into the form of humans: Portmann also stresses the intellectual dimension that cannot be separated from the human way of life. The

F. Jaroš (✉)
University of Hradec Králové, Hradec Králové, Czech Republic
e-mail: filip.jaros@uhk.cz

F. Jaroš, J. Klouda (eds.), *Adolf Portmann*, Biosemiotics 23,
https://doi.org/10.1007/978-3-030-67810-4_7

originality of his picture of humans is grounded in the way these modes are combined.

To characterize the position of humans among vertebrates, Portmann pays special attention to early ontogenetic processes intertwining the sensory, motoric, and social conditions of a new-born animal. Portmann (1935a, 1938) distinguishes two basic types of birth-condition among mammals and birds: *Nesthocker* ("nest squatters", altricial, nidicolous) and *Nestflüchter* ("nest fleers", precocial, nidifugous). In the case of mammals, a litter of a species of altricial type (e.g. most rodents and insectivores) is large (5–22), newborns are helpless, and their sense organs (ears and eyes) are developed to their full function only later. In contrast, a litter of a species of precocial young is small (usually 1–2), newborns are basically able to freely move and are partially independent (e.g. hoofed animals, whales). Newborn human babies show mixed features: usually one child is born, it is fully dependent on its mother in locomotion, even though its sense organs are already functional (for greater detail on the typology of early ontogeny – altricial vs. precocial – see Kurismaa, Chap. 6, in this volume). In contrast to young apes, which can (a few months after birth) actively grasp the hair of their mothers and cling on to their moving bodies, human babies are passively carried and transported.[1]

When the development of human infants during the first year of life is compared to that of other apes, we can find several differences. Firstly, the ratio of limbs to body weight does not allow a human newborn to stand on its extremities. The growth of legs is accelerated in the 6th month after birth. It is around the end of the first year that the legs can support a human baby, but the proportion of the head to body size is reduced even afterwards. Secondly, the brain of a human newborn has 3 times more weight compared to anthropoid apes (the overall body weight is approximately 2 times larger in the case of human newborns). Contrasting an adult to a newborn, the factor of brain growth is 4.3 for humans, 1.5–2.5 for apes (the number is similar for macaques, deer, and zebra, see Portmann 1990c: 130). In species other than humans, such rapid growth is characteristic for development during the embryonal phase. This observation, coupled with the classification of humans as secondarily altricial (*sekundärer Nesthocker*, Portmann 1941a: 921, 1941b: 512), leads Portmann to label the first year of a human newborn an "extra-uterine year" (*extrauterine Frühjahr*, Portmann 1941a: 924, 1941b: 517; *extraembryonale frühe Jahr*, 1942: 22). Human babies seem to be born prematurely, reaching the distinctive form of the human species only at the end of the first year, with the primary feature of an upright posture. Another formulation is given by Portmann

[1] Portmann does not comment on the fact that human babies (at least in some cultures) are left unattended by their mothers. According to Matsuzawa (2012: 300), the so-called supine posture of human neonates has important implications for their cognitive development (e.g. face-to-face communication, object manipulation). In contrast, postnatal chimpanzee infants constantly struggle to cling by lifting one arm and the contra-lateral leg. However, Matsuzawa's findings cannot be generalized to all ape species: a gorilla infant called Goma was provably able to take a supine posture until the age of two months (Lang 1961: 24).

(1965: 41) when he metaphorically talks of two uterine phases that take approximately the same time:

> We can thus say that the growing human being is born out of the mother's body into a second uterus in which he traverses the second half of his embryonic life: this is the *social uterus* (*der soziale Mutterschoß* in orig. Portmann 1965).

The resemblance between the speed of growth of a human fetus and an infant in its first year was discovered before Portmann,[2] but it was the Swiss researcher who pointed out the significance of the extra-uterine year for the formation of humanity. According to him, three distinctive elements of the human species are acquired between the first 7 and 12 months in the life of an infant: an upright posture, speech using words, and the basics of rational thought (Portmann 1990c: 132). Significantly, these capabilities can only be mastered with the help of a human social group, where the mother occupies a central position from the perspective of the infant. A human child is encouraged to make his first steps and praised for achievements in articulating basic words. It is the social group which attentively fosters his bodily and mental development after birth. There is a striking harmony between the developmental phase of a child and the readiness of a community for her training. The fusion of maturation, experience, and adaptation to social life is a distinctive mark of the human species (Portmann 1968b: 267).

The coordination of distinct developmental processes enables the acquisition of the most characteristic of human skills: speech. From a physiological perspective, the necessary condition for word articulation is the formation of a peculiar cavity in the throat, which is gradually created by the descent of the larynx. This process is both ontogenetically and phylogenetically connected with the formation of an upright position of the body. The ability to speak also requires a matching level of brain development. From an evolutionary perspective, the most important prerequisite for human speech is the complex social relationships among a group of primates (Portmann 1990b: 103). Great apes who have learned sign systems designed by humans use them primarily to achieve social contact with them. Here, it is important to understand that chimpanzees are more sensitive to visual than auditory cues: given the different structure of their vocal organs, they are not capable of producing articulated sounds but manage to use a number of gestures. On top of its evolutionary origin, Portmann (1990b: 98) stresses the unique capacity of human speech in constructing representations: he regards words to be fully arbitrary signs (i.e. symbols). To sum up, speech is the *expression* of the human condition, and the articulated exchange of meaningful words characterizes our mode of existence:

> Our speech is, presumably, the mark which most strikingly characterizes the human way of existence as being the expression of a special relationship to the world (Portmann 1990c: 126).

---

[2] Portmann (1968b: 266) mentions the work of E. von Lange and R. E. Scammon in this respect and talks about the phenomenon of humans as prematurely born as well recognized.

The Reception of Portmann's Biological Concepts in the International Scientific Community

Among the authors sympathetic to the ideas of philosophical anthropology, Portmann had a special position: he was a biologist by his education and professional career. Although his thinking was very much indebted to the Continental tradition (emphasizing form over function, holism over reductionism), Grene (1968) simplifies the state of affairs when she evaluates Portmann's work in a strict opposition to neo-Darwinism. His position is more balanced, and his life-long motivation surely was not to overthrow the Darwinian paradigm, but rather to open a space for an alternative view of living organisms stemming from a better sense for morphological and behavioural specifics and their meaning in the social life of individual species. Such a second path of biology posits a counterpart to the neo-Darwinian theories and approaches focused on the level of genes. The validity of mainstream research is not disputed; Portmann advocates an alternative point of view, which was threatened by the largely technological emphasis of laboratory research already during the peak of his career.

Already as a young researcher, Portmann was in contact with leading neo-Darwinian biologists from the University of Oxford (Julian Huxley, Gavin de Beer; Illies 1976: 82–3). Portmann participated at the Darwin Centennial in Chicago in November 1959 (cf. Portmann 1962: 8) which testifies to his contacts with leading protagonists of the modern synthesis. To illustrate some positive opinions regarding the major postulates of the modern synthesis, we can mention his agreement with Theodosius Dobzhansky on the importance of contingent mutations for the evolution of species, even if genetic heredity is outperformed by cultural evolution in the case of humans (Portmann 1962: 10).[3] Hence, Portmann's position is better characterized as critical of the so-called "hardened synthesis" (the phrase used by Gould 2010), the mainstream of Darwinian theory that, since the late 1940's, had been washing off other than genocentric views of evolution. Alongside, we find Portmann in opposition to the predominant belief in the omnipotence of natural selection that results in the origin of new species by continuous processes. On the top of that, "the hardened synthesis" insisted on an analysis of every animal and human organ through the optics of adaptation.[4]

---

[3] In the English edition of a popular pictorial publication, *The Beauty of Butterflies*, Portmann (1945: 10) envisages the importance of natural selection for establishing mimicry complexes among butterflies. This contrasts with the formulation in the German original, *Falterschönheit* (Portmann 1935b: 20), where the usefulness of mimicry for survival is contested. Since the English edition has an introduction by Julian Huxley, a stark proponent of natural selection, it is possible that Portmann's own opinion on mimicry was negated to serve the theoretical goals of the new synthesis.

[4] One could also argue that Portmann, "deeply in his heart", held consistent doubts against Darwinian ontology as such. For example, he openly criticized Darwin's metaphor of the struggle for life and highlights various cases of cooperation among animals, asserting that "attraction to other members of the species precedes all hostility and repulsion" (Portmann 1964a: 70).

The fate of Portmann's theoretical biology was very much connected with sociological factors in the biology of the second half of the 20th century. Gould (2010) is correct that the focus on the causal unity of explanations reigned mainly in English-speaking countries and culminated in the decades after World War II. With the English language becoming *lingua franca* of the world of science, other "national styles" of biological research and thought have backed off into the shadows (Gould 2010: 90; Komárek 2003). This development was amplified by the shift of research paradigms in German-speaking areas: the school of morphology and comparative development was superseded by phylogenetic systematics. More specifically, the tradition of idealistic (structuralist) morphology was in decline even before the onset of World War II, and Adolf Naef, Wilhelm Troll, and Adolf Portmann were the last outstanding representatives of this approach (cf. Rieppel 2016: 106; Monastra 2000). It is questionable if the label of idealistic morphologist is suitable to Portmann: the research tradition is known also as systematic or comparative morphology and does not subsume a unified theoretical and methodological approach (Levit et al. 2014). Yet, if we take Goethe's doctrine of form (*Gestalt* transcends pure spatial and mechanical accounts of organismal bodies) as a characteristic feature of idealistic morphology, two crucial works of Portmann (1948, 1960a) are indebted to this way of thought. Portmann also participated in the so-called "Colloquium Palatinum" (Gestalt-colloquium) co-organized by Troll; however, the quality of their relations is not known (cf. Nickel 1996: 182, 190).[5]

In his crucial theoretical-biological writings, Portmann (1948, 1953, 1960a) became not just reluctant to match his theories with genocentrism and adaptationism, but he also increasingly contrasted his approach to the technological and mechanical view of life as such, which left him more and more isolated in the field of zoology towards the end of his career (cf. Koechlin 2004; Rieppel 2016: 100–101). Nevertheless, the 1950s and 1960s bring success to Portmann in terms of publishing English translations of his main books on theoretical biology. However, it is important to distinguish impact on a broader intellectual audience and acceptance from the leading professionals in the field. To provide a balanced picture, it is informative to look at the journal reviews: what we find is basically incomprehension of the philosophical and theoretical background of Portmann's biology.

All of Portmann's main theoretical-biological writings were translated into English in the 1950s and 1960s (Portmann 1952, 1961a, 1964b). *Animal Forms and Patterns* (Portmann 1952) was negatively reviewed by Varley (1953) and ambivalently by Amadon (1954). A leading British ethologist, William H. Thorpe (1962), was dissatisfied to find that due to the lapse in time since the German original was issued (Portmann 1953), the content of the second English edition of *Animals as*

Portmann's respect to the results of neo-Darwinian genetic research was paid rather to the direct outcomes than the theoretical underpinnings. To get a sincere portrait of his relations to the proponents of the modern synthesis, we also have to take into account the conciliatory and respectful traits in his character.

[5] A slightly different approach which might be dubbed *phenomenological ethology* is characteristic for Portmann (1953) (I am indebted to Jiří Klouda for discussions on this issue).

Social Beings (Portmann 1961a) was at least 7 years old, and he therefore could not recommend the book to any professional in the field.[6] Especially informative here is the discussion of Gavin de Beer (1966) and John W. Yolton (1966) over *New Paths in Biology* (Portmann 1964b) in *The New York Review of Books*. Sir Gavin de Beer, a director of the Natural History Museum in London and a president of the Linnean Society of London, criticized Portmann for not providing experiments and articulated hypotheses that could confirm his theoretical programme. Moreover, Portmann's reserved opinion on investigations into genetics is portrayed as a retreat to vitalism and giving up the ambition to find a unifying principle in biology. In his response to de Beer (1966), John W. Yolton (1966) emphasized the philosophical nature of Portmann's writings and offered that his masterpiece should have been called "New Paths from Biology". Nevertheless, he rightly remarks that it is difficult to understand Portmann without a basic orientation in the Continental tradition (which Yolton seems to possess).[7]

Portmann's writing was nevertheless acknowledged by some Anglophone scholars. Incidentally, the first congenial interpretation of Portmann's *biological* work written in English came from the discipline of *philosophy*. It could be said that the positive evaluation by the philosopher Marjorie Grene (1965, 1968) signified the transfer of Portmann's biological thought into the realm of philosophical biology and philosophical anthropology, areas dominated by scholars with a philosophical background.[8] It is instructive that even Steven J. Gould (1977a: 74), who made his reputation as a fierce opponent of "the hardened synthesis" and who had sympathies with the Continental tradition of biology (and had read Portmann in German), labelled Portmann's approach a "spiritual view of evolution" (or alternatively as non-Darwinian, "psychogenic view"; Gould 1977b: 349, 369). Gould supposedly targeted seemingly teleological moments in Portmann's writing: as an example, the early birth of a human infant was seen as a function of his "openness to the world" (Portmann 1965: 40).[9] We can conclude that with the exception of the students who graduated under Portmann, the theoretical-biological framework peculiar to

---

[6]Also, Roger A. Stamm criticizes Portmann for not following new approaches in ethology (personal comm., 14.6.2017). It is fair to say that the second German edition of *Das Tier als soziales Wesen* (Portmann 1964c) does show some updates regarding the ethological literature. Moreover, Portmann does not seek a classic ethological programme here, but promotes a *phenomenological* perspective to animal behaviour (Portmann 1954: 178). W. G. van der Kloot (1962) also found Portmann (1961a) outdated due to the time lapse in translation, but his short review in *Science* is otherwise positive.

[7]Virtually no English translations paid respect to the quality of Portmann's original titles. This fact was reflected in another critical review of *New Paths in Biology* (1964b) by Oppenheimer (1965).

[8]It is interesting to read Grene's critical stance toward her earlier appreciation of philosophical anthropologists, as found in Grene (1968, 1974). She even refused to include Portmann's work among significant influences on her thought (Grene 2002: 80–81).

[9]Portmann asserts that the ontogeny of human beings does not make sense without an acknowledgment of the full (adult) form, which is characterized by elaborate cultural life. This need not be viewed as a teleological argument, as it is clear that the specificity of *any* species is grounded in early developmental phases.

Portmann's work (concepts of *Innerlichkeit* and *Rank*) was not followed by any professional biologist of international significance.

It might be considered a misfortune for the spread of Portmann's ideas that he did not become closely connected to the predominantly German school of ethology. Despite the rich contacts of Portmann with leading bird ethologist Otto Koehler, the latter turned hostile towards the theoretical framework of the former after reading *Animal Forms and Patterns* (Die Tiergestalt, Portmann 1948) and *A Zoologist Looks at Humankind* (Biologische Fragmente zu einer Lehre vom Menschen, Portmann 1944b) (Ritter 2000: 249). Konrad Lorenz, who is repeatedly cited in Portmann's writings, met him in Basel in the 1950s[10] and later made a critical reference to his concept of self-presentation (*Selbstdarstellung*) (Lorenz 1983).[11] Adolf Remane, as president of the German Zoological Society in 1963/4, strongly criticized Portmann's theoretical biology as metaphysical speculation (*metaphysische Schreibtischgedanken*, cited in Ritter 2000: 207 and Illies 1976: 167). Yet, from a different perspective, the teachings of Portmann and Remane were similar in their consideration of *Gestalt* and anti-reductionist epistemological approaches (Rieppel 2016: 105).[12] To conclude, the main reason for Portmann's growing isolation from leading centres of biological research from the 1950s on was not just the reluctance to embrace neo-Darwinian theory. As can be seen on the example of Gould, a science of form and corresponding anti-selectionist theories were regarded at least as a valuable dialectic partner for most of the leading protagonists of the modern synthesis. Furthermore, the consideration of *Gestalt* has heavily influenced at least the German tradition of ethological research, which was still influential several decades after WWII. Portmann simply invested more effort to tread new paths for his idiosyncratic biology than to relate to the research agenda of his contemporaries. In accordance with this goal, many of his theoretical writings were published as popular scientific books, aimed at persons interested in biological problematics rather than professionals in the field.

In contrast with Portmann's theoretical views, some of his empirical concepts and data proved relevant for the next generations of biologists, and this applies also beyond the group of his doctoral students (from whom Katharina Wirz has been widely cited – see Wirz 1950; Mangold-Wirz 1966). Portmann's distinction between altricial and precocial types of newborn birds and mammals was coupled with the

---

[10] Personal comm. with Roger A. Stamm (7.3.2019). The author did research at the personal library of Lorenz, which is accessible in the Konrad Lorenz Institute. There are copies of several books by Portmann (Portmann 1944a, b, 1956, 1960a, 1967a, 1970) and at least one (Portmann 1967a) has marks of reading by Lorenz. While Portmann's estimation for Lorenz's findings in ethology is visible, he expresses a critical stance toward his uncritical but catchy parallels between humans and animals (Portmann and Illies 1976: 293–297). Furthermore, he openly criticizes Lorenz's support for the universal impact of the selection theory in the section called "Against the simplifiers" (*Gegen die Simplifikateure*) (Portmann 1974: 39–41). I thank Jiří Klouda for the last reference.

[11] I thank Pietro Conte for the reference. For the concept of self-presentation and its opposition to Darwinian functionalism see Toepfer (2011).

[12] Portmann's former doctoral student Roger A. Stamm portrays the mutual relations of Portmann to Lorenz, Remane and Koehler as respectful rather than bitter (personal comm., 7.3. 2019).

hypothesis of change in ontogenetic type as a prerequisite of higher cerebralization (dubbed "Portmann's rule of precedence", but disputed by Gould 1977b: 349; positively tested on strepsirrhines by Martin 1975 and on birds by Mlíkovský 1985). Stemming from these investigations, Portmann labelled the ontogenetic state of humans "secondarily altricial" (for the suitability of this term and concept from the current point of view see Rosenberg and Trevathan 2002; Trevathan and Rosenberg 2016; see also Kurismaa, Chap. 6, in this volume).

Portmann's interpretation of human babies as somewhat prematurely born (at least in comparison with apes) has proved to be the most influential of his findings. The related concept of the social uterus has never been forgotten in both the philosophical and scientific communities and was repeatedly echoed also among Anglophone anthropologists (Montagu 1983). Having provided a specific biological characteristic of human ontogeny, Portmann did not seek to subsume humankind under the domain of biological research. In the evaluation of his treatment of human specificity, we need to be more cautious than Müller (1988: 139), who views Portmann's affirmation of human uniqueness as an inference from his morphological and physiological studies. There was surely also a strong theoretical motivation bolstered by the cultural milieu: Portmann basically worked for the emancipation of the picture of humans from the tenets of a universal evolutionary theory based on a simplified reading of Darwin. In the interwar period, one-sided interpretations had become increasingly connected with a political ideology and gave rise to vulgar Darwinism in Germany.[13] Conceptually, Portmann forged such an understanding of humans that would overcome traditional boundaries between the sciences and humanities; not, however, by favouring one set of methodologies over the other, but by developing an interpretative framework that would start off from the phenomena on the boundary of traditional concepts of *soma* and *psyche* (social uterus, physiology of emotions). Since anthropology takes an intermediate position between life sciences, social sciences and humanities, such a wide conception of human nature had better chances to survive as a scholarly concept than, for example, phenomenologically inspired views on animal morphology. Correspondingly, if labels like philosophical biology are extremely rare to come upon, philosophical anthropology has been formed into a standard programme in many (mostly Continental) curriculums of philosophy. Thus, Portmann is internationally recognized rather as an anthropologist than a biologist nowadays.

Portmann's Work in the Light of Philosophical Anthropology

Portmann's view on the special position of humans within the forms of life is influenced by the tradition of philosophical and physiological anthropology (mainly Plessner and Buytendijk). Fischer (2009: 155) characterizes the whole movement of

---

[13] For the link between political motivations and the frame of his theories, as well as Portmann's own reflections on the topic, see Ritter (2000: 252).

philosophical anthropology as an attempt to counterbalance Darwinian reduction of the human mind to non-human cases on the one hand, and a teleological linkage between the human body and mind (as in German idealism) on the other hand. It is likely due to his biological education that the evolutionary view is more visible in Portmann's writing in comparison to the thinkers creating the core of philosophical anthropology (Scheler, Plessner, Gehlen). The evolutionary approach brings a special emphasis on the homology of anatomical, morphological, and behavioural features between humans, apes, and other forms of life. In the tradition of Darwin's interpretation, this position highlights the continuity of mental faculties among animals and humans. Distinctions between the respective forms of existence are a matter of accentuation rather than kind (Portmann 1990c: 136). This is the basis which must be remembered before any attempt to draw a line separating humans from animals can be made.

> In line with these common hereditary processes [concentrated in the chromosomes], the propagation and preservation of the species among humans follow the same rules as those which govern animal life and the preservation of species among plants. Nor must we ever forget the correspondence that links our whole structure, in all its tissues and organs, to the large family circle of the higher vertebrates. (Portmann 1965: 39)

Nevertheless, there is also a view coming from an opposite direction which substantially informs Portmann's opinion on the nature of humans. Within the movement of philosophical anthropology, Brentari (2018) identifies a strong motivation to find a special position of humans within the organic world (*Sonderstellung des Menschen*, see Scheler 1928). Indeed, we can also find passages written by Portmann which acknowledge the specific nature of the human mind.

> Biology does not seek the 'spirit' which can be separated from soul and body. It seeks the organic materials of a special world-relation which, because of historical considerations, has here been termed 'intellectual and spiritual' (Portmann 1990c: 137).

The preceding quotations delimit the frame of Portmann's thought about the relations between behavioural and cognitive capacities of humans and animals. Throughout his work, we repeatedly come upon a strategy where human characteristics are initially matched to those of animals; then, in a second step, comes differentiation, where a specific (and positively valued) human feature is acknowledged. Generally speaking, humans are shown to have a richer mental world and behavioural opportunities compared to other forms of life, but they are not portrayed as separated from them. Portmann does not seem to be captivated by Scheler's concept of spirit (*Geist*) that separates human beings from animals. Humans are not just another sort of animal, but nor are they possessors of a substance or principle *opposing* to the world of nature (Portmann 1968c: 36–37; cf. Scheler 1928).

Another leitmotif of philosophical anthropology is the Herderian figure that portrays humans as vulnerable beings who overcome their deficiencies by the means of intellect and culture (Halák and Klouda 2018). In a classic interpretation of Gehlen (1966), humans are portrayed as "deficient beings" (*Mängelwesen*) whose instincts are undetermined and whose dispositions are open to behavioural plasticity and have to be further regulated by the formation of habits. Portmann (1962: 8–9, 1968a, 1970b) openly criticizes this position. The openness of human traits does not mark the vulnerability of humans, it is rather the asset of the species;

some instincts are replaced by the advanced organization of the cerebral cortex and its synapses, which eventually enable various means of cultural transmissions and the production of technological tools. Moreover, the openness is only relative; human babies are endowed with several instinct-based behavioural patterns, for example sensitiveness to the appearance of the characteristic pattern of the human face, expressed in a child's smile. In accordance with the postulates of his theoretical biology, Portmann highlights the self-presentative function of technological products. In contrast to Gehlen's interpretation, technological tools and appliances are not portrayed as the substitutes of imperfect body organs, but are produced as the results of the never-ending creative attempts of humans to improve and enhance their capacities together with their presentation (Portmann 1968a: 299–300). Similarly, the ontogeny of a human child cannot be comprehended as the retardation of the growth rates typical for apes (a critique of the concept of L. Bolk), but as a period which correlates with the opportunity to integrate into a social group (see also Kurismaa, Chap. 6, in this volume).

Portmann also develops Plessner's approach to *conditio humana* as "natural artificiality". In his interpretation, there is not any natural state of society, but it is characteristic of human life to form societies that are opposed to the world of nature. The state of civilization is our "artificial second nature" (Portmann 1964a: 70–75). Accordingly, Grene (1974: 285) has formulated the maxim of Portmann's anthropological stance, stating that "we are *biologically* formed to be *cultural* animals".

Another characteristic problem of philosophical anthropology is related to Umwelt–Welt distinction. In the interpretation of the leading thinkers of the movement, animals are bound to their environments according to their species-specific physiological and sensorial dispositions (*umweltgebunden*), whereas humans are open to the world (*weltoffen*) (Scheler's distinction). There are generic limitations of a behavioural set for every single species of animal. Further, human interactions with other people are to a large degree regulated by conscious will, whereas the life of animals is characterized by abrupt, involuntary changes in mood. As an example, humans are able to control their expressions according to social norm (laughter, weeping), whereas truly spontaneous manifestations (screaming, threatening gestures, and, partially, flushing) are rare (Portmann 1967b: 197). On the other hand, Portmann (1970a: 18, 1990c: 136) stresses the fact that many human acts are bound to their environments, whereas the capacity to learn is present among all vertebrates (e.g. the finch has to learn its song through imitation), and at least higher animals are able to perform new types of behaviours. Directly building on the Plessnerian motive, Portmann concludes that humans are special due to their competence to reflect on the natural limitations posed by the specificity of their Umwelten. Such a reflecting capacity subsequently enables humans to take a vantage point of view and reach beyond the initial limitations (see Plessner's concept of eccentric positionality in Tolone 2015).

To conclude, we argue that Portmann's anthropology gives a *transformative* account of human nature (see Jaroš and Maran 2019). It is a counterpart of the additive model of human nature where, to put it simply, humans are viewed as animals gifted with some special abilities. Under the transformative view, the human spirit is not a non-material entity added to the faculties already available to animals. Rather, the position can be characterized by a motto of Merleau-Ponty (2003: 208): "before being reason, humanity is another corporeity". The peculiarity of human organic form signifies a transformation from the lineage of the anthropoid apes, both in corporeality and the possibility to frame mental faculties culturally. As far as the developmental-evolutionary perspective is considered, this view must be contrasted with a postulate known as biogenetic law (ascribed to Ernst Haeckel) which asserts that during embryogenesis, novelties are superimposed on repetitions of earlier evolutionary phases. Giving the example of the early formation of the human curve of the spine in the 2nd month of uterine life, Portmann (1990c: 136) explicitly opposes biogenetic law and demonstrates that the human embryo does not start its development like the fetuses of apes and only gradually takes on specific human features, but is distinct right from the initial stage. In the most radical interpretation, our organic form prefigures the complexity of social structure and intellectual life, which drags us out of the world of other living beings:

> The more comprehensively anthropology tries to see its subject, the more manifest becomes Man's separateness and loneliness in the whole of living creatures (Portmann 1961a: 158–159).

Portmann's Programme of Basal Anthropology

As we have seen in the previous section, the question of the physiological background of humans' relation to the world was an important source for philosophical anthropologists. Apart from the core of this intellectual movement formed by philosophers, two thinkers educated in physiology and medicine (Buytendijk) and biology (Portmann) made significant contributions to the resulting picture of humans (and animals as their counterpart). It is also possible to subsume the work of von Uexküll, Buytendijk, and Portmann into a distinct class of anthropological physiology (Fagot-Largeault 2009). This approach is characteristic of the empirical accounts of phenomena that persuasively demonstrate the physiological background of human emotional and mental acts (e.g. turning red as a result of the feeling of shame). It was especially Buytendijk who provided innovative comprehension of the relations between *soma* and *psyche* in humans and animals, and Portmann acknowledges this influence on his own theoretical concepts (cf. Thinès 1996). Portmann (1968a: 301) considers the distinction between *soma* and *psyche* to be the outcome of pre-scientific thought, constituting different names for two polarities of a single unity. The psychosomatic view of human nature refuses either to a) reduce mind to physiological processes and behaviours (a path taken by some ethologists and psychologists), or to b) interpret mental processes independently and claim that the connection to physiological structures remain unknown (social sciences, also some traditions within cognitive science). Yet, the psychosomatic approach paradoxically tends to keep the differentiation of *soma* and *psyche* alive if it seeks to understand the connection between entities requiring separate names. Portmann sets foot on a different path, characterized by a "simultaneous consideration of both the structural and mental realms as being two aspects of a single being" (Portmann 1990a: 115). Although anthropological physiology does not form a separate ontology within evolutionary views,[14] Portmann's overt criticism of mechanistic accounts of life, his developmental perspective acknowledging different stages in the life of a human individual, his appreciation of the life form of humans as distinct from animals, and the ambition to provide a holistic account of humans uniting biology, physiology, anthropology, and humanities led him to use a distinct name for the programme: basal anthropology (*basale Anthropologie*: Portmann 1962: 1968a).[15]

Portmann's motivations for the project of basal anthropology as a new integrative science of humans were manifold, and some of them resulted in contradictions, at least in retrospective glance. Portmann was in strict opposition to attempts to subsume the question of human nature under a special biological, sociological, or religious theory (such as Marxism, Buddhism, or Christianity). In the characterization of humans as beings open to the world, Portmann has aligned himself to the thinkers stressing the historical dimension of humankind. At the same time, he claims universal validity for the programme of basal anthropology, which has aspirations to form "a genuine science of man" (Portmann 1962: 26). How to account for this tension between the ever-changing historical image of humans and the proclaimed universality of this new anthropology? Is not Portmann falling under the self-delusive spell of prophets, unconsciously excluding his own approach from the tenets of time?

In fact, Portmann acknowledges that his anthropology is also rooted in a specific intellectual milieu (Ritter 2000: 252). From the beginning of his scientific carrier, he showed great interest in the historical dimension of scientific theories, namely Darwin's evolutionary theory. However, Portmann does not make any relativistic arguments about the validity of scientific theories. The word "science" maintains a high esteem in his writings, related mostly to the solidity and validity of empirical findings. However, there is another connotation of scientific enterprise for Portmann, the one which is closely related to phenomenological worldview: it aspires to give primarily sincere descriptions, not theoretically rounded facts.

---

[14] Darwin's original writings are surprisingly "psychosomatic", especially consider Darwin (2009) – we can actually portray Darwin, Buytendijk, and Portmann as the thinkers transgressing the opposition between soma and psyche: the unity of both polarities is visible in behavioural acts of animals and humans. For another view on anthropological physiology see Kurismaa, Chap. 6, in this volume.

[15] In the only English translation of Portmann's texts dedicated to basal anthropology (from French), H. Kaal uses the term "basic anthropology". M. Landmann, who characterizes Portmann's approach as "anthropological monism" (Landmann 1985: 36) and put it among the cornerstones of his own synthetic approach, uses the term "fundamental anthropology" for both Portmann's and his programme (Landmann 1985: 262). In this volume, we prefer to talk about "basal anthropology", which better reflects a primary character of Portmann's considerations, rather than some basic axioms of a theoretically unified system.

Often, the aim of such descriptions is to show the particularity of every formal theory; in connection to the special theories of evolution, Portmann does not believe in their validity to account for the diversity and richness of life forms, and does not hesitate to talk about the mystery which resides in natural processes. As a result, Portmann's use of "science" is basically unpalatable for natural scientists and philosophers at the same time. Natural scientists miss formal and positively formulated postulates in Portmann's programme, whereas philosophers of the post-modern era must look with deep scepticism at its ambition to sketch a universal picture of human beings.

It is also characteristic that basal anthropology proclaims itself to be in the position of a servant (Portmann 1962: 24): if "terminate" (that is, ideologically and conceptually rounded) forms of anthropology give a positive programme for the characterization of human nature, and subsume individual diversities in favour of theoretical unity, basal anthropology warns against the partiality of similar worldviews. It does so at the price of its own compactness and the urge to intermingle with ruling scientific, cultural and ideological streams. It is not coincidental that Portmann (1970) dedicated strong attention to ethical issues, and the conservative character of basal anthropology foreshadows its usefulness rather as a regulative or prohibitive teaching than a ruling idea for scientific research. Nonetheless, it has great potential in education, as is envisaged by the call for individual responsibility and self-reflection by later environmental movements (see also Stamm, Chap. 4, in this volume).[16]

The united interpretation of the human form and its mental life has already been illustrated by the phenomenon of the social uterus. The ontogenetic view of human psychological and mental abilities is also applied for the life stages occurring after the first year of life: here, Portmann (1990a) repeatedly borrow facts from developmental psychologists R. A. Spitz and H. Wiesener. Apart from such widely received results, some of the most original, even if surely problematic moments in Portmman's view on the development of a human being are linked to thoughts peculiar to the teaching of *Typenlehre* (constitution approach). Portmann (1990a: 117–120) asserts that intellectual and bodily transformation during puberty is the result of the transition of the *form* of a human child and calls this stage of ontogeny the "second metamorphosis" (the first metamorphosis is related to age of six; Portmann 1990a: 111). Whereas a child is characterized by an emotional, dreamy inner life and pyknomorphic (i.e. thickset) somatotype up to ages 8–10, this transforms to rational thought and a leptosome (i.e. slender) figure during the later phase of puberty (ages 10–16). Portmann also envisages the peculiarity of human senescence and its significance for our social life. The lifespan of humans is twice as long as that of apes and surpasses any mammalian cases. Although senescence is usually portrayed by biologists as the decline of vitality and fitness, Portmann (1990a: 146) highlights instances of great individuals who have reached their

[16] If we understand Portmann's programme of anthropology more specifically, some of its basic empirical findings are echoed in the current fields of human evolutionary – and neuro-developmental science – see Kurismaa, Chap. 6, in this volume.

brightest accomplishments after the age of 80 (Sophocles, Radetzky, Titian). In conclusion, advanced age is seen as a quantitatively distinct period of human life when the mind of an individual can reach a balanced state and positively influence the cultural life of the society.

Existential considerations make another distinguishing feature of Portmann's basal anthropology. The problem of an individual life as an open task is transferred also onto a collective level: where is humanity as such heading? Portmann's train of thought shows that not merely the mental abilities of humans quickly evolve in contemporary cultural and technological evolution, but also the human form as such. We are far beyond the control of an evolutionary process, working on the organic structure of our species and influencing our mental life also through the unconscious (Portmann 1990a: 7–10). What humanity will mean in the upcoming era is a puzzling riddle (cf. Portmann 1970).

Portmann (1990a: 118–138) interprets data signalling an increase in the height of Western citizens in the twentieth century as the progressive occurrence of slender leptosomes (as opposed to sturdy pyknomorphs) in highly urbanized environments. Portmann believes that the leptosome type is statistically more attuned to intellectually demanding professions, whereas the pyknic type is prevalent in agricultural regions. Thus, a higher prevalence of leptosomes in urban areas can be accounted for by the migration of intellectual career-seekers into cities ("professional and academic selection", Portmann 1990a: 130).[17] Moreover, there are also hypothetical epigenetic influences increasing the growth factor (optic and acoustic stimuli) which are particularly ubiquitous in cities.

Portmann also links these shifts with an earlier onset of sexual maturity — both trends basically overlap in the twentieth century. At the same time, emotional and intellectual maturity takes place several years after the end of puberty and occurs later in subsequent generations. The antithetic shifts in sexual and mental metamorphosis results in a long period of imbalance, which Portmann (1990a: 128) evaluated as an important problem for contemporary society.

It is good to point out that similar worries had led Konrad Lorenz in the 1940s to a sharp critique of the self-domestication process in Western countries (cf. Lorenz 1983). Portmann (1964a: 73–74) also seems to be disturbed by the challenges of life in large cities facing the ancient biological dispositions of our species. As a classic example, we can name the calibration of human personal bonding to life in small groups. However, increasing urbanization is regarded as an inevitable fact of the modern era by Portmann and is seen as a prompt to search for alternative forms of social relations leading to a new balance. Accordingly, Portmann does not succumb either to a romantic view of healthy and unbounded life in rural settings, or techno-optimistic views largely linked to the social engineering of future generations (eugenics). Before WWII, Portmann's critique of eugenics was surely related to his social-democratic political orientation. Even before the discrediting of *Typenlehre*

---

[17] Portmann does not comment on possible differences in reproduction rates between urban and rural populations.

by the deeds of German Nazis, he had been very critical of any social order suppressing individuality in favour of groups (cf. Rieppel 2016: 98–101). His restrained position holds consistent also in the post-war era in the face of promises of engineering the human genome: Portmann warns against any unidirectional attempts to "improve" the evolution of humans.[18] Additionally, his balanced view on the problem of human types can be illustrated by his appreciation of the "mediocre" constitution (between leptosome and pyknomorphic) and the vulnerability of intellectually-oriented persons (Portmann 1970c: 274–277).

Portmann's emphasis on the dynamic evolution of humankind and its specific competence to encompass novel phenomena in the world and transfer them to new generations leads him to differentiate two types of evolution: organic *evolution* and human *history*. As history is broadly equated to the products of human intellectual life (Portmann 1990a: 11), we can say that this distinction basically relates to the distinction between natural selection and cultural evolution. Portmann agrees with the mainstream of neo-Darwinism (see the reference to Dobzhansky in Portmann 1962: 13) in the sense that cultural evolution has surpassed the influence of genetic changes on the formation of humankind. But in sharp contrast to the tendencies of the late 1960s that led to the formation of sociobiology in the mid 1970s, he is reluctant to subsume the process of cultural evolution under the theoretical domain of pristine biology. To illustrate his view, he refuses to intermingle the concepts of archaeology (e.g. migration, trade with artefacts) with the framework of theoretical biology (mainly working with natural selection).

Portmann's reluctance to subsume human culture and its history under the single interpretative framework of biology has profound theoretical implications. Moreover, in accordance with our previous analysis, we must emphasize that Portmann does not want to separate cultural processes from a layer that is often portrayed as our biological nature. Thus, any cultural anthropology that is not interested in our biological inheritance cannot be viewed as a legitimate heir to Portmann's programme. At the same time, reductive programmes of many traditions within human ethology and all of sociobiology underestimate the variability and plasticity of human behaviour. Crucially, human social life is only partially regulated by hereditary factors: it is the very form of the human family which has always been a subject to ongoing changes. Every new generation finds itself in a new social and cultural situation, so a human individual has to find its own way how to navigate through social structures. Complementarily, it is the state of constant friction within a society which stimulates some individuals to attempt to change it, and the very phenomenon of human responsibility is connected to reflective thinking about the state of our social life. According to Portmann (1965: 40), even the most simple man is capable of directing his interest at will and observing himself from an external vantage point. The individual human life is unfulfilled if it is rid of

[18] Portmann (1970d: 378) exclaims: "Hands off any life form we did not create ourselves". Similar moments in Portmann's work are also significant for the appreciation of his thought by theologian Karl Barth and philosopher Karl Jaspers, who were his colleagues at the University of Basel (Barth resided in the same building as Portmann, see Rieppel 2016: 102).

the question of its purpose. Since similar existential problems have the character of a task, basal anthropology can merely acknowledge its importance for the human form of life, but not give a binding answer.

Conclusion: Portmann's Answer to the Problem of Anthropological Difference

In our analysis of Portmann's anthropology and the problem of anthropological difference specifically, we must also take into account the ontological background that supplements the factual basis of the problem. Authors commenting on the relation of humans to animals always navigate between the appreciation of human uniqueness (difference in kind) and the setting of humans within the evolution of animal forms (difference in degree) (cf. Jaroš and Pudil 2020). Which position in the continuity/discontinuity debate is characteristic of Portmann's basal anthropology? What is the ontological stance informing the anthropological research of Portmann?

To answer these questions, we have first to take into account that there are two traditions with profoundly distinct aims and concerns in which Portmann's writings are embedded. Portmann the biologist marks conspicuous moments from the social life of apes (e.g. the social factor in the quality of the mothering behaviour of apes) as prerequisites to our form of life. Portmann the anthropologist asserts that the human form of existence is set apart as something quite special, due to its richness of world-relation.

It is our conclusion that the language of *discontinuity* is more characteristic of Portmann's writing about organic creatures. Portmann the biologist is compelled to the structuralist approach to morphology which regards living species as transformations from one form to another one, without the need to find intermediary forms that is typical of the Darwinian vision of evolutionary continuity. Portmann the anthropologist aims to provide a biologically informed picture of the special position of humans (*Sonderstellung*), not to elevate the level of consciousness and reflexivity among animals, or alternatively decrease specific capabilities of humans (Portmann 1968a: 303). In comparison with other zoologists writing about the biological features of humans in his time (Konrad Lorenz, Desmond Morris), Portmann refuses to encompass human behaviours and cultural assets with purely biological terminology. Quite the contrary: the determination of common features of the social life of higher animals and humans is not perceived as the final goal of reductive models, but as the base from which the specificity of the social life of every species can be exposed. In our interpretation of Portmann's work, even the inherently cultural life of humans does not lend human beings another ontological status, since dispositions for this are preformed from the very beginning of individual ontogeny and therefore ingrained into human organic nature. Human culture can be viewed as a species-specific form of social life with such a high degree of cognitive and instrumental compartmentalization that we need to *model* human intra-specific (and also human–animal) interactions by other than purely biological modes.

The previous view can be supported by dialectics characteristic of Portmann's thought. To depict a discontinuity between the human and animal relation to the world, a phenomenon highly characteristic of the human way of life is exposed, put in the context of human social and intellectual life (culture), and given a specific name ubiquitous in common language. For example, Portmann (1968a: 298) keeps a separate label for the performance of work (*Arbeiten*), related to our historic mode of existence, as opposed to animal behaviour *(Verhalten)*. Speech is not just communication among members of the same species, but the expression of the special human condition.

On the broad theoretical level, Portmann distinguishes two evolutionary processes: organic *evolution* and human *history*. As an example, animal societies evolve through the random change of hereditary factors, whereas human history is subject to formative decisions and crises (Portmann 1964a: 75). The following quotation puts the discontinuity in an expressive, poetic way:

> We know swifts as creatures hunting in summer during their whistling flights in the sky of our towns. They have been nesting on human buildings for thousands of years though their life form has not been touched in any sense. Sometimes they build their nests on Greek temples, sometimes on Gothic cathedrals or in cracks of modern viaducts and power plants, they do not care for the difference. However, the change in human buildings, where the birds ever again find their species-specific roosts, expresses the rapid transformations in the style and value-orientation of human life. (Portmann 1968a: 298)

This moment is suitable for reflections on Portmann's position towards the central problem of philosophical anthropology: the ontological relation of humans and animals. Although the quotation might be interpreted as downgrading the cognitive and social world of birds, this is not the goal. Portmann the biologist regards awe and admiration as strong emotional impulses serving to comprehend animal forms. In contrast to the core movement of philosophical anthropology, animals are not portrayed as a negative image of the symbolic capacities of humans, with a goal to find a hiatus separating humans from animals (cf. Brentari 2018). It is highlighted by Portmann (1968b: 270) that to find a proper characterization of our kind, we need to counterbalance between methods concentrating on parallels (comparative biology) and differences (not specified, probably some areas of humanities) between human and animal species. Portmann (1964a: 233) asserts that all higher animals are fundamentally social creatures, and it is just through the acknowledgment of this fact that one could reflect on the secret connection of humans and animals. Portmann highlights the ability of several species of birds to learn various songs and mimic even industrial noises (e.g. the lyrebird, see the last chapter of Portmann 1964a). Further, advancements in comprehending the basics of human symbolic systems are acknowledged in the case of experiments on the linguistic abilities of apes during the 1960s and 1970s. In fact, the symbolic capacities of communication systems among species of cetaceans (mainly dolphins) are believed even to surpass the case of apes. When Portmann (1990b: 119) finally finds humans superior in the richness of their relations to the world, it is not due to a single trait, but as the result of the combination of advanced mental abilities, the faculty of speech, and the special organs of relation to the world (i.e. hands).[19]

Moreover, Portmann's theoretical aim to establish the special position of humans in the realm of living beings is also recognizable in his biology. Here, it is important to mention the thesis of Ritter (2000) that both anthropological and biological theoretical programmes of Portmann were formulated during the period 1937–1944. In his studies on the early ontogeny of mammals during this time, Portmann contrasts the behavioural plasticity of humans and the relative rigidness of behaviours and movements in apes. From the perspective of contemporary primatology, he opened the gap too far. Chimpanzees can master techniques and habits which can become part of the traditions of their group and which are linked to the ecological circumstances. The set of behaviours which is transferred by imitation is so rich that most ethologists do not hesitate to talk about chimpanzee cultures (e.g. cracking nuts on stones reminiscent of anvils, rain dances). Crucially, newborn chimpanzees are dependent on the net of social relations in their group and the quality of their attachment to mother greatly influences their socio-cognitive abilities (cf. Bard et al. 2014). Moreover, in the light of experiments on interspecific symbolic communication with humans, one has to oppose the assertion that "in the well-protected, newborn ape, no essentially new possibilities for behavior and movement or for means of communication arise" (Portmann 1990a: 57). In other terms, the Umwelt of apes is less fixed than had been believed in the tradition of philosophical anthropology, and it can undergo significant transitions through socio-cognitive and ecological influences (cf. Tønnessen 2009).[20]

Nevertheless, to give a balanced picture, Portmann formed his opinion on the differences between humans and apes chiefly in the 1940s when virtually no studies of the social life of apes had been performed. He had an opportunity to supplement his stance in 1959, when the birth of the female gorilla Goma took place in the Basel Zoo. She was reared by the director of the zoo, Ernst M. Lang, and his wife, who carefully observed her early development (Lang 1961). Interestingly, Lang, Schenkel, and Siegrist (1965: 20) follow Portmann's typology of early mammalian development, but based on their observations of the development of Goma and her

---

[19] It is interesting to find Portmann (1990b: 113) writing about the negative influence of solely linguistic expressivity on creativity in art and science. Further, Polanyi's tacit knowledge is viewed not only as a prerequisite (or substituent) of linguistic communication, but as a crucial constituent of animal orientation in the world. Similar notions demonstrate that Portmann is aware of zoosemiotic (*sensu* Sebeok 1990) capacities in human individuals.

[20] Even so, it is still possible to hold the distinction between the Umwelt of ape species and Welt of humans, i.e. to find a qualitative difference in the semiotic relations of apes and humans to their environment. In the tradition of biosemiotics, the uniqueness of human language for both communication and thinking is usually seen as the benchmark of human *superior* cognitive faculties. Nevertheless, one can also highlight the complexity of the cognitive systems of higher animals and mark important differences between the Umwelts of distinct species. This seems to be a viable option for the contemporary Zeitgeist: to acknowledge the differences without making evaluative comparisons (cf. the model of diffuse discontinuities in Brentari 2018). For a comprehensive account about the narratives of anthropological difference see Jaroš and Maran (2019).

brother Jambo, they subsume gorillas under the type of secondary altriciality. Portmann (1961b) writes his commentary on Goma's family but incorporates the case into his framework of the major differences between apes and humans. Later, he incorporates pictures of Goma's mother and her infant into subsequent editions of *A Zoologist Looks at Humankind* (Portmann 1969). He confirms the tight bond between the infant and its mother but remains conclusive about the special phenomenon of the social uterus and its crucial significance in the development and acculturation of human infants.

Further, it might be argued that the holistic element in Portmann's consideration of human form comes at the expense of theoretical conceptuality. How do historical and evolutionary processes interfere in the transformations of the human form of life? Portmann criticizes the naïve belief that separate consideration of the processes can bring a trustworthy image of a prehistorical human (Portmann 1990a: 13–14): what a biologist seeks to explain as a result of ecological processes (or "organic advancement"), a historian interprets as a result of migration and trade (see also Tolone, Chap. 10, in this volume). While Portmann might be correct in identifying problems in the study of prehistoric humans (he also comments on the ideological bias in palaeoanthropology, see Portmann 1960b), he fails to make any comments on a proper theoretical stance or feasible methodologies combining the approaches of palaeoanthropology, evolutionary biology, and archaeology. That is surprising if we consider the degree to which basal anthropology needs a conceptualization of the relation between the domains heuristically viewed as body and mind, where the constituents lie at the intersections of evolutionary and historic processes.[21] Similarly to other contexts, Portmann concentrates on sketching counterexamples to dominant views but does not provide a theoretically compact alternative.

Even if Portmann magnified the difference between humans and apes, it does not mean that the narrative of discontinuity is outdated as such. Specification of the anthropological difference is an open problem, and the differential relations of animals and humans to the world are ongoing matters of research. Many comparative psychologists seek to find a cognitive trait that is unique to humans and that, since it appeared in the evolution of humanity, has transformed the human way of relating to the world. Tomasello (2014) links the ability to share attention with the interdependence of human individuals: further, rapid evolution of human culture is enabled by the will of parents and other group members to teach infants cultural innovations. Similar inclinations of contemporary researchers can be seen as the continuation of the theoretical programme of philosophical anthropology, supplemented with careful laboratory experiments.[22]

To conclude, Portmann has provided a basic interpretative framework for capturing both similarities and differences between humans and other forms of life. His profoundly developmental and evolutionary perspective on the human condition

---

[21] As an example, within neo-Darwinian conceptualizations of the relations between nature and culture, we can name dual inheritance theory.

[22] Indeed, as a pioneer of philosophical anthropology in the 21st century, Tomasello was awarded the Wiesbadener Helmuth-Plessner Award in 2014.

aligns him with current theories informing comparative studies of the cognitive and behavioural capacities of humans and great apes. Social womb is a strong metaphor, where biological, social, and cultural perspectives collapse into a single phenomenon, linking individual development with an overarching evolutionary and historic perspective. Portmann's assertion that the abilities pertinent for our species (a triad of upright posture, linguistic utterances, and rational thought) are acquired only due to co-occurrence of the genetic dispositions of a newborn infant and the intensive care it receives from a social group retains its validity for any future anthropological theory.

Although one has to agree that cultural processes form the human species to unmatched extent, it remains an open question if the conceptual distinction between organic inheritance and cultural evolution has to be singled out just for our species. The last decades boast numerous studies on animal cultures, mainly in the case of apes (cf. Laland and Galef 2009). There remains an option to broaden the idea of the social womb to non-human species and link the existence and transfer of group traditions to a sensitive period of early postnatal development. Under such a view, every form of social life in higher animals would have a contingent element, undetermined by the species-specific genetic information alone. After all it is probably a sign of great empirical validity if the concept of the social womb can serve as a point of departure for theories of either the specific plasticity of humans or the embeddedness of humankind in general evolutionary processes.

References

Amadon, D. (1954). Animal forms and patterns by Adolf Portmann. *Evolution, 8*(4), 405.

Bard, K. A., Bakeman, R., Boysen, S. T., & Leavens, D. A. (2014). Emotional engagements predict and enhance social cognition in young chimpanzees. *Developmental Science, 17*(5), 682–696.

Brentari, C. (2018). From the hiatus model to the diffuse discontinuities: A turning point in human-animal studies. *Biosemiotics, 11*(3), 331–345.

Darwin, C. (2009). *The expression of the emotions in man and animals*. New York: Oxford University Press.

De Beer, G. (1966, April 14). In the genes. New paths in biology by Adolf Portmann. *The New York Review of Books*. https://www.nybooks.com/articles/1966/04/14/in-the-genes/. Accessed 1 Jan 2019.

Fagot-Largeault, A. (2009). Anthropological physiology: von Uexküll, Portmann, Buytendijk. In A. Berthoz & Y. Christen (Eds.), *Neurobiology of "umwelt": How living beings perceive the world* (pp. 1–7). Berlin, Heidelberg: Springer.

Fischer, J. (2009). Exploring the core identity of philosophical anthropology through the works of Max Scheler, Helmuth Plessner, and Arnold Gehlen. *Iris, 1*(1), 153–170.

Gehlen, A. (1966). *Der Mensch. Seine Natur und Seine Stellung in der Welt*. Frankfurt am Main, Bonn: Athenäum.

Gould, S. J. (1977a). Human babies as embryos. In S. J. Gould (Ed.), *Ever since Darwin* (pp. 70–76). New York: Penguin Books.

Gould, S. J. (1977b). *Ontogeny and phylogeny*. Cambridge, MA/London: The Belknap Press.

Gould, S. J. (2010). The hardening of the modern synthesis. In M. G. Grene (Ed.), *Dimensions of Darwinism* (pp. 71–96). Cambridge: Cambridge University Press.

Grene, M. G. (1965). Beyond Darwinism: Portmann's thought. *Commentary, 40*(5), 31–38.
Grene, M. G. (1968). *Approaches to a philosophical biology*. New York: Basic Books.
Grene, M. G. (1974). *The understanding of nature. Essays in the philosophy of biology*. Dordrecht: D. Reidel.
Grene, M. G. (2002). Reply to Jacquelyn Ann K. Kegley. In R. E. Auxier & L. E. Hahn (Eds.), *The Philosophy of Marjorie Grene* (pp. 79–82). Chicago and La Salle IL: Open Court
Halák, J., & Klouda, J. (2018). The institution of life in Gehlen and Merleau-Ponty: Searching for the common ground for the anthropological difference. *Human Studies, 41*(3), 371–394.
Illies, J. (1976). *Das Geheimnis des Lebendigen*. München: Kindler.
Jaroš, F., & Maran, T. (2019). Humans on top, humans among the other animals: Narratives of anthropological difference. *Biosemiotics, 12*(3), 381–403.
Jaroš, F., & Pudil, M. (2020). Cognitive systems of human and non-human animals: At the crossroads of phenomenology, ethology and biosemiotics. *Biosemiotics, 13*(2), 155–177.
van der Kloot, W. G. (1962). Animals as social beings. Adolph Portmann. *Science, 135*(3503), 523.
Koechlin, F. (2004, June 10). Don Quijote der Laboratorien. *Die Wochenzeitung* (Basel).
Komárek, S. (2003). *Mimicry, aposematism and related phenomena. Mimetism in nature and the history of its study*. München: Lincom Europa.
Laland, K., & Galef, B. G. (Eds.). (2009). *The question of animal culture*. Cambridge, MA: Harvard University Press.
Landmann, M. (1985). *Fundamental anthropology* (D. J. Parent, Trans.). Washington DC: The Center for Advanced Research in Phenomenology & University Press of America.
Lang, E. M. (1961). *Goma, das Gorillakind*. Stuttgart, Wien: Albert Müller, Rüschlikon-Zürich.
Lang, E. M., Schenkel, R., & Siegrist, E. (1965). *Gorilla, Mutter und Kind*. Basel, Hamburg, Wien: Basilius Presse.
Levit, G. S., Hossfeld, U., & Olsson, L. (2014). The Darwinian revolution in Germany: From evolutionary morphology to the modern synthesis. *Endeavour, 38*(3–4), 268–279.
Lorenz, K. (1983). *Der Abbau des Menschlichen*. München: Piper.
Mangold-Wirz, K. (1966). Cerebralisation und Ontogenesemodus bei Eutheria. *Acta Anatomica, 63*(4), 449–508.
Martin, R. D. (1975). The bearing of reproductive behavior and ontogeny on strepsirhine phylogeny. In W. P. Luckett & F. S. Szalay (Eds.), *Phylogeny of the primates. A multidisciplinary approach* (pp. 265–298). New York, London: Plenum Press.
Matsuzawa, T. (2012). What is uniquely human? A view from comparative cognitive development in humans and chimpanzees. In F. B. M. de Waal & P. F. Ferrari (Eds.), *The primate mind: Built to connect with other minds* (pp. 288–305). Cambridge, MA, London: Harvard University Press.
Merleau-Ponty, M. (2003) *Nature. Course notes from the Collège de France* (R. Vallier, Trans.). Evanston: Northwestern University Press.
Mlíkovský, J. (1985). Portmann's rule of precedence and the evolution of encephalization in birds (Aves). In J. Mlíkovský & V. J. A. Novák (Eds.), *Evolution and morphogenesis* (pp. 559–566). Praha: Academia.
Monastra, G. (2000). Adolf Portmann, Wilhelm Troll: due biologi della morfologia idealistica. *Rivista di Biologia, 93*(2), 1000–1004.
Montagu, A. (1983). *Growing young*. New York: McGraw-Hill.
Müller, H. (1988). *Philosophische Grundlagen der Anthropologie Adolf Portmanns*. Weinheim: VCH, Acta Humaniora.
Nickel, G. (1996). *Wilhelm Troll (1897–1978). Eine Biographie*. Halle (Saale): Deutsche Akademie der Naturforscher Leopoldina.
Oppenheimer, J. (1965). New paths in biology. Ruth Nanda Anshen, Adolf Portmann, Arnold J. Pomerans. *The Quarterly Review of Biology, 40*(3), 281–282.
Portmann, A. (1935a). Die Ontogenese der Vögel als Evolutionsproblem. *Acta Biotheoretica, 1*(1–2), 59–90.

Portmann, A. (1935b). Tropische Falter. In *Falterschönheit. Exotische Schmetterlinge in farbigen Naturaufnahmen* (pp. 13–20). Introduction by H. Hesse. Bern: Iris Books.
Portmann, A. (1938). Die Ontogenese der Säugetiere als Evolutionsproblem. *Biomorphosis, 1*(1), 49–66; *1*(2), 109–126.
Portmann, A. (1941a). Die biologische Bedeutung des ersten Lebensjahres beim Menschen. *Sweizerische medizinische Wochenschrift, 71*(32), 921–924.
Portmann, A. (1941b). Die Tragzeiten der Primaten und die Dauer der Schwangerschaft beim Menschen: ein Problem der vergleichen Biologie. *Revue Suisse de Zoologie, 48*(3), 511–518.
Portmann, A. (1942). *Die Biologie und das neue Menschenbild.* Bern: Herbert Lang & Cie.
Portmann, A. (1944a). *Vom Ursprung des Menschen.* Basel: Reinhardt.
Portmann, A. (1944b). *Biologische Fragmente zu einer Lehre vom Menschen.* Basel: Schwabe.
Portmann, A. (1945). Tropical butterflies. In *The beauty of butterflies* (pp. 7–10). Introduction by J. Huxley). New York, Toronto: Iris Books.
Portmann, A. (1948). *Die Tiergestalt. Studien über die Bedeutung der tierischen Erscheinung.* Basel: Friedrich Reinhardt. [English translation: *Animal forms and patterns*, Portmann 1952.]
Portmann, A. (1952). *Animal forms and patterns* (H. Czech, Trans.). London: Faber and Faber.
Portmann, A. (1953). *Das Tier als soziales Wesen.* Zürich: Rhein-Verlag. [English translation: *Animals as social beings*, Portmann 1961.]
Portmann, A. (1954). Biologie auf neuen Wegen. In J. Moras & H. Paeschke (Eds.), *Deutscher Geist zwischen gestern und morgen. Bilanz der kulturellen Entwicklung seit 1945* (pp. 172–188). Stuttgart: Deutsche Verlagsanstalt.
Portmann, A. (1956). *Zoologie und das neue Bild vom Menschen.* Hamburg: Rowohlt.
Portmann, A. (1960a). *Neue Wege der Biologie.* München: Piper. [English translation: *New paths in biology*, Portmann 1964b.]
Portmann, A. (1960b). *Der Pfeil des Humanen. Über P. Teilhard de Chardin.* Freiburg im Breisgau, München: Alber.
Portmann, A. (1961a). *Animals as social beings* (O. Coburn, Trans.). New York: Viking.
Portmann, A. (1961b). "Gomas" erste Wachstumsphase. In *Goma – das Basler Gorillakind* (Bulletin Nr. 8). Basel: Documenta Geigy.
Portmann, A. (1962). The preface to a science of man. Translated by H. Kaal. *Diogenes, 10*(40), 1–26.
Portmann, A. (1964a). *Animals as social beings* (O. Coburn, Trans.). New York, Evanston: Harper & Row.
Portmann, A. (1964b). *New paths in biology* (A. J. Pomerans, Trans.). New York: Harper & Row.
Portmann, A. (1964c). *Das Tier als soziales Wesen* (2nd ed.). Freiburg im Breisgau: Herder.
Portmann, A. (1965). The special position of man in the realm of the living. *Commentary, 40*(5), 38–41. [German original: Portmann, A. (1957). Die Sonderstellung des Menschen im Reich des Lebendigen. *Universitas, 12*(4), 337–344.]
Portmann, A. (1967a). *Zoologie aus vier Jahrzehnten.* München: Piper.
Portmann, A. (1967b). *Animal forms and patterns* (H. Czech, Trans.) (2nd ed.). New York: Schocken Books.
Portmann, A. (1968a). Um eine basale Anthropologie. In A. Portmann (Ed.), *Biologie und Geist* (pp. 293–308). Frankfurt am Main: Suhrkamp.
Portmann, A. (1968b). Metamorphose der Tiere. In A. Portmann (Ed.), *Biologie und Geist* (pp. 233–272). Frankfurt am Main: Suhrkamp.
Portmann, A. (1968c). Die Biologie und das Phänomen des Geistigen. In A. Portmann (Ed.), *Biologie und Geist* (pp. 9–49). Frankfurt am Main: Suhrkamp.
Portmann, A. (1969). *Biologische Fragmente zu einer Lehre vom Menschen* (3rd, enlarged ed.). Basel: Schwabe. [English translation: *A zoologist looks at humankind*, Portmann 1990a].
Portmann, A. (1970). *Entläßt die Natur den Menschen? Gesammelte Aufsätze zur Biologie und Anthropologie.* München: Piper.
Portmann, A. (1970a). Neue Fronten der biologischen Arbeit. In A. Portmann (Ed.), *Entläßt die Natur den Menschen?* (pp. 13–28). München: Piper.

Portmann, A. (1970b). Der Mensch – ein Mängelwesen? In A. Portmann (Ed.), *Entläßt die Natur den Menschen?* (pp. 200–209). München: Piper.
Portmann, A. (1970c). Die Entwicklungsbeschleunigung der Jugend ais biologisches und soziales Problem. In A. Portmann (Ed.), *Entläßt die Natur den Menschen?* (pp. 253–281). München: Piper.
Portmann, A. (1970d). Biologie als technische Weltmacht. In A. Portmann (Ed.), *Entläßt die Natur den Menschen?* (pp. 364–379). München: Piper.
Portmann, A. (1974). *An den Grenzen des Wissens*. Wien, Düsseldorf: Econ.
Portmann, A. (1990a). *A zoologist looks at humankind* (J. Schaefer, Trans.). New York: Columbia University Press.
Portmann, A. (1990b). The path leading to words: Levels of living communication. In *Essays in philosophical zoology by Adolf Portmann: The living form and the seeing eye* (pp. 97–124) (R. Carter, Trans.). Lewiston: The Edwin Mellen Press.
Portmann, A. (1990c). Human beings in the perspective of the theory of evolution. In *Essays in philosophical zoology by Adolf Portmann: The living form and the seeing eye* (pp. 125–144) (R. Carter, Trans.). Lewiston: The Edwin Mellen Press.
Portmann, A., & Illies, J. (1976). Ein Grenzgespräch. In J. Illies (Ed.), *Das Geheimnis des Lebendigen* (pp. 291–312). München: Kindler.
Rieppel, O. (2016). *Phylogenetic systematics. Haeckel to Hennig*. Boca Raton: CRC Press.
Ritter, M. (2000). Die Biologie Adolf Portmanns in zeitgeschichtlichem Kontext. *Basler Zeitschrift für Geschichte und Altertumskunde, 100*, 207–254.
Rosenberg, K., & Trevathan, W. (2002). Birth, obstetrics and human evolution. *BJOG : An International Journal of Obstetrics and Gynaecology, 109*(11), 1199–1206.
Scheler, M. (1928). *Die Stellung des Menschen im Kosmos*. Darmstadt: Reichl.
Sebeok, T. A. (1990). *Essays in zoosemiotics*. Toronto: Toronto Semiotic Circle.
Thinès, G. (1996). La forme animale selon Buytendijk et Portmann. *Études Phénoménologiques, 12*(23/24), 195–207.
Thorpe, W. H. (1962). Co-operation: Review of animals as social beings by Adolf Portmann. *Journal of Animal Ecology, 31*(1), 169–170.
Toepfer, G. (2011). Selbstdarstellung. In G. Toepfer (Ed.), *Historisches Wörterbuch der Biologie* (pp. 246–253). Stuttgart, Weimar: J. B. Metzler.
Tolone, O. (2015). Plessner und Adolf Portmann. Zur philosophischen Bestimmung des Menschen durch Exzentrizität und Frühgeburt. In K. Köchy & F. Michelini (Eds.), *Zwischen den Kulturen. Plessners "Stufen des Organischen" im zeithistorischen Kontext* (pp. 141–160). Freiburg im Breisgau: Alber.
Tomasello, M. (2014). *A natural history of human thinking*. Cambridge, MA: Harvard University Press.
Tønnessen, M. (2009). Umwelt transitions: Uexküll and environmental change. *Biosemiotics, 2*(1), 47–64.
Trevathan, W. R., & Rosenberg, K. R. (2016). Human evolution and the helpless infant. In W. R. Trevathan & K. R. Rosenberg (Eds.), *Costly and cute: Helpless infants and human evolution* (pp. 1–28). Santa Fe, Albuquerque: School for Advanced Research Press, University of New Mexico Press.
Varley, G. C. (1953). Animal forms and patterns - a study of the appearance of animals. By Adolf Portmann, Hella Czech and Sabine Baur. *Journal of Ecology, 41*(2), 401–402.
Wirz, K. (1950). Zur quantitativen Bestimmung der Rangordnung bei Säugetieren. *Acta Anatomica, 9*(1–2), 134–196.
Yolton, J. W. (1966, June 9). New paths in biology. *The New York Review of Books*. https://www.nybooks.com/articles/1966/06/09/new-paths-in-biology/. Accessed 1 Jan 2019.

Filip Jaroš is an assistant professor at the Department of Philosophy and Social Sciences, University of Hradec Králové. He spent one year (2017/8) as a postdoc researcher at the Department of Semiotics, University of Tartu. He obtained his PhD in theoretical and evolutionary biology at Charles University in Prague. His main scholarly interests include the biosocial philosophy of

Adolf Portmann, human-animal communication and the problem of anthropological difference. Filip Jaroš is an Editor-in-Chief of the journal *Biosemiotics* and co-edited a book about animal aesthetics, *Krása a zvíře. Studie o vztahu estetických a etických hodnot zvířat* [Beauty and the beast: essays on aesthetic and ethical values of animals, in Czech].

Part III
Philosophical Aspects of Portmann's Work

Chapter 8
Portmann, Goethe and Modern Biology: Two and a Half Ways of Looking at Nature

Markus Wild

Abstract A fundamental and bold claim of Portmann's philosophy of biology is a thesis about the autonomy of self-representation (*Selbstdarstellung*) of all living beings: "Self-presentation has to be understood as a basic fact of life, on a par with self-maintenance and the preservation of the species." In other words, the perceivable appearance of organisms cannot be reduced to its chemical, physiological, morphological or functional causes, but must be understood as a phenomenon in its own right. The aim of the following contribution is threefold. First, it aims at establishing the relevant context for the bold claim by embedding the claim in Portmann's broader view of biology. Secondly, it aims at reconstructing Portmann's rather scattered argument for the bold claim by recasting it as an appeal to what Portmann calls "unaddressed authentic appearances" (*unadressierte eigentliche Erscheinungen*). Finally, it presents Portmann's partly phenomenological approach to biological phenomena as a reluctant relative of Goethe's approach, thereby claiming that Portmann is no advocate of a genuine Goethean science of nature.

Portmann, Goethe and Modern Biology

Adolf Portmann characterized his own approach to biology in close proximity to what he took Goethe's approach to be, namely a phenomenological approach based on morphological research. It is safe to say that he accepted something close to Fritz Heinemann's account of Goethe's phenomenological method: "It begins with phenomena, proceeds through them, and ends with them, returning at the last from the Ur-phenomena to the particulars whose claims have not at any point been abrogated." (Heinemann 1934: 79).

However, Portmann was rather reluctant in accepting the "Ur-phenomena" as something in the reach of natural science, whether in the detached, 'Platonic' interpretation as ideas ante rem or in the non-detached, 'Aristotelian' interpretation as ideas in re. Rather, he follows Goethe's advice to start with a specific empirical

M. Wild (✉)
University of Basel, Basel, Switzerland
e-mail: markus.wild@unibas.ch

F. Jaroš, J. Klouda (eds.), *Adolf Portmann*, Biosemiotics 23,
https://doi.org/10.1007/978-3-030-67810-4_8

phenomenon in nature, then convert it into a scientific phenomenon by means of a controlled sequence of experiments that keeps the phenomenon in its integrity, yet under conditions different from the first encounters, and thereby finally establish the pure phenomenon beyond the contingencies of the specific empirical encounters. In contrast to those upholding Goethean science as an alternative to modern natural science, Portmann saw Goethe's stance as an alternative within it (cf. Amrine and Zucker 1987), counterbalancing the (as Portmann had it) one-sided, atomistic, mechanical, and manipulative approach of modern biology to the living world.

Portmann dedicated three of his essays to Goethe's work on natural science (Portmann 1953a, b, 1973). His late essay "Goethe and the Concept of Metamorphosis" (Portmann 1973, English translation 1987) certainly constitutes the pinnacle of his life-long engagement with Goethe's natural science. In these essays Portmann makes use of an analogy which illustrates two fundamentally different ways of looking at nature. If a grand drama is performed, there are at least two very different points of view, the view from the front, and the view from behind the stage:

> Behind the stage I find all the paraphernalia, the technical underpinnings that make a performance possible: the wind and thunder machines, the provisions that make even possible the appearance of ghosts. In front of the stage completely different rules prevail: here it is a matter of relationship between human beings. I must know the language; I must grasp the meaning of words and gestures; must give myself over entirely to the sensory appearance [...] In contemporary scientific research the work behind the stage has developed in a powerful and extremely specialized way. For many, it represents true research, the actual goal. And yet there are many tasks, e.g. ethology, which require a biologist, an observer of the drama, in front of the stage [...]. That which I experience before the stage is not 'more correct' than that which the physiologist, the microbiologist, the molecular biologist fathoms in his or her study behind the stage of life. Research proceeding from each of the two starting-points has meaning. (Portmann 1987: 138, cf. 1973: 139)

Portmann's aim as a researcher was to insist on the necessity and autonomy of the view from the front on the spectacle of life, especially animal life. He was defending not so much Goethe's specific point of view on plant or animal morphology as the need to adopt the general point of view on living phenomena that Goethe himself occupies in his research.[1] In order to see Portmann's point of view, I will follow his hint that zoology, physiology, morphology, ethology, and anthropology[2] require "an observer of the drama [...] in front of the stage."

After World War II, there had been dramatic advances in molecular biology, most notably in molecular genetics and code-cracking. To this day, this kind of research

---

[1] Portmann's general point of view on biology is laid out in Portmann (1961). This relatively small book constitutes the *summa* of Portmann's thinking on biology. The philosophical importance of Portmann's reflection on the nature of biology is best acknowledged in Grene (1968: 3–54, 1974:s 254–293).

[2] Portmann worked on these topics during his very productive career. For morphology cf. Portmann (1948a, last edition Senn 2013) and Portmann (1948b), which deals with the "self-representation" (*Selbstdarstellung*) of the animal Gestalt; for ethology cf. Portmann (1953c); for anthropology cf. Portmann (1944, 1956).

has opened up innumerable possibilities of changing, manipulating, controlling, and creating living forms. A case in point is the new methods for targeted genome editing such as the tools developed from the user-friendly CRISPR-Cas system. Following Max Scheler's Sociology of Knowledge (Scheler 1926, English translation Scheler 1980), Portmann calls the type of knowledge that the kind biological research behind the stage is driving at "knowledge for mastery" (*Herrschaftswissen* or *Leistungswissen*). This kind of knowledge is useful as a technological tool for mastering the material world. Both Scheler and Portmann were concerned about the one-sided pragmatic and technological attitude of modern science directed toward the control and mastery of nature. While Scheler thought that knowledge of mastery should be united with the learning of the cultured human being (knowledge of erudition, *Bildungswissen*) and, finally, set under the guidance of knowledge of salvation (*Erlösungswissen*), Portmann shies away from such grandiose programs of unification and opts for a more modest integration of the Goethean point of view into the scientific exploration of living beings.

The examples of backstage machinery in Portmann's theatre analogy clearly suggest a mechanical interpretation of the invisible workings of living beings, and this interpretation aims at the manipulation of living organisms as we would expect from "knowledge for mastery." According to the mechanical view, organisms are complex machines with functional parts designed by evolution to facilitate survival and reproduction. I think that Portmann considers this view a superficial hermeneutical tool, legitimate as far as it goes, yet according to him it betrays not only a one-sided point of view (the view from backstage only) but also reflects a serious misunderstanding of the metabolic processes going on in living beings. Portmann (1961) acknowledges the complexity of life-processes even on the macromolecular level, and this view has been supported by some contemporary researchers. As the chemist Peter Moore puts it: "Macromolecular devices are not machines" (Moore 2012: 8). In the medium and at the scale of the cell, matter behaves very differently from how it behaves elsewhere. In the cell, as opposed to what we find in machines, roars a constant "molecular storm": there is constant spontaneous motion not driven by anything external; larger molecules rearrange themselves spontaneously and vibrate; everything is constantly bombarded by water-molecules; electrical charge is an ever-present factor; the chemical processes show an astonishing number of random side effects (cf. Godfrey-Smith 2016).

After World War II, however, the almost equally new and fascinating science of animal behavior – ethology and its more recent development, cognitive ethology (Allen and Bekoff 1997; Shettleworth 2010) – started the study of individual and collective action patterns of animals and the cognitive skills and capacities giving rise to those patterns. Ethologists must watch the animals they are studying literally hour by hour, preferably in species-specific, natural conditions. Ethologists certainly perform sophisticated experiments, interfering with the environments of the animals under observation, and they elaborate abstract, theoretical, and even mathematical models and explanations of the behavioral patterns encountered. However, in contrast to their colleagues in molecular biology working backstage, ethologists never escape their position in front of the drama of life: they observe what the animals do. To be sure, many ethologists think that, ideally, systematic observation and experimentation in the wild would be reproduced by controlled experiments in the laboratory in order to isolate the cognitive mechanisms responsible for the relevant action patterns; the cognitive mechanisms should eventually be reduced to neurological processes in the brain. For the time being, however, ethologists have to observe what the animals they are talking about do, and some of them refrain from the neurological reductionism for theoretical and/or ethical reasons.

Portmann's Bold Claim: The Autonomy of Self-presentation

While Portmann's original and main area of research was comparative morphology (Portmann 1948a), he also wrote a comprehensive overview of ethological research (Portmann 1953c), since both areas of research, morphology and ethology, are directed at the macroscopic, visible, and observable features of the animal world, whereas the structures in molecular biology are accessible only by highly indirect and analytical techniques. As Portmann puts it in *Neue Wege der Biologie*: the aim of backstage biological research is "the exploration of the living substance beyond the frontiers of the visible" (*die Ergründung der lebenden Substanz jenseits der Grenze der Sichtbarkeit*) (Portmann 1961: 9). In contrast, the macroscopic view in front of the drama of life is directed at the whole living being. In our ordinary encounters with animals and plants, we observe and experience both their appearance (*Erscheinung*) and their "centricity" (*Innerlichkeit*).

When ethologists watch individual or collective animal behavior, they watch it – as Marjorie Grene puts it – "in a manner analogous to the way in which we might watch another human being" (Grene 1974: 256). Thus, the animal is viewed as a center of activities, a conscious or non-conscious center of movements, drives, perceptions, social interactions, achievements, and failures. What Portmann calls "centricity" (*Innerlichkeit*) is this center of activity to be found in every living being:

> Inwardness [centricity] is a reality without location, it is seated neither in a spatial interior nor in some headquarters, rather it is an active center present in the whole animal. (*Die Innerlichkeit ist eine undimensionale Wirklichkeit, die nicht im räumlichen Inneren, nicht in einem Zentrum ihren Sitz hat, sondern die eine tätige Mitte ist, die im ganzen Tier gegenwärtig ist.*) (Portmann 1961: 46)

How is this notion of centricity to be understood? As I have indicated, Portmann was rather reluctant to accept great metaphysical ideas such as Goethe's "Ur-phenomena" or grandiose philosophical programs such as Scheler's search for "knowledge of salvation." We have thus reason to doubt whether Portmann took "centricity" to be a name for, or pointer to, some inner active principle or life force distinguishing all living beings from non-living entities. As I see it, Portmann emphasizes a logical or mereological point by highlighting the fact that we have to take the whole organism as the subject of the ascription of activities typical for living beings. Thus, the whole animal eats, hides, moves, perceives, interacts, etc., and not but a part of the animal; the whole plant grows, blooms, converts light energy, communicates, etc., not but a part of it. To be sure, there are parts of an organism that are necessary in order to sustain some activity of the whole living being. Neither will the cat be able to visually spot a bird without the eyes, nor will the elm be able to engage in photosynthesis without chloroplasts, yet both organisms as such do the perceiving of potential food and the converting of light energy. Therefore, the expression "centricity" points at the fact that living beings are centers of their own activities. According to Portmann, the higher the organism, the more its activities are under central control. This amounts to a clearer articulation of centricity in, say, felines than in deciduous trees. With the concept of centricity Portmann criticizes the exclusive ascription of biological and psychological predicates (such as "perceives" or "converts") to parts of an organism that apply to the organism as a whole. As he sees it, the view from backstage is nothing but the tendency to ascribe such predicates to an organism's part, while the view from the front refuses this kind of illicit predication, since it invites invalid inferences and arguments. The so-called "mereological fallacy" claims that it is a mistake to ascribe predicates to parts that belong to a whole. In contrast to contemporary discussions for and against the mereological fallacy in the life-sciences (Bennett and Hacker 2003; Smit and Hacker 2014), Portmann holds that the legitimacy or illegitimacy of the ascription of predicates to parts of an organism that apply to the organism as a whole only holds within the perspectives from behind and before the stage. However, since there is a clear sense in which some biological and psychological predicates have to be ascribed to the whole organism as opposed to its enabling parts, the Goethean view from the front gains additional justification from this mereological argument.

In what follows I will concentrate on Portmann's account of appearance (*Erscheinung*), more specifically on self-presentation (*Selbstdarstellung*) as a fundamental and autonomous kind of appearance. Despite the fact that both appearance (*Erscheinung*) and centricity (*Innerlichkeit*) are correlated and essential aspects of life from the perspective in front of the drama, of the macroscopic, visible, and observable manifestation of living organisms, this chapter will focus on appearance, not only for the sake of brevity, but also because of the central role appearance plays in Portmann's work. Now, the fundamental and bold claim of Portmann's work across all domains of biology is the following:

> Self-presentation [or "display"] has to be understood as a basic fact of life, on a par with self-maintenance and the preservation of the species. (*Selbstdarstellung muss als eine der Selbsterhaltung und Arterhaltung gleichzusetzende Grundtatsache des Lebendigen aufgefasst werden.*) (Portmann 1957: 40)

Self-presentation takes center stage in a science that takes the phenomenal side of living beings seriously. Moreover, self-presentation is a fundamental fact of life, on a par with self-preservation. The main goal of the remainder of this chapter is the reconstruction of Portmann's reasons for this bold claim.

Portmann's View of Biology

According to my view of Portmann, the following chart lays out the bare bones of his system of biology or "*Lebenswissenschaft*" in a very schematic form:

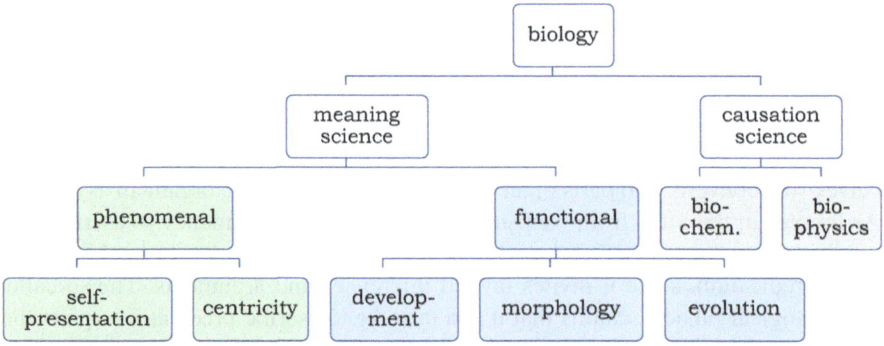

A few comments are in order. The realm of causation science is only poorly represented by biochemistry and biophysics; all research in genetics, molecular biology and microbiology – all research aiming at "the exploration of the living substance beyond the frontiers of the visible" (Portmann 1961: 9) – might be subsumed under the heading of causation science in biology and (partly) under the heading of functional meaning science, but certainly not under the heading of phenomenal meaning science. The chart is more nuanced than the analogy with the drama performance. Clearly, what I call the "science of causes" or "causal explanation" (Portmann 1953a) aims at a causal explanation of biological phenomena and is represented by the view from backstage; what I suggest to call the "science of meanings" aims at the investigation of the meaning of biological phenomena for the organisms themselves and is represented by the view from the front. However, the functional parts of meaning science which explain the function of biological structure according to development, morphology, and evolution lie half-way between the two perspectives represented in the theatre analogy. The functional parts of meaning science are underpinned by biochemical processes; they are, however, not reducible to these very processes. Portmann's reasons for this claim are not the concern of this chapter; however, they basically take the shape of what has come to be acknowledged as the thesis of multiple realization (Polger and Shapiro 2016). In short, the same function can be fulfilled by diverging kinds of biochemical processes or by different physiological and morphological structures. Therefore, the function of a biological trait is not reducible to the material constitution of this trait. From the point of view of the etiological theory of functions (cf. Millikan 1989; Neander 1991; Griffiths 1993; Godfrey-Smith 1994; Mitchell 1995; Neander and Rosenberg 2012; Kramer 2014), an additional reason for the multiple realization and, therefore, relative autonomy[3] of the function of a structure in relation to its material constitution is the idea that history is essential for function. Simply put, functions are selected effects in recent history, and different material structures can be selected for the same purpose. Thus, function is relatively autonomous not only because of the thesis of multiple realization, but also because history is essential for function.

On the one hand, it is not hard to see why Portmann thinks of functional biology as part of the "science of meanings," since functional structures do have a point for the organism – they aim at something, they are there because they fulfill a certain role. In this weak teleological sense, functional structures are meaningful structures. On the other hand, since function can be accounted for by natural selection – biological functions are naturally selected effects in recent evolutionary history – the functional departments of meaning science are at home in a broadly Darwinian picture of life structured by evolution by means of natural selection, inclusive fitness, self-maintenance and the preservation of the species. Thus, Portmann's bold claim about self-presentation stems primarily from the phenomenal parts of meaning science. Portmann is not only opposed to the view that the functional parts will ultimately be reduced to causation science, he is also opposed to the view that the phenomenal parts of biology can be reduced to the functional parts (and ultimately to causation science).

Kinds of Appearances

Portmann distinguishes between Gestalt and form in the following manner:

> We call the visible body, the total appearance of the animal, its 'Gestalt,' and we will distinguish from it the 'form' of the visible and the hidden parts of the animal. (*Wir wollen den sichtbaren Leib, die gesamte Erscheinung des Tiers seine 'Gestalt' nennen und von ihr absondern die ‚Form', die den einzelnen Gliedern zukommt, den sichtbaren ebenso wie den verborgenen.*) (Portmann 1948b: 33, my translation, cf. 1967: 33)

In some passages of *Animal Forms and Patterns* (Die Tiergestalt), Portmann makes the following (implicit) argument: Forms (e.g. feathers) are parts of other forms (e.g. wings); every form is part of a Gestalt (e.g. the visible bird); forms have functions, yet the Gestalt (the whole perceptible animal) does not have a function; since only items having functions fall under genuine functional explanations (under functional meaning or causation science), the animal Gestalt does not fall under genuine functional explanations. Since the Gestalt of the whole animal is part of phenomenal meaning science, this is a first attempt to establish the autonomy of self-presentation in the animal kingdom.

---

[3] By expressing this thought as a "relative autonomy" I take criticisms of very liberal formulations of the thesis of multiple realization into account (Shapiro 2004); to be sure, there are material and biological constraints for function-carrying structures, yet the idea of multiple realization still holds within those constraints.

More important than the distinction between Gestalt and form is Portmann's distinction between different kinds of appearances. He distinguishes between authentic and inauthentic appearances (or phenomena) and, within the set of authentic appearances, he distinguishes between addressed appearances and unaddressed appearances. Portmann holds that there are "unaddressed authentic appearances" (*unadressierte eigentliche Erscheinungen*) and that they are part of the self-presentation of living beings.

The whole set of perceptible (morphological, behavioral, acoustic, etc.) patterns of living beings in space and time Portmann calls "authentic appearances" (*eigentliche Erscheinungen*). On the other hand, the subject matters of molecular biology, physiology, anatomy, etc., are called "inauthentic appearances" (*uneigentliche Erscheinungen*). Marjorie Grene thinks that this distinction is an attack on the modern Galilean view that the book of the world is written in mathematical code and that the sensory qualities (the so-called secondary qualities) are not part of the structure of the objective world, but rather of the subjective experience of the world. According to Grene, all appearances that exist because there are living beings – the world of color and taste and smell and sound – are authentic appearances in Portmann's sense; they are part of the Umwelt or Welt of every living organism (Grene 1974).

Grene's interpretation is certainly on the right track, but she misses the morphological point of Portmann's concept of authentic appearance. Portmann thinks that organisms are essentially a "pre-arranged relationship" (*vorbereitete Beziehungen*) to the world (Portmann 1965: 13–32). The way the world appears, or rather can appear, to an organism is predetermined by the organism's biological endowment, which prepares the possible relations the organism will be able to entertain with the world. Portmann's main example is the human world of color, and he thinks that Goethe's Farbenlehre is primarily an investigation into the laws of how color appears to creatures like us. On the one hand, Goethe's color circle, for example, presents an objective sequence of perceptible hues with their own intrinsic order and lawful relations. On the other hand, Goethe's work on color is the objective expression of a subjective phenomenon, namely of the way creatures like us – according to their pre-arranged relations to the world – see the world. The visible world of other creatures – Portmann refers on several occasions to the perception of color in honeybees (e.g. Portmann 1972) – almost certainly contains a different circle of visible hues and, accordingly, a different kind of intrinsic order and some distinct lawful relations. According to the ecological view, colors are partly dependent on the biological nature of the perceiving animal (Matthen 2000, 2005). Colors correspond to surface properties as visually perceived or classified by certain species of animals. Portmann shares the view that color perception and color spaces cannot be identified with human color perception and human color spaces. His reference to comparative color perception (as in honeybees) indicates that an adequate theory of color vision should refrain from taking human color experience as the only source for color information (cf. Portmann 1965: 123 ff.). This view does not imply that Goethe's Farbenlehre is not a serious and objective enterprise, but it limits the extension of the theory to the ecological space occupied by human color vision. Or rather, he views with suspicion the Goethean idea that human color vision allows access to the very nature of color because human beings possess a special cognitive capacity – intuitive judgement – responsible for the very nature of natural species such as color, clouds, plants, or animals.[4]

Now, what about the morphological point of Portmann's concept of authentic appearances? According to Grene, the whole perceptible world contains authentic appearances in Portmann's sense, but Portmann wanted to make a specific point about the appearance of living organisms. The visible world of, say, the giant cuttlefish contains living beings and non-living entities; moreover, its body is itself an appearance for other living beings. In Grene's interpretation, there are simply too many authentic appearances beyond the realm of the senses. Portmann's target is much more specific. In *Die Tiergestalt*, he establishes with great care a distinction between transparent animals and semi-transparent animals mostly living in water – jellyfish or fish larvae are salient examples – and opaque animals. Whereas in the case of transparent and semi-transparent animals the inner organs are visible from the outside, this is not the case with opaque animals such as bees, sharks, toads, snakes, birds of prey or rodents. Opaque animals have developed an organ that clearly hides certain phenomena (e.g. inner organs and bones) from sight and clearly exposes other phenomena (e.g. hair and feathers). From the perspective of the whole and intact animal, the inauthentic appearances have to be hidden from sight, otherwise the animal would be neither whole nor intact (e.g. seeing the entrails of a mouse is basically in contradiction with its preservation); authentic appearances, on the other hand, are visible if the animal is in a whole and healthy state. Thus, the distinction between authentic and inauthentic appearances nicely corresponds to two important biological features of animals: the opaque skin and the state of health. According to Portmann, the development of opaque skins – or more generally, opaque boundaries – amounts to the development of an organ for self-presentation.

Addressed authentic appearances are meant to be seen and/or do have a function for the animal's self-maintenance; they are either other-related (semantic, e.g. signals, or cryptic, e.g. mimicry) or self-related (e.g. movement or thermic protection). Addressed authentic appearances clearly have biological functions; they therefore fall under the reign of functional meaning and causation science. Signaling, camouflage, or thermoregulation carry their functional role on their sleeves – they are more or less easily assimilated under the Darwinian paradigm of self-preservation. By contrast, unaddressed authentic appearances are not meant to be seen and do not have a function for the animal. (They are expressions of the animal's centricity). Since they do not lend themselves to functional explanations within the Darwinian paradigm of self-preservation, the reality of unaddressed authentic appearances constitutes the purest case of self-presentation (see also Conte, Chap. 9, in this volume). So, do unaddressed authentic appearances exist?

---

[4] For the notion of intuitive judgement in Goethe's view of natural philosophy or natural science cf. Förster (2012), Hindrichs (2011). For elaborations on Förster's view see Haag und Wild (2013), for an elaboration of different views of intuition see Wild (2012).

The Case for Unaddressed Authentic Appearances

Portmann holds that nature is rife with patterns that exist without any possibility of direct natural selection. The genetic endowments of organisms thus include special systems directed at the development and display of complex visible patterns of appearance.

Portmann's favorite examples are an informal group of tiny marine snails, the so-called Opisthobranchia (Portmann 1948a: 93–115). The members of this group are richly ornamented and colored, resembling fantastic dragons and fancy spaceships. Is there any survival value to the Gestalt and forms of these snails? Opisthobranchia are loners – they do not communicate by sensory means, their sensory organs are only poorly developed, they reproduce by external means, they hide in case of danger. Moreover, they vary strongly in both form and color, yet the selective environment is the same or at least similar (so there seems to be no environmental selection pressure). Finally, during the development of these organisms, a surprising amount of energy is invested in the maintenance of their forms and colors. Therefore, the rich ornamentation and coloration of the Opisthobranchia is an instance of an unaddressed authentic appearance and thus a case of self-presentation.

Portmann thinks that this argument can be generalized. Since biological explanations are either functional or phenomenal, and since many authentic appearances are displayed in a systematic way throughout the animal kingdom and are part of natural development, produced and maintained with considerable costs and hence neither products of chance nor side effects, they cannot fall under causal explanations alone. Many authentic appearances have to fall under phenomenal explanations, since they are unaddressed authentic appearances. These are essential parts of the animal Gestalt. Consequently, there has to be a science of the animal Gestalt. The object of this science is self-presentation. Sometimes, Portmann even claims that self-presentation is more fundamental for living beings than self-preservation, because unaddressed authentic appearances are a condition of evolution through natural selection: "A pattern can only function as camouflage or as signal if it is already in existence." (*Denn als Tarnung oder Signal kann ja ein Muster erst wirken, wenn es bereits da ist*) (Portmann 1961: 125).

Da Capo: Portmann and Goethe

Hjalmar Hegge claimed that Portmann "uses the same methods" as Goethe (Hegge 1996: 1). Eckart Förster (2012) has provided us with a very convincing exposition of Goethe's methodology of the intuitive understanding. According to Förster, a complete series (either of phenomena or of experiments) and the attentive observation of the transitions between the elements of the series are to provide a basis for studying the idea underlying the whole. Since the transitions may not be perceptible (in contrast to the elements of the series), one has to be very careful in the intuitive take-up of the transitions between the (perceptible) elements.

What about Portmann's method? As we have seen, Portmann takes issue with modern biology's strong and one-sided tendency towards the mechanical interpretation of the invisible workings of living beings. Moreover, he thinks that the use of animal models is an expression of this trend, since reliance on animal models limits the scope of living beings that come into view. By contrast, Portmann's research relies heavily on extensive series of phenomena. Let me briefly point out two examples from Portmann's more empirical work, the early ontogenesis of birds, mammals, and especially humans.

In his work on the ontogenesis of birds, Portmann and his co-workers measured hundreds of bird brains in order to determine the encephalization quotient. The encephalization quotient not only determines the ratio of brain hemisphere size to body size, but also the ratio of brain hemisphere size to the sizes of other parts of the brain and, importantly, the relative growth of brain hemisphere size in relation to the growth of the sizes of other parts of brain in early development, i.e. the static and dynamical intracerebral indices (Portmann 1946, 1947).[5]

Portmann was very much interested in various species' degree of independence at birth, whether they are "nest-dwelling" (nidicolous or altricial) or "nest-fleeing" (nidifugous or precocial) species. Concerning mammals, the young animal's independence or precociality increases with the degree of development (compare hedgehogs and orangutans). In humans, as in other primates, one would thus expect a very strong expression of precociality, but this is not the case. Human babies are born in an immature and helpless condition so as to require intensive care, yet some aspects of the human baby are precocial: we are born with our eyes and ears wide open. Humans have thus retained the sensory aspects of the precociality of higher mammals, but have recovered decidedly altricial aspects of lower mammals. We are – as Portmann puts it – "secondarily altricial" (see also Kurismaa and Jaroš, Chaps. 1, 6 and 7, in this volume).[6]

In both cases Portmann produces, there are very detailed series of developments (avian brain development, development of precociality in mammals). The point of

[5] Portmann's extensive data collection has recently played a role in the establishment of a new nomenclature of the bird brain. Primates have been taken to be the most cognitively advanced creatures in the animal kingdom. Recently, birds have been shown to be cognitively very sophisticated as well. The New Caledonian crow, the blue jay or the grey parrot are just the more celebrated examples. In primates and other mammals, the exercise of cognitive skills is controlled by the neocortex, a brain structure birds lack. The brains of mammals and birds have very different structures; in addition, birds have much smaller brains than apes. However, both types of brain are wired in similar ways. This, however, raises two questions: Are avian and mammalian brains comparable? Do we have to redraw the atlas of the avian brain? Simply put, the answer to both questions is: Yes.

[6] Portmann's additional thesis accounts for an "extra-uterine year" (*extrauterinäres Frühjahr*) during the first year after birth, allowing the human infant to develop cognitive and motor-neuronal skills in a sociocultural environment, which is also an inference from a comparative series. For a recent discussion of Portmann's hypothesis see Dunsworth et al. (2012).

both series is to become aware of transitions – transitions in static and dynamical cerebral development over time and transitions in development in precociality and altriciality. However, the first series and the second series point in a very different way to the idea underlying the whole. In the first (avian) series, the "dynamic type" sought for in the transitions of the series is the avian type underlying the series; in the second (mammalian) series, however, the "dynamic type" showing itself is the human Gestalt contrasting with the mammalian series (or as Portmann puts it: the human type "breaks the rule that applies to mammals"). Thus, for Portmann, as for Goethe, Gestalt is a moving, a becoming, a passing thing. The doctrine of forms is the doctrine of transformation. Yet how do the quantitative data of the first series relate to Portmann's promotion of phenomenal meaning science, centricity (*Innerlichkeit*) and self-presentation (*Selbstdarstellung*)? The key concept here is "rank" (*Rang*). Portmann holds that our natural way of talking of "lower" and "higher" animals is central to our understanding of life. Criteria for higher rank are cerebralization (Portmann 1948a: 137–152), the marked difference between adult animal Gestalt and embryonic Gestalt in its early stage (Portmann 1948b: 41), the complexity and differentiation of the sensory apparatus (Portmann 1948b: 69, 1961: 90), and spatial autonomy (Portmann 1961: 88). In other words, the criteria for higher rank closely correspond to the differentiation of the centricity of the animal. The Gestalt of animals – their self-presentation – is an expression of their centricity. Therefore, there have to be laws of Gestalt relating centricity and self-presentation according to rank. For example: "The higher in rank the inner organization of an animal the more the animal's head is marked as the leading pole in the animal's Gestalt and emphasized by its distinctive form" (Portmann 1948b: 76). In this sense, the form of the head is a distinctive mark of higher rank. The head unites three vital functions: perception, nourishment, orientation.

Hegge (1996) holds that Portmann's research into early human ontogenesis is evidence for a strongly teleological perspective on life in general, since the explanation of the different developmental stages is to be found in something that appears clearly only later in development. Hegge may be right from the developmental perspective, even from the perspective of functional biology; however, I very much doubt that the inference to the teleological structure of life in general is justified. For instance, the breaking of the rule that applies to mammals at the early stages of the human form in no way makes retrospective sense of lower mammalian ranks; rather, the contrast with the mammalian series points to the specifics of the human Gestalt. According to Portmann, the facts about the (early) human Gestalt – "secondary altriciality" and "extra-uterine year" – make it evident that human beings are biologically formed to be cultural animals – that is our pre-arranged relationship (*vorbereitete Beziehung*) to the world. Moreover, Portmann does not hold – as Goethe does, according to Förster (2012) – that human beings are endowed with a special cognitive capacity – intuitive judgement – capable of grasping the idea, or essence or "Ur-Phenomena" behind or in natural phenomena such as colors, plants, or animals. Rather, using Goethe's phenomenological method (Heinemann 1934) without its metaphysical underpinning, Portmann counter-balances the one-sided approach of contemporary biology in order to emphasize the topics of meaning, centricity, self-presentation and rank within modern life science.

References

Allen, C., & Bekoff, M. (1997). *Species of mind. The philosophy and biology of cognitive ethology*. Cambridge: MIT Press.

Amrine, F., & Zucker, F. J. (1987). Goethe's science. An alternative to modern science or within it – Or no alternative at all? In F. Amrine & F. J. Zucker (Eds.), *Goethe and the sciences. A reappraisal* (pp. 373–388). Boston: D. Riedel.

Bennett, M. R., & Hacker, P. M. S. (2003). *Philosophical foundations of neuroscience*. Oxford: Blackwell Publishing.

Dunsworth, H. M., et al. (2012). Metabolic hypothesis for human altriciality. *Proceedings of the National Academy of Sciences of the United States of America, 109*(38), 15212–15216.

Förster, E. (2012). *The twenty-five years of philosophy. A systematic reconstruction*. Cambridge: Harvard University Press.

Godfrey-Smith, P. (1994). A modern history theory of functions. *Noûs, 28*(3), 344–362.

Godfrey-Smith, P. (2016). Mind, matter, and metabolism. *The Journal of Philosophy, 113*(10), 481–506.

Grene, M. (1968). *Approaches to a philosophical biology*. New York: Basic Books.

Grene, M. (1974). *The understanding of nature. Essays in the philosophy of biology*. Dordrecht: Springer.

Griffiths, P. (1993). Functional analysis and proper functions. *British Journal for the Philosophy of Science, 44*, 409–422.

Haag, J., & Wild, M. (Eds.). (2013). *Übergänge - diskursiv oder intuitiv? Essays zu Eckart Försters Die 25 Jahre der Philosophie*. Frankfurt/M: Klostermann.

Hegge, H. (1996). Transcending darwinism in the spirit of goethe's science: A philosophical perspective on the works of Adolf Portmann. *Newsletter of the Society for the Evolution of Science, 12*(2), 1–26.

Heinemann, F. (1934). Goethe's phenomenological method. *Philosophy, 9*(33), 67–81.

Hindrichs, G. (2011). Goethe's notion of an intuitive power of judgment. *Goethe Yearbook, 18*, 51–65.

Krämer, D. M. (2014). Revisiting recent etiological theories of functions. *Biology and Philosophy, 29*(5), 747–759.

Matthen, M. (2000). The disunity of color. *Philosophical Review, 108*, 47–84.

Matthen, M. (2005). *Seeing, doing and knowing*. Oxford: Oxford University Press.

Millikan, R. (1989). In defense of proper functions. *Philosophy of Science, 56*(2), 288–302.

Mitchell, S. D. (1995). Functions, fitness, and disposition. *Biology and Philosophy, 10*, 39–54.

Moore, P. (2012). How should we think about the ribosome? *Annual Review of Biophysics, 41*, 1–19.

Neander, K. (1991). Functions as selected effects. *Philosophy of Science, 58*, 168–184.

Neander, K., & Rosenberg, A. (2012). Solving the circularity problem for functions: A response to Nanay. *The Journal of Philosophy, 109*(10), 613–622.

Polger, T. W., & Shapiro, L. A. (2016). *The multiple realization book*. Oxford: Oxford University Press.

Portmann, A. (1944). *Biologische Fragmente zu einer Lehre vom Menschen*. Basel: Schwabe.

Portmann, A. (1946). Étude sur la cérébralisation chez les oiseaux I. *Alauda, 14*, 2–20.

Portmann, A. (1947). Étude sur la cérébralisation chez les oiseaux II. *Les indices intra-cérébraux. Alauda, 15*, 1–15.

Portmann, A. (1948a). *Einführung in die vergleichende Morphologie der Wirbeltiere*. Basel: Schwabe.
Portmann, A. (1948b). *Die Tiergestalt. Studien über die Bedeutung der tierischen Erscheinung*. Basel: Friedrich Reinhardt. (Translation into English: *Animal Forms and Patterns: A Study of the Appearance of Animals*, 1967).
Portmann, A. (1953a). Goethes Naturforschung. *Neue Schweizer Rundschau, 21*, 406–422.
Portmann, A. (1953b). *Biologie und Geist*. Zürich: Rhein-Verlag.
Portmann, A. (1953c). *Das Tier als soziales Wesen*. Zürich: Rhein-Verlag. (Translation into English: *Animals as Social Beings*, 1961).
Portmann, A. (1956). *Zoologie und das neue Bild des Menschen*. Hamburg: Rowohlt. (Partial translation into English: *A Zoologist Looks at Humankind*, 1990).
Portmann, A. (1957). Die Erscheinung der lebendigen Gestalten im Lichtfelde. In K. Ziegler (Ed.), *Wesen und Wirklichkeit des Menschen. Festschrift für Helmut Plessner* (pp. 29–41). Göttingen: Vandenhoeck & Ruprecht.
Portmann, A. (1961). *Neue Wege der Biologie*. München: Piper.
Portmann, A. (1965). *Aufbruch der Lebensforschung*. Zürich: Rhein-Verlag.
Portmann, A. (1972). Farbensinn und Bedeutungen der Farben aus biologischer Sicht. *Eranos Jahrbuch, 41*, 465–491.
Portmann, A. (1973). Goethe und der Begriff der Metamorphose. *Goethe Jahrbuch, 90*, 11–21.
Portmann, A. (1987). Goethe and the concept of metamorphosis. In F. Amrine, F. J. Zucker, & H. Wheeler (Eds.), *Goethe and the sciences. A reappraisal* (pp. 133–145). Dordrecht: Springer.
Scheler, M. (1926). *Die Wissensformen und die Gesellschaft. Probleme einer Soziologie des Wissens*. Leipzig: Der Neue Geist Verlag.
Scheler, M. (1980). *Problems of a sociology of knowledge*. London: Routledge & Kegan Press.
Senn, D. G. (2013). *Einführung in die Morphologie der Wirbeltiere* (begründet von Adolf Portmann). Basel: Schwabe.
Shapiro, L. A. (2004). *The mind incarnate*. Cambridge: MIT Press.
Shettleworth, S. J. (2010). *Cognition, evolution and behavior*. Oxford: Oxford University Press.
Smit, H., & Hacker, P. M. S. (2014). Seven misconceptions about the mereological fallacy: A compilation for the perplexed. *Erkenntnis, 79*(5), 1077–1097.
Wild, M. (2012). Intuitionen, intuitiver Verstand, Intuition. *Deutsche Zeitschrift für Philosophie, 60*(6), 1011–1018.

Markus Wild is Professor of Theoretical Philosophy at the University of Basel (Switzerland). His main research interests include early modern philosophy, philosophy of mind, and animal philosophy. Among his most recent publications are "Intentionality and flexibility in animal communication" (with C. Sievers and T. Gruber), in *The Routledge Handbook of Philosophy of Animal Minds* (ed. K. Andrews and J. Beck, 2018); "What is Biological about Aristotelian Naturalism?", in *Aristotelian Naturalism: A Research Companion* (ed. M. Hähnel, Springer 2019); and "Conceptual Thought Without Language? The Case from Animal Cognition," in *Concepts in Thought, Action, and Emotion: New Essays* (ed. C. Demmerling and D. Schröder, Routledge 2021).

Chapter 9
Playing Appearances: On Some Aspects of Portmann's Contribution to Philosophical Aesthetics

Pietro Conte

Abstract Widely regarded as Adolf Portmann's most important (and controversial) work, *Animal Forms and Patterns* has been the object of extensive analysis and criticism from many different perspectives, such as evolutionary and theoretical biology, psychology, philosophy, ethology, and environmental studies. Yet, very little attention has been paid so far to the significant differences between the first edition of the book, dating from 1948, and the second edition, which only appeared 12 years later, in 1960. After highlighting the crucial role that such differences assumed in the evolution of the Portmannian way of understanding the "spectacle" of nature and its meaning, I shall concentrate on the topicality of Portmann's thought in the specific field of morphological aesthetics, particularly focussing on the notion of play as a possible bridging concept between science and art.

The Deepest Is the Surface

It was 1948 when Adolf Portmann published *Animal Forms and Patterns* (Die Tiergestalt), a masterwork that, although representing the ripe fruit of a lifetime of research, appeared at an extremely untimely moment. A professional biologist, the author had already gained international attention and respect thanks to numerous scientific papers that in 1931 had earned him – at the age of barely 34 – the chair of Zoology at the University of Basel. But these studies also left him with a deep dissatisfaction in terms of well-established scientific paradigms that seemed to leave no room for doubt. Precisely this dissatisfaction manifested in *Animal Forms and Patterns* and became even more pronounced in 1960, as the second edition of the work was published, expanded and revised in many important aspects.[1]

[1] All the existing translations of *Die Tiergestalt* are based on the first German edition of the book, with the exception of the Italian *La forma degli animali. Studi sul significato dell'apparenza*

P. Conte (✉)
Ca' Foscari University of Venice, Venice, Italy
e-mail: pietro.conte@unive.it

F. Jaroš, J. Klouda (eds.), *Adolf Portmann*, Biosemiotics 23,
https://doi.org/10.1007/978-3-030-67810-4_9

To clarify what was at stake, it would be convenient to begin with the initial words of the first chapter, which represent a precise statement of intention:

> Man's experience, from very early times, has led him to realize that, as regards much of what we see around us, appearances are deceptive and conceal from us the true nature of things. Natural science itself has confirmed the opinion that only beneath many and varied skins do we find the essential core of reality. And indeed, discoveries in the realm of the invisible, and about what is not readily accessible, have produced notable achievements", thus reinforcing the view that the "great riddles of life" are to be sought only "in the invisible. (Portmann 1967: 17, orig. Portmann 1948)

The story recalled here is very old and calls into question the scheme based on the opposition between *eidos* and *eidolon*, being and appearing, knowledge and opinion – a scheme that found its first monumental systematization in Plato's thought.[2] Within this theoretical framework, what appears – the phenomenon – is but a degraded image of something (no matter how you name it: original, model, reality) that, in turn, is not an image. Yet, along with modern scientific development this paradigm seems to undergo a drastic change: objectivity and truth are not being chased anymore in the metaphysical regions of Hyperouranion but rather in the microscopic and submicroscopic space of atoms and molecules. If Plato took sensible forms as merely partial and unsuccessful imitations of the ideal Forms, modern physics, biology, and physiology consider instead those forms as representing a secondary and accessory manifestation of absolutely concrete processes that remain nonetheless hidden under (or if you prefer, behind) deceptive appearance. From both points of view, the essential is precluded to the senses. To catch it, we need different eyes: the eyes of the mind freed from the bonds of corporeality in one case, the artificial eyes of sophisticated technical tools in the other.

The existence of a true world beyond the perceptible world is therefore a *vexata quaestio* against which Goethe – whose scientific studies were a source of great inspiration for Portmann – had already protested. In his famous and highly controversial studies on botany and zoology, the author of *The Metamorphosis of Plants* left no shortage of criticism of the indiscriminate use of lenses and microscopes, which he accused of confusing "the unaided human senses" (Goethe 1998: 502).[3] These words certainly did not aim to create a polemic against technical progress or the use of appliances meant to unveil new fields of research and exploration which would otherwise remain completely inaccessible. Rather, Goethe's concern related to the possible disparagement of the sensible dimension and the sphere of appearance. The real risk is to lose sight – literally – of the world around us, that "life-world" (as Husserl defines it) which is both the precondition and the horizon for every possible experience.

---

fenomenica degli animali. Milan: Raffaello Cortina, 2011 (not to be confused with *Le forme degli animali*, published in 1960 and based, again, on the first edition of Portmann's work).

[2] This systematisation is not at all devoid of ambiguities, and even inconsistencies, in Plato himself, as masterfully shown by Ernst Cassirer (Cassirer 2013).

[3] On Portmann's reading of Goethe's scientific studies see Hegge (1996) and Wild, Chap. 8, in this volume.

As in Goethe, so also in Portmann the discussion concerns a theoretical model that makes of truth a nucleus hidden under more or less numerous deceptive cover-ups. With regard to zoology, the idea that everything immediately given to the senses would be nothing but an accidental consequence of much more essential structures leads to the view that the outward appearance of an animal exists "just in order to preserve the essential internal mechanism, by means of acquiring food and by moving, by avoiding enemies and by finding the opposite sex" (Portmann 1967: 204, orig. 1948).

A crucial point is reached here in Portmannian reflection: that of the relationship between outer and inner, between what is perceptible with the naked eye and what remains inaccessible to the direct gaze. The criticism that runs through every single page of *Animal Forms and Patterns* concerns the reduction of outer appearances to mere by-products of the internal processes or to features that are useful in the struggle for life. From this perspective, understanding the shape of an organism only means identifying its survival value, that is, its ability to facilitate the survival of individuals and the species they belong to.

Reference is made here to Darwin, but it is worth clarifying immediately that Portmann's target is by no means the English naturalist, whose work is unequivocally solemnized as "pioneering" and "extraordinary" (Portmann 1976: 78), but rather his unspecified "popularisers" (cf. Portmann 1958a), who chose the path of a radical reductionism meant to interpret the multiple aspects of the phenotype either as functional features for the preservation of life or as insignificant epiphenomena of fundamental physiological processes. There is no middle ground: either outer appearances are useful in the struggle for survival and could be explained on the basis of a strict selection programme, or they have no meaning at all and must be considered mere side effects of the vital functions of the organisms. In the first case, outer appearances are condemned to eternal subordination to the principle of utility; in the second, "superficial" turns out to be but a byword for "superfluous", and the visible form is considered a negligible manifestation of priority functional circuits invisible to the naked eye.

In Portmann's view, instead, "surface" should not necessarily be understood as the opposite of "deepness"; on the contrary: one need only pay attention to the role appearances play within human society (Carnevali 2012)[4] to understand that "nothing is deeper in man than his skin" – an adage from Paul Valéry (1971: 33) taken up afterwards, and slightly modified, by Gilles Deleuze (1990: 7: "The deepest is the skin"). To put it briefly, there does indeed exist a superficial deepness, a deepness *in* and *of* the surface, and there is no need to refer to Didier Anzieu's "skin ego" (Anzieu 1989) to get acquainted with the fact that the skin delimits our body and protects our own individual identity, while at the same time it clothes us and opens us to the outside world, introducing us into the company of other people. The skin is "an agent of display and membership", like "the palimpsest of our biography"

[4] Not coincidentally, Carnevali's book refers directly and from the very first pages to Portmann's notion of "appearance".

(Cappelletto 2012: 59): it exposes us, as Jean-Luc Nancy (2008: 32) used to say, playing on the homophony between French words *exposition* and *ex-peau-sition*.

For Portmann, however, this holds true not only for human skin, but also for the fur, the feathers, the scales or the slough of non-human animals. Even more: it also holds true for all vegetables, fungi, and even those humble slime moulds that Portmann describes in some crucial pages purposely added to the second edition of *Animal Forms and Patterns* in 1960. A genuine "form value"[5] is to be ascribed to even the simplest protist.

Ostending Form

Thus, Portmann proposes to completely rework the notion of "organism" and to look for a redefinition that he described as the "crux of a most significant change towards comprehensive biology" (Portmann 1970b: 14). It concerns reinterpreting, from a biological perspective, the basic principle of Gestalt psychology that the whole is greater than the sum of its parts, so as to show that organisms cannot be reduced to the sum of their organs. As Portmann's former doctoral student Roger Alfred Stamm recalls, the starting point of the work of the Swiss biologist lies in the assumption that "whoever considers the organism only as a series of particular functions loses the sense of totality: our direct experience of life is an encounter with a living being (plant, animal, human being) in its immediate manifestation, i.e. as a *totality*" (Stamm 1999: 55).

This very same theoretical principle had been set out several years before Portmann by Frederik Buytendijk. In a short essay dating from 1928, "Intuitively Accessible Features of the Organic" (Anschauliche Kennzeichen des Organischen), the Dutch physiologist had emphasized that the usual definition of "totality made of heterogeneous parts" or "complex unity" does not capture the essence of the organic, which also implies (as a distinct Gestalt principle) a clean-cut delimitation of its surface with regard to a non-organized background. And this delimitation means freedom: "Precisely because the background is *behind* the organic form, the latter remains free, independent: rather than a functional value (*Wirkungswert*), it displays a value of being (*Seinswert*)" (Buytendijk 1958: 10). To be sure, this does not mean that the organisms have no functional relationship with the surrounding environment, with which, on the contrary, they form an inseparable unity (as Jakob von Uexküll, whose scientific work was greatly admired by both Buytendijk and

[5]The term is used frequently in Portmann (1967, orig. Portmann 1948), for instance where it expresses "the valency of a living thing in the same way as the nuclear number of an atom represents its position in a system. The form value does not give any information about the performance of an organ; it merely indicates the place occupied by this part of the body in a classification of organic forms" (46–47).

Portmann, had already argued).[6] But all of this is secondary compared to the primary essence of the organic, which consists in the displaying of its value of being: "If a totality exhibits primarily a functional value, it can only be an organ, never an organism" (Buytendijk 1958: 10).

There is no doubt that all organisms are constituted by structures that follow physical-chemical laws and function mechanically, hence giving the impression that there is nothing superfluous, useless, or excessive in them. But this is not enough, according to Buytendijk, to define the essence of an *organism*:

> In nature, the machines are beautifully painted, they are made with elegance and refined taste [...]. One has the impression that the whole rational machine exists only to display that splendour with a boundless waste of energy. Briefly, the organic is the vehicle of an ostensive value of being, and its essence consists in wealth, luxury. (Buytendijk 1958: 12)

From this perspective, exhibition of the value of being appears as an inalienable trait of life itself, an *essential* characteristic of the organic: the living is not merely what survives, but what expresses itself through the exhibition of its own singularity – exhibition that should be regarded as even more original and stronger than the desire to preserve biological existence. Portmann makes this idea his own, directly referring to Buytendijk's essay and endorsing the critique of the functionalist one-sided considerations and the longing for a new "phenomenology of animal morphology" (Thinès 1996: 202) that, while not ignoring the importance of functional elements, could bring out a broader meaning of phenomenal appearance:

> We shall have to form a very different conception of the whole organism. We shall have to realize that not only the eye, heart and brain arise by an ordered plan of developmental processes, but also the way in which outward appearance is formed, the ornamental shapes and the designs as a whole. (Portmann 1967, 142, orig. 1948)

Thus, morphology. In the wake of a centuries-old tradition based on the Goethean principle that "everything that is must also manifest and show itself" (Goethe 1987: 349), Portmann regards the study of the external form not really as an alternative to functionalism, but rather as its balancing counterweight: his criticism does not concern the theory of natural selection, the concept of adaptation or that of fitness, but how these concepts are *used* and the reductionism they can possibly (but by no means necessarily) lead to.[7] The attempt to explain organic reality as solely a chain of physical-chemical reactions can easily be made subservient to a worldview based on the desire to dominate nature and on an impoverished concept of the organic, as if everything could be reduced to a combination of physiologically determined and purposefully oriented factors. To be sure, for Portmann the organism is certainly this, but at the same time it is much more than this: interpreting it as a mere

[6] Portmann, in particular, penned a long foreword to the 1956 edition of Uexküll's masterpiece (originally published in 1934) *Streifzüge durch die Umwelten von Tieren und Menschen. Ein Bilderbuch unsichtbarer Welten* (Uexküll 1956: 7–17).

[7] As highlighted by Salvatore Tedesco (2010: 29), Portmann's thought – particularly of late years – is to be regarded as clearly embedded "in an evolutionary framework".

Zweckgestalt, as a simple "purposeful form", means losing sight of characteristics as fundamental as those associated with vital functions.

Morphology teaches us to study form as an element with its own specific value, with an irreducible meaning to which we have only access through intuition, but which does not, for this reason, escape genuinely scientific consideration. By consciously employing such an ambiguous and equivocal term as *Anschauung*, Portmann intends to emphasize that relying on "intuition" and on the "naked-eye vision" does not mean falling into mysticism, but on the contrary sharpening one's gaze to disclose a kind of knowledge otherwise unachievable: a knowledge *of the image*, an *aesthetic* knowledge. In order to understand an organism, it is certainly necessary to take into account the functionalistic value of its organs, without at the same time leaving aside their aesthetic value (as it deserves equal consideration). An organism cannot be reduced to a well-thought-out complex of structures aimed at the preservation of existence, for it manifests instead – albeit within certain boundaries – unlimited possibilities of "freer configuration" (Portmann 1964: 84).[8] Purposeless (*zweckfrei*) is not the same as meaningless (*sinnlos*): many aspects of animal forms and patterns seem to have no immediate survival purpose or functional value, but this does not mean that they are meaningless or play no role in the existence of the organism (Portmann 1970b: 27).

The analysis of these peculiar phenomenal characteristics is defined by Portmann as *phanerology*. Phaneric (from the ancient Greek word *phaneros*, "evident, manifest") are all the formal features perceivable with the naked eye; phanero-logy is the attempt to find out the *logos* (the logic) of these formal features – thus, the *logic of aesthetics*. These elements have too often been associated exclusively with the functions of preservation based on a lock-and-key model: "Characteristic colours or movements, particularly significant calls were interpreted to trigger certain actions. These dispositives were technically defined as *releasers* [...]. However, we want to put aside the functional interpretation of the phaneric traits." (Portmann 1965: 49).

What is at stake here is the well-known problem of the peacock's tail and, more generally, of all the multiple attempts to attribute to outer appearances – without exception – a peculiar adaptive function, such as attracting the conspecific individuals of the opposite sex, intimidating potential enemies, or blending in to deceive predators. The aim is to establish a biunique correlation between the sensory organs (interpreted as receiving devices) and the phenotypic characteristics (considered as transmitting ones), assuming that this connection is functional to life preservation. For Portmann these attempts, based on the identification of specific "organs to be seen" (Portmann 1967: 110, orig. 1948, transl. modified), are certainly of great importance and often correct, but they do not cover all the meanings of the phenomenal appearance of animals: certain structures are made simply *to be (even though potentially) gazed at*, and it is precisely in their being aimed at a *possible* gaze that

[8]This quote is taken directly from the original German edition of Portmann's *Neue Wege der Biologie* (Piper, München 1960: 84), as it has been inexplicably expunged from the English translation by Arnold J. Pomerans (Portmann 1964: 52).

their ultimate meaning is to be found: an *aesthetic* meaning that cannot be further explained.

Form Does Not Follow Function

It is a matter of exteriority, a matter of *façade*, a term which is not to be meant as a "deceptive appearance" hiding something more genuine and essential, but rather as the manifestation of characteristics that have in themselves – in *pure appearing* – their irreducible meaning. Portmann's reflection focuses here on the concept of "ornament", which offers an important point of contact between aesthetics and biology.[9] Is it possible to consider an ornamental element – no matter if it appears on the pediment of a temple, on the ledge of a building, or on the body of an animal – as endowed with a specific value? Is it legitimate to attribute to it a specific prowess, or even a purpose? Or, on the contrary, would it be more appropriate to speak of superfluous decoration?

Both in architecture and in biology, the core problem lies in the relationship between form and function, and it is certainly no coincidence that Portmann knows and quotes the famous formula coined by Louis Sullivan, *form follows function*[10]:

> A physiologically oriented thinking in architecture began to prevail which was intended as an extreme case of the idea of function. According to this thesis, form should be the strict expression of a performance – as it is in nature. […] The formula: "Form follows function" became the catch-word for many of the reformers among the architects. This physiological concept led, above all else, to the devaluation of ornamentation. As a consequence, it gradually led to the rejection of all the expressive value of the outer appearance of buildings as well. "A façade": since that time this term has come to refer to a negative subject, to the gross, forbidding example of empty ostentation, to senseless veiling of the living function. This physiological thinking then led on to the further broadened implication that the act of building must itself be determined completely in the same way as it is for the organism – from within outward, beginning from the support functions. The strict realization of this principle requires that the form of the outside be presented as coming from inside, from itself – as a pure being rather than as a false appearance. Today we must realize how the one-sidedness of this principle misinterprets natural forms and how widely it threatens to divert us from a real understanding of living form. (Portmann 1990b: 151)

To criticize the theory that decoration would be nothing more than a superfluous element of architectural structure means, in biology, to criticize the idea that the outer appearance of animals is nothing but an inessential addition to physiological processes, only suitable – at best – for providing information on the proper functioning of internal organs. From this point of view, Portmann would have certainly endorsed Sullivan's assertion that a decorated structure "cannot be stripped of its system of ornament without destroying its individuality" (Sullivan 1979: 188).

---

[9] On the relationship between animal and human ornaments with direct reference to Portmann's thought see Prévost (2009, 2010, 2012).

[10] Cf. Portmann (1976: 32).

It is basically the ancient problem of "hypertelic" structures, forms and actions, which seem to go beyond (*hyper*) any rationally conceivable purpose (*telos*). Think of the classic example of the "extended phenotype" provided by the unusually complex courtship ritual of the bowerbird. The male gets a bit of twigs, builds a small nest and decorates it using feathers, shells, small flowers, coloured berries and even pieces of glass or plastic. While entering the "flat", the female witnesses a choreographic parade consisting of a complex dance accompanied by calls and songs, until she eventually decides whether to couple with the candidate partner. Evolutionary biology provides various explanations of this mating ritual, the most famous of which refers to the notion of "costly signing": the luxury of decor implies a considerable cost in terms of time and energy, and this cost, in turn, would reveal the high genetic quality and excellent fitness of the candidate. However, when faced with such a complexity, one is reminded of Stephen Jay Gould and Richard Lewontin's reference to Darwin's "pluralistic approach" and their warning against a Panglossian paradigm grounded just on the principle of adaptation: "Often, evolutionists use *consistency* with natural selection as the sole criterion and consider their work done when they concoct a plausible story. But plausible stories can always be told" (Gould and Lewontin 1979: 588).

There is more. To bring all phenomenal traits back to a series of useful functions for survival does not explain their *specificity*, i.e. their *being so and not otherwise*. Stating that a particularly striking colouring serves to attract potential partners, or that a cryptic pattern is useful to confuse predators, is certainly correct (no coincidence that Portmann himself devotes a thorough study to mimicry).[11] With this, however, nothing is said about the variety and morphological specificity of phenomenal appearance. At stake is the question of the particular, of the quality, of the aesthetic:

> To state [the] function does not in the least touch upon or explain the special manner in which a task is carried through by a certain form: nothing has been said about the particular pattern which characterizes a bird's feathery crest, or the wings of a butterfly; the specific shape of antlers or horns, the pattern on a zebra's head, are by no means explained. (Portmann 1967, 208–209, orig. 1948)

Neither the concepts of utility and fitness, nor the (alleged) survival instinct, can account for the fanciful diversity of appearances produced by organisms. Moreover, the forms and colours of an animal's fur or those of a flower's corolla *can*, indeed, produce a defined optical effect: they are destined for, and perfectly adapted to, the eye of a particular receiver – an eye that is therefore to be regarded as having a selective role. Yet Portmann insists on the idea that through this way of associating a particular phenomenon to a particular function the *origin* of the formal traits is not in the slightest clarified. On the contrary, it is presupposed:

> In order that an optical characteristic could act on the eye, it must first *exist*: this obvious observation shows that the chromatic structures and all the other elements of the skin surface could not have been *originated* from the selection, which at most may have *favoured*

[11] Portmann (1959).

> certain variants instead of other ones, consolidating certain formal patterns instead of other ones. (Portmann 1970a: 80, emphasis mine)

As a consequence, the functional cycle between seer and seen must be considered as insufficient to account for the extraordinary complexity of the issues at stake, and this not only because all explanations based solely on the role of selection processes explain neither the origin nor the specificity of outer appearance, but also because the organs to be seen are but a particular case among much more numerous "optically indifferent" surface patterns. In a crucial passage added to the second edition of *Animal Forms and Patterns*, Portmann clarifies:

> The contrast between outer and inner configuration is already manifested at levels in which no cryptic or semantic meaning can be attributed to the external form and to the skin pattern. In these cases, the formal characteristics are optically indifferent, but nevertheless they follow specific development rules. The optically indifferent configuration represents the starting point for the development of optically acting forms, which also have specific roles in the game of life (*Lebensspiel*). (Portmann 1960a: 241, translation mine)

In this context, Portmann makes explicit reference to another important name of twentieth-century theoretical biology, Raymond Ruyer, who in 1958 published *The Genesis of Living Forms* (La Genèse des formes vivantes). The French scholar, in turn, directly refers to Portmann's work, particularly agreeing with him on the existence of phenomenal characteristics which are not explained by the selective pressures exerted by the observer's eye. Beyond that, even for Ruyer the organs to be seen are, indeed, very numerous, but they still represent an exceptional case within the extraordinary complexity of the phenomenal appearance of animals. The ability to perceive images is, in fact, a goal reached only at a very late stage in the evolutionary history of organisms, which means that the forms that presuppose this ability must have developed *after* those optically indifferent.

A veritable change of perspective is then needed: primary are not the forms that appeal to an eye looking at them and that can therefore be explained on the basis of a certain functional role. On the contrary, what is primary is morphogenesis, the production of forms *tout court*, of which the allaesthetic forms (that is, the forms only conceivable in relation to the distance receptors) are but a particular case. All organisms are not merely living machines whose primary function is metabolism and who only aim to survive for as long as possible. They are, above all, beings which display themselves in their singularity, without this self-presentation being primarily, solely, or necessarily related to sense organs.

A great deal of confusion seems to have arisen on this crucial (and, to be sure, highly problematic) aspect of Portmann's thought. In a famous, unfinished work published posthumously in 1978, *The Life of the Mind*, Hannah Arendt makes enthusiastic reference to many different essays of the Swiss biologist, including *Animal Forms and Patterns*. The first chapter, which is devoted to the meaning of outer appearance and its truth value, begins with the assumption that "the world men are born into contains many things, natural and artificial, living and dead, transient and sempiternal, all of which have in common that they *appear* and hence are meant to be seen, heard, touched, tasted, and smelled, to be perceived by sentient creatures endowed with the appropriate sense organs. Nothing could appear, the word 'appearance' would make no sense, if recipients of appearances did not exist" (Arendt 1978: 19). This, however, is a macroscopic simplification of Portmann's argument, according to which the organs to be seen are a particular characteristic of only higher animals:

> Even if all human beings, actually, all the organisms equipped with eyes, suddenly disappeared off the face of the Earth, we would still have no other word than 'appearance' to indicate that even after all human beings and animals capable of seeing vanished, there would still be organisms characterized by inwardness and self-presentation to possible senses. (Portmann 1964: 221)[12]

Self-presentation (*Selbstdarstellung*) is the key term, and notably it is not mentioned even once in the first edition of *Animal Forms and Patterns*; it appears instead in the second, and precisely in that concluding section devoted to moulds which was not included in the 1948 edition. It seems as if Portmann decided to rework his *opus magnum* and to republish it in 1960 in order to add this last chapter and to introduce the concept of "self-presentation" – a concept that in the meantime had already been studied in depth in a fundamental essay dated 1958, "Self-presentation as the Ground for Form Configuration of Living Beings" (Selbstdarstellung als Motiv der lebendigen Formbildung; Portmann 1958b).

Self-presentation is the presentation of one's being this way and not otherwise. As such, it is not limited to human beings, but characterizes all animals, plants, fungi. Portmann is keen to point out that self-presentation is a name for the displaying of a self (*Selbst-Darstellung*), not of an ego (*Ich-Darstellung*), which means that self-presentation is not necessarily related to consciousness, but to the inwardness (*Innerlichkeit*) which distinguishes – although in a so-far incomprehensible, hence unexplainable way – even vegetal organisms. The outer appearance of a plant or animal is therefore "a coat which raises its wearer to a place apart" (Portmann 1967: 25, orig. 1948), a dress that identifies an organism not only in its belonging to a particular species, but also in its unique and unrepeatable individuality. Living beings present themselves, put themselves on display as in a shop window (Portmann 1970c: 131) or on a theatre stage:

> Genetics allows us to look behind the scenes of the theatre. We may watch the way in which the actors get ready, how the machinery produces the effects of thunder and rain; how everything works together so that, by the complicated action of many invisible helpers, a play having an intelligible sequence is finally unfolded before the spectator. But such a glimpse behind the scenes tells us neither the gist of the play nor its significance. [...] Whether we are able to understand the play which is being enacted before our eyes depends upon other requisites than a grasp of the technique of the performance. (Portmann 1967: 161–162, orig. Portmann 1948)

[12] This quote is taken directly from the original German edition of Portmann's *Neue Wege der Biologie* (Piper, München 1960: 221), as it has been inexplicably expunged from the English translation by Arnold J. Pomerans (Portmann 1964: 154).

The Play of Appearances

The metaphors of the display window and the theatrical stage should not be misinterpreted as if self-presentation would be necessarily correlated to a receptive organ. The second edition of *Animal Forms and Patterns* presents an afterthought to the concepts of "form" and "morphology", a crucial rethinking aimed at highlighting precisely the autonomy of formal value and the fact that self-presentation is *not* necessarily intended for receptive organs. The existence of phenomenal manifestations that find in themselves – *in pure appearing* – their meaning implies the need to reformulate the very concept of "function", without limiting it solely to what appears to be useful in terms of the survival of the individual or species. From this perspective, the priority is not the spectator but the spectacle: *contra* Berkeley, "to be" is *not* the same as "to be perceived", for there also exists – to quote Ruyer once again (1966: 15) – a "spectacle without spectators". In the second edition of *Animal Forms and Patterns*, Portmann (1960a: 253) introduces the notion of "unaddressed phenomena (*unadressierten Erscheinungen*)",[13] that is, outer appearances which are not headed to any particular recipient, formal traits whose function consists simply in *appearing*: "The colours and forms in which these structures manifest themselves have a meaning that is never numbered among the functions necessary for life preservation: the meaning of representation (*Repräsentation*), the meaning of self-presentation of the species" (Portmann 1960a: 251). In many cases, it is impossible to couple a visual phenomenon with a single specific receiver and a clearly identifiable function; the animal's appearance seems to be "sent 'into the blue' (*'ins Blaue' gesendet*)" (Portmann 1960a: 254), headed to innumerable *possible* receivers. The meaning of the "signal" becomes open, ambiguous, virtually multiple.

Within this theoretical framework, even the Merleau-Pontian interpretation of Portmann's work turns out to be rather equivocal. On the one hand, in the famous lectures given at the Collège de France on the concept of "nature", the French philosopher takes up the formula that "Being cannot be defined outside of perceived being", arguing that Portmann's work "seems to confirm this way of seeing things" (Merleau-Ponty 2003: 189). The course was held during the academic year of 1957–1958, when neither the essay on self-presentation (Portmann 1958b), nor that on "The Evaluation of the Role of Appearance in Organisms" (Die Beurteilung der Erscheinung im Organischen; Portmann 1970a), nor the second edition of *Animal Forms and Patterns* (1960) had yet been published. This can certainly explain the lack of reference to "unaddressed appearances", but not the fact that Merleau-Ponty does not even take into account "optically indifferent" structures – a theme that also goes in a diametrically opposite direction with respect to the Berkeleyan principle *esse est percipi*, and that it was already very present in the edition of 1948.

On the other hand, however, Merleau-Ponty himself seems to have noticed the problematic nature of his own interpretation. In the working notes for *The Visible and the Invisible*, which date from April 1960 – shortly before the publication of the

[13] On this, see Dewitte (2001).

second edition of *Animal Forms and Patterns* (whose foreword dates from the autumn of 1960), but 2 years after the publication of Portmann's essay on self-presentation (which however, as explained above, Merleau-Ponty does not mention)[14] – a significant change takes place. Once again, the French phenomenologist sets out from the "organs to be seen", but he then underlines that the very fact that these organs do imply someone looking at them "would not itself be possible if antecedently the part of the body in question were not *visible*, if there were not, around each part of the body, a halo of *visibility*". The emphasis here is on the *virtuality* of perception, on the *possibility* of vision that precedes actual vision and paves the way to it: "This visible not actually seen [...] is a presence of the imminent, the latent, or the hidden" (Merleau-Ponty 1968: 244–245).[15] What comes forth is the *potential* nature of perception: the fact that ornamental traits are to be found in even the lowest forms of organic life, where seeing eyes play no role at all, can only be explained by arguing that these very same traits have in themselves "the *possibility* of being-looked-at (*die Möglichkeit des Angeschautseins*)" (Portmann 1990a: 25).[16]

This openness to the possible is the openness to the dimension of multiple meanings, virtuality, the imaginary, creativity, pretense and making as-if, in a word: to the dimension of *play*. In Portmann's view, appearing is, at the most elementary level, playing; and playing, in turn, cannot be degraded either to a waste of time or to an incidental, superfluous break between two "serious" activities. Quite the contrary, it is – like appearing – "a basic property of life (*eine basale Lebenseigenschaft*)", as Portmann maintains, once again, in the last chapter purposely added to the second edition of *Animal Forms and Patterns* (Portmann 1960a: 252). Life would serve the art of appearing and not the contrary: in other words, living beings would appear, and would tirelessly continue to produce new appearances, "for the sake of appearing and playing with appearances" (Dufourcq 2016: 68).

From this perspective, appearing seems to be autotelic and self-satisfactory like play. It is therefore no coincidence that Portmann became more and more interested in the enigmatic question of ludic activities. Although this aspect is key to understanding some central issues of his work, it has been scarcely investigated until a very recent time, when it began to be discussed by the scientific community, and in particular by Italian researchers.

Portmann's way of approaching the phenomenon of play found its first clear formulation in the essay "Play and Life" (Spiel und Leben), which dates from 1960 (that is, not coincidentally, the same year of the second edition of *Animal Forms and*

[14] Significantly, the French translation of this text was published in the journal *Études Phénoménologiques*, devoted to phenomenology and phenomenological aesthetics (Portmann 1996).

[15] On this, see Gambazzi (1999), where the author argues that the potential character of being-seen extends far beyond the "realm of the mere empirical" and is not limited to "the simple fact of someone actually seeing".

[16] Translation modified; emphasis mine. On the convergence of Portmann's and Merleau-Ponty's views on this specific point, see Dewitte (1998) and Amodio (2012).

Patterns). In it, the criticism concerning all scientific paradigms based just on the principles of adaptation, fitness, and function, is developed into a peculiar concept of playful behaviours carried out not only by human beings, but also by highly-developed non-human animals such as crows. From the very beginning, Portmann is firm in asserting that play cannot be reduced – as in Karl Groos's famous "pre-exercise theory" – to a youthful necessary practice of skills that will later be of aid in the struggle for life (Groos 1898). To be sure, this vision contains an element of truth, and it can often provide an explanation of some specific behaviours, but it is also undoubtedly partial:

> We must consider play from a different perspective, focussing on the wide room-for-freedom (*Raum der Freiheit*) and on playing as the expression of a free giving-form to an inexhaustible supply of drives and impulses headed towards activity (*Tun*), not towards exercise (*Üben*). (Portmann 1960b: 13)

This idea is also at the core of a later essay also specifically devoted to the investigation of the meaning of playful activities: "Play as Structuralized Time" (Das Spiel als gestaltete Zeit; Portmann 1975). By adopting a strictly phenomenological approach, Portmann maintains that play makes it possible to transform the meaningless flow of "dead" time into meaningfully *lived* time. Commenting on this point, Oreste Tolone points out: "It is as if, at a certain point of organic evolution, it became fundamental to guarantee the survival of organisms, the continuation of life, but also, or especially, for life to represent itself, displaying itself in all its superabundance of forms, in its euphoria and complexity". The more complex the organism, the more open to infinite new possibilities and, thus, the more free to play. The evolution from simpler organisms to higher developed animals goes hand in hand with an "increasingly clear distinction between functional structures and aesthetic-representative structures" (Tolone 2012: 233). Other Italian scholars, like for instance Salvatore Tedesco and Valentina Flak, also have made reference to Portmann (as well as to Buytendijk and Caillois) in order to highlight the role ludic activities play in exhibiting (in dis-playing) a genuine "value of being" (Tedesco 2008: 89; Flak 2003).

Now, the two essays that Portmann specifically devotes to the question of play are based on the very same presumption, namely, that in order for animals to play, they need to be able to count on: "1) A complex relation to the environment, provided by highly-developed organisation of both the body and the psyche; 2) freedom from immediate feeding needs; 3) a condition of security and protection within the group they belong to" (Portmann 1975: 839). Yet this reasoning only concerns animal *behaviour*, which after all is what is commonly understood when talking about "play". What about animal *appearance*? Can we legitimately speak not only of play *behaviour*, but also of a play of *appearances*?

Let's recapitulate some important points we have clarified so far. The surface display is part of the presentation of the self of a living being. Even when no perceiving eye can be said to play a role in shaping patterns and designs, the *possibility* of perceived beings already exists, "and that potentiality leads to the further promise of an enrichment of living relationships which grows out of the existence of those patterns and designs which uniquely characterize distinct individuals" (Portmann 1990a: 25). In other words, play contributes to "inwardness": "What we call inwardness is the specific mode of existence of living beings. We know it best from our own experience, but find evidence of it in other organisms, though decreasingly as we move farther away from man" (Portmann 1954: 345).

This observation is very significant: on the one side, it is only with higher animals' play that the evidence for interiority first becomes compelling; on the other side, however, we know from the closing remarks of the second edition of *Animal Forms and Patterns* that Portmann attributed inwardness – even if only an enigmatic and so far inexplicable form of inwardness – not only to *lower* animal forms, but also to vegetables, fungi, and even slime moulds. If so, these very simple organisms, too, are to be thought of as playing with their own appearances. They too display

> unmistakable characteristics which help to express a picture altogether peculiar to that particular species. In all these particular points, they far exceed the requirements of structure formation which is merely adequate for organic preservation. This expressive display of the picture belonging to a species permits us to transgress a significant frontier: namely, it may still be correct to make definitive statements concerning the elementary functions once we have crossed that frontier, but those statements are not sufficient to explain what it is which we in fact meet with on the other side of that frontier. *Every* organic form-structure transcends elementary necessities. (Portmann 1990c: 140, emphasis mine)

From this point of view, play proves to be a truly universal characteristic of organisms as such, an essential trait of life itself. And it is perhaps no coincidence that in the very same year Portmann published both the second edition of *Animal Forms and Patterns* and his first essay on play, another German-speaking scholar, Eugen Fink – he too, like Merleau-Ponty, a phenomenologist – published his most famous masterpiece, *Play as Symbol of the World* (Spiel als Weltsymbol; 1960). To some readers, the conclusion of this work might well sound very Portmannian:

> The play-metaphor fails as a cosmic likeness if we fixate on the personality of a player and on the character of appearance belonging to the playworldly scene. [...] The play of the world is no one's play, because only in it are there first someones, persons, human beings, and gods. And the playworld of the play of the world is not an "appearance" (*Schein*), but rather the coming to appearance (*die Erscheinung*). Coming to appearance is the universal emergence of all beings, things, and events into a common presence that integrates everything individuated into a presence – in our midst. What we ordinarily already call the world is the world-dimension of presence, the dimension of coming to appearance wherein things are really separated from each other, but also still integrated in spatial and temporal vicinity and connected to each other by strict rules. But the world is also the nameless realm of absence, from which things come forth into appearance and into which they again vanish [...]. All beings are cosmic playthings but all players, too, are themselves only played. Coming to appearance is a mask, behind which "no one," behind which nothing, is – precisely as the nothing. (Fink 2016: 214–215)

What Fink suggests here is that the act of bringing to appearance is the essence of existence *tout court*, a phenomenal mask behind which there is no one and no thing at all: the *Erscheinung* is the *Wesen* (appearance is the substance in itself). Portmann shares this anti-Platonic, strictly phenomenological paradigm as he argues that playing with appearances is at the very core of life itself in its uninterrupted metamorphosis. This does not undermine Darwinism itself but the most radically mechanical interpretations of Darwinian natural selection. Portmann is, indeed, a Darwinist in that he fully recognizes the role of natural selection in the stabilization of populations (though loudly opposing – as it has been shown above – that this force is responsible for the *creation* of novel structures and species); at the same time, he does not share the reductionist aspects of neo-Darwinism in that he gives priority to creativity (even if only an *unconscious* form of creativity), to the production of forms, to the play of appearances. From this perspective, play has to be regarded not only as a behaviour but as a free production of forms that, once having come to light, are indeed subject to the action of selective pressures. The latter, however, can explain neither the specificity of a particular form nor its origin, which must be ascribed to such a fundamental quality of life itself as play. To exist means to express oneself (*Selbst-darstellen*), which in turn means to play with appearances. In this sense, the player is not only the individual or the species, but life itself: evolution is the action of selection pressures upon the still magmatic material provided by a continuous activity of playing with forms – and this is precisely where the ontology of play meets the biological significance of all ludic activities.

References

Amodio, P. (2012). Di profondità filosofiche e superficialità viventi tra Hannah Arendt e Adolf Portmann. In P. Amodio, C. Fuschetto, & F. Gambardella (Eds.), *Underscores* (pp. 81–122). Napoli: Giannini.

Anzieu, D. (1989). *The skin ego* (C. Turne, Trans.). London/New Haven: Yale University Press.

Arendt, H. (1978). *The life of the mind*. Orlando: Harcourt.

Buytendijk, F. J. J. (1958). Anschauliche Kennzeichen des Organischen. In *Das Menschliche. Wege zu seinem Verständnis* (pp. 1–13). Stuttgart: Koehler.

Cappelletto, C. (2012). Skin: Towards an aesthetic identity. In P. Bellasi & M. Mazzotta (Eds.), *Pelle di donna. Identità e bellezza tra arte e scienza* (pp. 50–60). Milan: Mazzotta.

Carnevali, B. (2012). *Le apparenze sociali. Una filosofia del prestigio*. Bologna: il Mulino.

Cassirer, E. (2013). Eidos and eidolon: The problem of beauty and art in the dialogues of Plato. In *The Warburg years (1919–1933): Essays on language, art, myth, and technology* (pp. 214–243). Translated and with an introduction by S. G. Lofts with A. Calcagno. New Haven: Yale University Press.

Deleuze, G. (1990). *The logic of sense* (M. Lester & C. Stivale, Trans.). New York: Columbia University Press.

Dewitte, J. (1998). L'interanimalité comme intercorporéité et intervisibilité: Merleau-Ponty lecteur de Portmann. In J. Gayon & P. F. Moreau (Eds.), *Corps et individuation. Annales doctorales 1* (pp. 89–117). Dijon: Editions Universitaires de Dijon.

Dewitte, J. (2001). Adolf Portmann et l'apparence inadressée. *Prétentaine, 14–15*, 207–223.

Dufourcq, A. (2016). A phenomenological approach to the imaginary of animals. In M. Tønnessen, K. Armstrong Oma, & S. Rattasepp (Eds.), *Thinking about animals in the age of the Anthropocene* (pp. 55–72). Lanham/Boulder/New York/London: Lexington.

Fink, E. (2016). Play as symbol of the world. In *Play as symbol of the world and other writings*. Translated by A. Moore and C. Turner (pp. 33–215). Bloomington/Indianapolis: Indiana University Press. [German original: Fink, E. (1960). *Spiel als Weltsymbol*. Stuttgart: Kohlhammer].

Flak, V. (2003). Per un'estetica animale fenomenologica. *Leitmotiv, 3*, 93–103.
Gambazzi, P. (1999). *L'occhio e il suo inconscio*. Milan: Raffaello Cortina.
Goethe, J. W. (1987). Morphologie. In *Sämtliche Werke* (Vol. 24, p. 349). Frankfurt a. M: Deutscher Klassiker Verlag.
Goethe, J. W. (1998). *Maxims and reflections*. Translated by E. Stopp, ed. with an introduction and notes by P. Hutchinson. London: Penguin.
Gould, S. J., & Lewontin, R. C. (1979). The spandrels of San Marco and the Panglossian paradigm: A critique of the adaptationist programme. *Proceedings of the Royal Society of London, 205*, 581–598.
Groos, K. (1898). *The play of animals*. Translated by E. L. Baldwin, with a preface and an appendix by J. M. Baldwin. New York: Appleton.
Hegge, H. (1996). Transcending Darwinism in the spirit of Goethe's science: A philosophical perspective on the works of Adolf Portmann. *Newsletter of the Society for the Evolution of Science, 12*(2), 1–26.
Merleau-Ponty, M. (1968). *The visible and the invisible: Followed by working notes*. Translated by A. Lingis, edited by C. Lefort. Evanston: Northwestern University Press.
Merleau-Ponty, M. (2003). *Nature: Course notes from the College de France*. Translated by R. Vallier, edited by D. Seglard. Evanston: Northwestern University Press.
Nancy, J.-L. (2008). *Corpus*. (R. A. Rand, Trans.). New York: Fordham University Press.
Portmann, A. (1954). Biology and the phenomenon of the spiritual. In J. Campbell (Ed.), *Spirit and nature: Papers from the Eranos Yearbook* (Vol. 1, pp. 342–370). Princeton: Princeton University Press.
Portmann, A. (1958a). Kampf und Frieden als biologisches Problem. *Eranos Jahrbuch, 27*, 453–483.
Portmann, A. (1958b). Selbstdarstellung als Motiv der lebendigen Formbildung. In *Geist und Werk. Aus der Werkstatt unserer Autoren. Zum 75. Geburtstag von D. Brody* (pp. 139–173). Zürich: Rhein-Verlag.
Portmann, A. (1959). *Animal camouflage*. (A. J. Pomerans, Trans.). Ann Arbor: University of Michigan Press.
Portmann, A. (1960a). *Die Tiergestalt. Studien über die Bedeutung der tierischen Erscheinung, zweite, neubearbeitete und erweiterte Auflage*. Basel: Reinhardt.
Portmann, A. (1960b). *Spiel und Leben*. In *Gärten, Menschen, Spiele* (pp. 11–18). Basel: Pharos.
Portmann, A. (1964). *New paths in biology*. (A. J. Pomerans, Trans.). New York/Evanston/London: Harper & Row.
Portmann, A. (1965). Erleuchtung und Erscheinung im Lebendigen. In *Aufbruch der Lebensforschung* (pp. 33–57). Rhein: Zürich.
Portmann, A. (1967). *Animal forms and patterns: A study of the appearance of animals*. Translated by H. Czech. London: Faber & Faber. [German original: Portmann, A. (1948). Die Tiergestalt: Studien über die Bedeutung der tierischen Erscheinung. Basel: Friedrich Reinhardt].
Portmann, A. (1970a). Die Beurteilung der Erscheinung im Organischen. In *Entläßt die Natur den Menschen? Gesammelte Aufsätze zur Biologie und Anthropologie* (pp. 58–87). Münich: Piper.
Portmann, A. (1970b). Neue Fronten der biologischen Arbeit. In *Entläßt die Natur den Menschen? Gesammelte Aufsätze zur Biologie und Anthropologie* (pp. 13–28). Münich: Piper.
Portmann, A. (1970c). Die Formen des Welterlebens und die Bedeutung der Gestaltforschung. In *Entläßt die Natur den Menschen? Gesammelte Aufsätze zur Biologie und Anthropologie* (pp. 121–136). Münich: Piper.
Portmann, A. (1975). Das Spiel als gestaltete Zeit. *Merkur. Deutsche Zeitschrift für Europäisches Denken, 29*(9), 832–843.
Portmann, A. (1976). *An den Grenzen des Wissens. Vom Beitrag der Biologie zu einem neuen Weltbild*. Frankfurt a. M: Fischer.
Portmann, A. (1990a). The living thing as a pre-arranged relationship. In *Essays in philosophical zoology by Adolf Portmann: The living form and the seeing eye* (pp. 21–43). Translated, with an interpretive essay, by R. B. Carter. Lampeter: The Edwin Mellen Press.

Portmann, A. (1990b). What does the living form mean to us? In *Essays in philosophical zoology by Adolf Portmann: The living form and the seeing eye* (pp. 147–159). Translated, with an interpretive essay, by R. B. Carter. Lampeter: The Edwin Mellen Press.

Portmann, A. (1990c). Human beings in the perspective of the theory of evolution. In *Essays in philosophical zoology by Adolf Portmann: The living form and the seeing eye* (pp. 125–144). Translated, with an interpretive essay, by R. B. Carter. Lampeter: The Edwin Mellen Press.

Portmann, A. (1996). L'autoprésentation, motif de l'elaboration des formes vivantes. *Études Phénoménologiques, 12*(23–24), 131–164.

Prévost, B. (2009). L'élégance animale. Esthétique et zoologie selon Adolf Portmann. *Images Re-vues, 6.* http://imagesrevues.revues.org/379

Prévost, B. (2010). Les apparences inadressées. Usages de Portmann (doutes sur le spectateur). http://www.fabula.org/atelier.php?Les_apparences_inadress%26eacute%3Bes

Prévost, B. (2012). Cosmique cosmétique. Pour une cosmologie de la parure. *Images Re-vues, 10.* http://imagesrevues.revues.org/2181

Ruyer, R. (1958). *La Genèse des formes vivantes*. Paris: Flammarion.

Ruyer, R. (1966). *Paradoxes de la conscience et limites de l'automatisme*. Paris: Albin Michel.

Stamm, R. A. (1999). L'intériorité, dimension fondamentale de la vie. *Revue européenne des sciences sociales, 37*(115), 55–73.

Sullivan, L. H. (1979). Ornament in architecture. In *Kindergarten chats and other writings* (pp. 187–190). New York: Dover.

Tedesco, S. (2008). *Forme viventi. Antropologia ed estetica dell'espressione*. Milan: Mimesis.

Tedesco, S. (2010). Morfologia estetica. In *Alcune relazioni fra estetica e scienza naturale*. Palermo: Aesthetica Preprint.

Thinès, G. (1996). La forme animale selon Buytendijk et Portmann. *Études phénoménologiques, 12*(23–24), 195–207.

Tolone, O. (2012). Adolf Portmann: The expressive function of technique. *Etica & Politica / Ethics & Politics, 14*(1), 230–243.

Valéry, P. (1971). Idée fixe. In *The collected works of Paul Valéry* (Vol. 5). Translated by D. Paul, with a preface by J. Mathews and an introduction by P. Wheelwright. Princeton: Princeton University Press.

von Uexküll, J. (1956). *Streifzüge durch die Umwelten von Tieren und Menschen. Ein Bilderbuch unsichtbarer Welten*. Reinbek bei Hamburg: Rowohlt.

Pietro Conte is an Assistant Professor of Aesthetics at Ca' Foscari University of Venice and team member of the ERC project "An-Iconology: History, Theory, and Practices of Environmental Images". His research focuses on illusion, hyperrealism, mimicry, and the multifarious practices of un-framing, a thematic cluster he first explored in the monographs *Unframing Aesthetics* and *In carne e cera. Estetica e fenomenologia dell'iperrealismo* (Flesh and Wax: Aesthetics and Phenomenology of Hyperrealism). He is the editor of the Italian translation of Adolf Portmann's *Animal Forms and Patterns*.

Chapter 10
Ptolemaic Man and Copernican Man: In Favour of "Depth Anthropology"

Oreste Tolone

Abstract This essay highlights how in the anthropology of Portmann, nature and culture, history and biology, coexist in man in a form of close collaboration, which has as its premise "world-openness" and the "extra-uterine year". This dual belonging, which Portmann reworks in his basal anthropology, sees man inhabit two worlds at the same time: a *primary world*, which he shares with all other living things, which he accepts as immediately endowed with meaning and which makes him part of a connected system of integrated signs and communications; and a rational, cultural *secondary world*. The latter has the obligation to *give* meaning to life, whilst safeguarding human health. The risk of a hypertrophy of the secondary world – a warning, in our opinion, to be found also in the thought of Morin and Lorenz – requires an integration of the two worlds, through what we have called "depth anthropology". In the last part of the essay, in fact, we have introduced this expression as a different declination of basal anthropology, in which attention is placed on the positive and indispensable role of the primary world.

Biology and History

In *A Zoologist Looks at Humankind* (Portmann 1990a, Biologische Fragmente zu einer Lehre vom Menschen), which appeared for the first time in 1944, Adolf Portmann lays the foundations for a basal philosophical anthropology. In order to achieve a balanced and not preconceived image of man, Portmann stresses in this book how fundamental it is to lay down a methodological premise: the biologist and the philosopher cannot be guided from the outset by the conviction that man can be explained exclusively from below – that is to say, from the simplest forms – or exclusively from above – by introducing a spirit unrelated to life, as in Scheler (Portmann 1968a: 11, 36).[1] On the contrary, we must remain open to this double

[1] Portmann's position is critical both towards Scheler, whose stance is determined by a "metaphysical presetting to the mind" (Portmann 1968a: 36), but also towards vitalism. Jakob von Uexküll's

O. Tolone (✉)
D'Annunzio University of Chieti–Pescara, Chieti–Pescara, Italy
e-mail: orestetolone@gmail.com

F. Jaroš, J. Klouda (eds.), *Adolf Portmann*, Biosemiotics 23,
https://doi.org/10.1007/978-3-030-67810-4_10

possibility, rejecting any form of a priori reductionism: "The vision of life looking down from above, from the point of view of the human being, is a necessary complement to the attempt at building from beneath, to proceeding from the simplest forms" (Portmann 1990a: 6).[2] In particular, reductionism from the bottom (more dangerous and current), must be challenged. It is Portmann's conviction that it is pointless to look at the bones that paleontologists bring to light to find a link between animal and man. Man, in fact, is a polarity – he is the union of Nature and History, of a double evolution (Portmann 1962: 9). This anthropological conviction – the counterpart to the methodological premise – is the fundamental assumption of Portmann's entire anthropology. In man, this organic evolution is supplemented by historical evolution. Man's evolution is as much a biological premise as it is the result of "migration, trade, miscegenation" (Portmann 1990a: 13): immaterial factors that leave no trace and that therefore, in the mists of time, make way purely for material investigation. For this reason:

> Every fact that the biologist working in this field would like to explain by ascribing it to the descent of one human type from another and to organic advancement, the historian understands, often more accurately, to be a result of migration, trade, miscegenation, and so on. In the field of prehistory, the two different concepts of evolution necessarily meet. And because prehistory, particularly its early phases, has been studied primarily through the methods of natural science, it has been especially easy for the biological concept of evolution to gain the upper hand. (Portmann 1990a: 13)

The truth is that the biology of man develops and is completed only in the field of his social life, of the group, of free decision and of history. "Even in the eyes of the biologist, man must appear to be a special being influenced by his history, and also a life form with a peculiar second nature, that is to say, culture" (Portmann 1951: 25). The biologist who focuses on the past, however, is not used to this exercise and reduces the historical dimension to a pure and late consequence of its biological counterpart.[3]

Planmässigkeit ("correspondence to a plan") is perceived by Portmann as a dangerous vitalist failure, to extrabiological factors (Portmann 1956: 16). Otherwise, Uexküll plays a significant role in Portmann's thinking.

[2] "When the biologist begins his work, the spiritual should not be seen in advance either as a raid from above or as an explosion from below" (Portmann 1968a: 11). This methodological approach has the merit of placing man in continuity with the animal kingdom, but does not preclude the possibility of his own *Sonderstellung* ("special position").

[3] The difficulty of communication and understanding between natural sciences and humanistic knowledge, which is highlighted in the study of man, fuels the idea of profound discontinuity between nature and culture: an idea that Portmann does not share. On the contrary, his attempt at integration places his reflection in potential dialogue with contemporary biosemiotics. See Bateson (1979); Morin (1974); Hoffmeyer (1996); Cobley (2010). For a reading of basal anthropology in a prevalent socio-genetic key see Kurismaa, Chap. 6, in this volume.

Portmann substantiates this important assumption by using the theory of "the first extra-uterine year" (*das extrauterine Frühjahr*) (Portmann 1990a: 80–98).[4] His physiological, early (by one year) birth places man in a vulnerable position in the world, which requires, from the very beginning, a social context of protection and guarantee: a social uterus (Portmann 1960b: 41, 1965f: 41, 1974: 63–89; Fischer 2008; Nestor 2017) within which he can develop, *culturally*, his own biological structure. In other words, not only does man add a cultural dimension to the biological, genetic dimension, but he also takes on the task of *historically shaping the biology of man*. Man is asked to decide on his own natural form. This makes him a structurally composite being, living at the same time in "two distinct worlds" (Portmann 1990b: 459) communicating with each other. The human paradox consists of the fact that nature seems to create the conditions for overcoming oneself in culture, or rather, it seems to rely upon the social dimension.[5] In the human mammal, which is considered "secondarily altricial" (Portmann 1960b: 43; DeSilva 2016: 70–73), nature interrupts its natural course, laying the foundations for a different evolution of nature as historical or, rather, as historical-natural.[6] From a biological point of view, this link between nature and history is identified in the first year of a child's life (Portmann 1990a: 50–62), a year in which the hereditary, genetic and natural dimension of man – open and malleable by virtue of his relative prematurity – allows the social dimension to define the contours of his nature. If we then consider that the "human triad" (*die menschliche Trias*) (Portmann 1968b: 72, 1960b: 44),[7] as Portmann defines it – his upright position, the development of language and the development of thought – is achieved outside the womb, and therefore culturally, we understand how the specificity of man is the result of a double dimension. Premature birth and the relaxation of his animal instincts make man a being "open to the world" (Portmann 1965f: 40, 1974: 181). This mysterious openness, however, does not appear unconditional and absolute; this openness projects the animal beyond itself, whilst positioning it within two distinct and connected worlds.

---

[4] The English translation "extra-uterine Spring", although suggestive, is not without misunderstandings and does not adequately render the complexity of the term "Früh-Jahr". For this reason, we suggest the expressions "extra-uterine premature year" or "extra-uterine prematurity".

[5] The human propensity to social cooperation – according to practices of shared intentionality (*we-intentionality*) – has been identified by Michael Tomasello as the assumption that explains the specificity – linguistic, religious and cognitive – of the human ape (Tomasello 2014).

[6] Portmann speaks in this regard of *Natürlichkeit des Geschichtlichen* ("Naturalness of the historical") (Portmann 1968b: 73).

[7] "The biological consideration reveals something else that is important: it reveals such a close ontogenetic relationship between all the gestalt characteristics and those of our behaviour, that we cannot understand the behaviour, the social structure, the idiosyncrasies of language and tradition as epiphenomena, as late acquisitions, as a late consequence of a purely bodily anthropogenesis. Instead, we must accept a unity of human characteristics in which we see them all as functionally connected members of a system" (Portmann 1968b: 72).

Man and His Doubles

The mystery of man consists precisely of this problematic coexistence: of being a member of two worlds, which are proposed as contiguous or alternative, identical or even opposite. The *polarity* between nature and history, however, carries with it many others; man appears to be as much conditioned by his own genetic heritage as freely able to shape his own nature. For him, remaining in the mother's womb is just as important as remaining in the social uterus; he still appears to be bound to innate behavioural patterns, sensitive to forms and to triggering objects, as much as his action seems freed from such mechanisms. This double aspect of man (*Doppelaspekt*) recalls the position developed by Helmuth Plessner, with whom Portmann had a heated exchange but whose theory of eccentricity seemed highly complementary to that of the extra-uterine prematurity (Plessner 1975: 81–105; Fischer 2008).[8] The "three anthropological laws" developed by Plessner (Plessner 1975: 309–346), in fact, seem to summarise this double belonging. The first and second laws, in particular, reaffirm that man is a naturally artificial animal, constantly tempted to flatten the artificial over the natural or the natural over the artificial, or to reabsorb mediacy in the immediacy of the natural response, and vice versa (Plessner 1975: 321–341). The idea of a "utopian standpoint" (*utopisches Standort*) (Plessner 1975: 341–346), identified in the third law, better expresses this intrinsic possibility of man living in other worlds, constantly projecting himself beyond the natural world from where he originates – as well as the cultural world in which he was born. This eccentricity, this ability to interrupt the natural flow of one's existence, places man in the uncomfortable position of having to *decide* constantly in which world to live. It forces him to act in constant "mediacy", by virtue of which each natural element assumes a very particular physiognomy. The "human triad", which Portmann identifies as the physical counterpart of temporal prematurity, is also structurally subject to a dual interpretation. The human triad is as much a form of natural and immediate expressiveness as it is a form of "controlled expression" (Portmann 1965g: 208, 1968a: 30).

The controlled means of expression creates, through tradition, the unprecedented possibility of a rapid change of living space and therefore a special kind of development, which is not a simple continuation of organic evolution, but a completely new form of facing the task of life and which we know as the sphere of history. This thereby produces a second nature created by us, which is not directed against nature and therefore unnatural, but is something new, a work for which we are responsible in a completely different way. (Portmann 1968b: 74).

Portmann reminds us several times that this clear distinction and separation between worlds is fictitious. In fact, in man, his standing, his speaking and his thinking are at the same time the expression of an archaic, immediate, natural, unconscious world, but also the expression of a free, controlled, mediated and artificial action. On one hand, he seems to belong to a primary world, with which he seems

[8] On the close relationship between Portmann's thoughts and those of Plessner, see Tolone (2015).

perfectly in tune. On the other, he acts in a secondary world, from which he is distanced thanks to his rational and technical capacity.

Primary World and "Ptolemaic" Man: The Aesthetic Function

The World of the Senses

The development of a world increasingly dominated by reason, in which everything is, in principle, calculable and technicalised, in which there is an irreversible "disenchantment of the world" (Portmann 1968d: 100), an absolutisation of scientific rationality, pushes Portmann to recover an understanding of the "primary" world. This recovery must not appear to be simply a nostalgic return to the past, to an accommodating unity with nature, or simply reminiscent of a superseded phase. As he states:

> [...] My stubborn, ever renewed emphasis on the role of our original worldview may be perceived here and there as a fad, as a prejudice bordering on obsession, or as reactionary. Well, this 'mania' is not 'reactionary' – but it is certainly a reaction, a protest against the contempt with which progressive research often pushes our most natural way of experiencing into the background. (Portmann 1974: 160)

The original wisdom with which man understands the world should neither be grasped as the final horizon of man, nor reabsorbed by the rational vision of reality. So, of what does this primary understanding of the world essentially consist?

Before achieving a reliable expression of everything through reason, man possesses an instinctive and natural understanding of the world, emerging from the intense and direct experience of our senses. The first human world, of which we are aware, is the world of the primitive, Ptolemaic man; that is to say, man immediately immersed in and surrounded by a reality to which he instinctively gives credit and which is based on the reliable work of the senses. It is evident that "the first great truth is always the result of the immediate experience of the senses on which every conception of the connection between the things of the world is based" (Portmann 1960a: 23). Ptolemaic man accepts the original experience as a direct and elementary source of knowledge and builds his truths as a result of his own hereditary structures and his complex sensitive experiences. Unlike rational knowledge, in which elementary experiences are learned, primary knowledge simply knows (*weiß*) this experience: it knows that it is so, and it cannot be questioned (Portmann 1968d: 104). The world opens up in the sign of belonging, of a hereditary and sensitive concordance with man.

Man as a Telluric Event

Before his conscious and complex interaction with the world, man moves within it naturally, with an innate wisdom. He knows how to orient himself in the world, recognises the earth as his natural homeland, in which he moves following routes and orientations that he finds within himself. He is primarily a "telluric event" (Portmann 1968g: 221), an individual who tramples on the surface as if it were his own homeland, the "earth as the house of life". Every natural movement, every insertion in the environment is based upon a partial "innate structure of world integration" (Portmann 1968g: 219), in which he is placed from that first moment. For example, he always moves within forces and constants, which act as a stable frame of reference, from which he is unable to escape. In particular, he acts within gravitational fields, as a vertical being with its own centre of gravity, which presupposes a horizontal earth to be walked on, a stable light source positioned "above" (Portmann 1968g: 218), in the sky, which persists over time. As a result, the organisation of man, as of all species, has structures that create relationships of every kind with the inanimate environment.

> These structures increase in size and complexity with the total differentiation of life forms and include those components related to large, planetary conditions: organised integration into the gravitational field and evaluation of gravity as a stimulating effect, organised integration into the light field of solar radiation and the rhythm given by changes from day to night. These structures are all given in an unconsciously running life, in complicated interactions, and also position us men, before and beyond all conscious experiences, in this telluric event. (Portmann 1968g: 212)[9]

The Mediocosmos

This primordial world in which we take our first steps can be defined as the *mediocosmos* (Portmann 1965e: 222–223, 1974: 160), namely, the intermediate cosmos in which everyday life takes place and in which man preserves his central and naive position: "the sphere which I call mediocosmos is the only space we know and really trust, in which we know the superior organisation of living" (Portmann 1974: 161). It differs from the *microcosm* and the *macrocosm*, with which we come into contact exclusively as a result of the Copernican revolution, that is, indirect and scientific understanding, which pushes man into contact with the realm of the "invisible". These areas project man more and more beyond himself and his own natural homeland up to a form of extraneousness, which replaces the intimate world of daily familiarity with the secondary world of science. It also differs from the inter-world (*Zwischenwelt*) (Portmann 1964: 81, 1965e: 204–210), or intermediate

---

[9] The consonance with the "mystical solidarity" of man towards the animal and plant world, recognised by Mircea Eliade, is striking (Portmann 1968g: 229; Görner 2016: 47).

world.[10] It represents that ever wider and intrusive sphere of artificial objects, created by man, which progressively isolate him and uproot him from the primary world.

The World of Childhood and Sleep

This structural anchorage to a geocentric, Ptolemaic, innate and senses-based world guarantees protection and safety: the same things that accompany man in his childhood. The primary world is similar to that of the child and is the first spiritual articulation by man, taking place under the auspices of trust, immediacy and beauty. Unlike the world of waking and rational awareness, this world emerges in an almost unconscious, dreamlike way and is marked by a hidden order, governed by mysterious, hereditary laws that preside over dreams (Portmann 1965d: 134).

The Pyknic and Feminine World

The possibility of this double approach to reality – more sensitive or rational – has its roots in the "structural bipotence" of life. Human life, in fact, always presents itself according to male and female sexual differences. There is no *tout court* human being, nor can we know the human being, except in its male-female articulation. This "structural bipotence remains a basic plasmatic fact for all life and it can emerge in many ways" (Portmann 1968a: 39), including a different way of experiencing the world (Portmann 1960a: 23–24). According to Portmann, the feminine dimension preserves a series of characteristics typical of the first phase of youth, which can be restricted to the first ten years. It thus appears to be closer to the primary world of senses, sleep, imagination and creativity.

Strength of feeling and an inclination towards a harmonious insertion in the primary world is found to be typical of the pyknic character type. According to Ernst Kretschmer's constitutional psychology, which attempts to study the relationship between physical structure, temperament and predisposition to mental pathologies, there are three somato-constitutional types: pyknic, asthenic or leptosome, and athletic.[11] The pyknic type – curved, roundish and stockier than the long-limbed

[10] "Modern man lives in three worldviews: he remains suspended lifelong in everyday life – according to our naive experience – in a mediocosmos, according to whose measure and centre we organise our feelings and actions. But we penetrate beyond the limits of this mediocosm into a submicroscopic world of molecular and submolecular forces, into a microcosm – we penetrate beyond the everyday zones into extraterrestrial spheres, into a macrocosm, which determines our life on earth, our spiritual creation, our economic efforts, even if we hardly take our first 'steps' in this world" (Portmann 1964: 48).

[11] The theory of personalities by Ernst Kretschmer (1888–1964) attempted to relate the biotypological structure to specific psychic imbalances; in the leptosome, schizotimic, the tendency towards schizophrenia prevails; in the pyknic, cyclothymic, the tendency towards manic-depres-

leptosome – is more inclined towards intense, cyclothymic, extrovert affectivity, decidedly oriented towards the social dimension, "open to colours, to the richness of the world and tied to a unitary interpretation of it" (Portmann 1960a: 27). In his 1958 essay entitled "*Don Quijote und Sancho Pansa*", Portmann even attempts to group these physical-characterial traits according to a double category: that of *Grundmensch* and *Kopfmensch* ("basic human" and "head human") (Portmann 1958: 34).[12]

From a cultural point of view, Portmann highlights, albeit with great circumspection, that such polarity is best embodied in the different approaches to the world and to life influenced by Oriental culture and art, in the sense that the pyknic-asthenic duality is reflected in Eastern and Western culture, respectively (Portmann 1968: 90–1, 109, 232, 313–333). This is not only due to the renunciation by Oriental art of means of perspective, chiaroscuro and anatomical studies in favour of other practices; it is not only due to the aversion to a typically Western form of pictorial naturalism; it is not only due to the extraordinary favour reserved to artistic forms such as calligraphy, in the Far East, or to spontaneous poetic compositions, in Japan (Portmann 1968h: 331). Rather, this typically Oriental sensibility towards imaginative, artistic thought reveals a different approach to nature, conceived as "helpful and friendly". It implies that man is disposed to perceive himself as a "member of the world and in particular of the earthly" (Portmann 1968g: 232).

The Mythical World and Imaginative Activity

This primary world, sensitive, feminine, dreamlike, terrestrial, oriental, pyknic, childish, is the result of an approach to the world that could be defined as "aesthetic". In Ptolemaic man, living in the primary world, it is the aesthetic function that determines the relationship with the objects in the world, leaving them intact and preserving that initial value, that comes from the immediate sensation of form, colour, sound, etc. Paradoxically, in so-called "primitive man" an attitude governed by amazement, wonder and order prevails: Aesthetic activity not only accepts sensory data, but relies upon it and seeks it out; it lives above all on the sensitivity that manifests itself in the first act. It is the way of quality, sounds and colours, palpable impressions and smells, which appear at each moment, and spontaneously, the

sive disorder predominates. This theory, expressed in particular in the work *Physique and Charachter* (1921, orig. Körperbau und Charakter), is now predominantly of historical interest, since these correspondences have not been confirmed. Portmann also refers to the work of the psychiatrist and neurologist Klaus Konrad, *Constitution-type. Theoretical Foundation and Practical Determination* (1941, orig. Der Konstitutionstypus. Theoretische Grundlegung und praktische Bestimmung), which reduces the basic typologies to two.

[12] The biologist Florianne Koechlin reminds us that this pair seemed to be plastically embodied by Portmann himself, thin, ascetic and slender, and the theologian Karl Barth, with his small constitution, whom he often accompanied. Even Portmann referred to himself as a: "Don Quixote in the technical world of laboratories" (Koechlin 2004).

whole field of consciousness, even in the forms of reflection that we see internally, even in the thoughts we hear within the silence of ourselves. This function is a deep source of emotion, a powerful means of persuasion and also of knowledge and is of inestimable value. Although it does not lead to the material conquest of the world, its action, in the order of quality, is as important as that of logic and science. This aesthetic behaviour appears today, compared to the effects of the theoretical attitude, to be an experience without brutal intervention, a more tender approach, dominated by a love of things and by respect for mystery that only appears in the sensitive appearance (Portmann 1948: 137).

This aesthetic function is characterised by the prevailing role of fantasy, of a primary imaginative thought (*das imaginierende Denken*) (Portmann 1968d: 109), capable of creating an image of the world of marvellous power and unity (Portmann 1965d: 143). Within it, our senses find confirmation of their validity: we have grown with imaginative power to create a unitary image of the world, a world "in which the senses are fully developed and in which we experience the action of these senses as meaningful and well-founded" (Portmann 1960a: 29; Pinotti Tedesco 2013). This aesthetic function is absolutely dominant in prehistory, among primitive peoples, as testified by rock paintings; it determined religious life, collective action, social organisation (Portmann 1968h: 313–314). It is exercised in the prevalence of an imaginative, "visual" language (Portmann 1965d: 138),[13] of a thought through images, a metaphorical one which proceeds through analogies, preferably inferred from contemplation and from sensitive experience.[14] This thought draws on knowledge, on a richness of innate and unconscious meaning, which possesses the power of "vision".

This unitary vision of the whole is a "mythical", poetic vision, which does not simply represent an epistemological obstacle to the correct scientific understanding of reality, but plays a constructive role – as Gaston Bachelard points out in *Earth and Reveries of Will* (1948, orig. La Terre et les rêveries de la volonté), cited by Portmann himself. On the contrary, the history of science is full of examples which testify to the proactive value of imaginative thought, considered "a powerful fact of our nature" (Portmann 1968d: 117).

[13] "It is the choice of the West between two different methods of overcoming the world, between two ways of the spirit working, which in the human being should act in full harmony" (Portmann 1968c: 90).

[14] The importance in Portmann of the metaphorical cognitive approach, compared to the abstract one, reveals the capacity of allegorical thinking to grasp, in an analogical way, the unity and communicative creativity in which it is "aesthetically" immersed from the beginning. The extension of this metaphorical process to the whole living world is carried out by W. Wheeler (2010).

Isomorphism, Supra-Individuality, World of Images

A human, therefore, is able to insert himself into reality, to elaborate intuitively a coherent image of the world, in which there seems to be complete correspondence between all living beings. Only the use of aesthetic and imaginative intuition of the surrounding world, only recovered attention to the self-expression of nature can allow us to understand the intimate consonance that exists between creatures: the intimate harmony of the parts, having a moderating effect on the arrogance of reason. Pollination, for example, presupposes that the flower is seen by the butterfly and that the scent is perceived; the chromatic nuances of living things presuppose the appearance of sentient eyes, or in any case organs of sight, in order to appear. It is as if there were a correspondence between an innate structure, present in the interiority of the organism, and reality. It is as if the way of being of two distinct organisms acts according to a pre-established harmony, or rather, as if it were merged into a single, superindividual entity (*überindividueller Zusammenhang*) (Portmann 1965b: 63). The orientation of birds – during the migration period – in relation to constellations presupposes that the interiority of the animal is connected, from the beginning, with the vision of the sky. It is as if, Kant would say, there is a sense to which everything contributes. Wolfgang Koehler calls this intimate correspondence "isomorphism" (Portmann 1968d: 121). In the human experience of the world there are also isomorphic structural elements, such as, for example, the smile of a newborn child and the model of the human face (Portmann 1968d: 125).

From the outset, there have been triggers (*Gestaltreizen*) (Portmann 1968f: 183–184) – generally simple, conspicuous, improbable – specific to each species, which favour the development of relationships with the things of the environment.[15] Rather than sounds, scents, etc., images trigger instinctive reactions, which reveal an original connection, of which the external aspect of things becomes in some way the bearer. In particular, an animal's coat and human skin seem to specialise in this complex expressive capacity. In man, however, the relaxation of his instinctive connections (Portmann 1968f: 196) favours a multiplication of sign-stimuli, a free and cultural expression, which can give any image or sound an evocative value (*Evokator-Wirkung*) (Portmann 1968f: 202). The "word" thus takes on a magical value, becoming a "transformer" (Portmann 1968f: 180), a kind of human biocatalyser (*Biokatalysator*) (Portmann 1968f: 197), freely able to activate the behaviour of men.

Man, like any other plant or animal, fits within a living unit, a Ptolemaic world, whose understanding concerns the *ordre du coeur* (Portmann 1968c: 86).[16] Not even

---

[15] In particular, "only this energetic transformation of Gestalt stimuli into images, and further into actions catalysed by these images – which is characteristic of the animal realm – creates the super-individual relationships on which all higher forms of social life are based" (Portmann 1968f: 185).

[16] In this sense, the Ptolemaic (geocentric) position of Portmann differs from that outlined by Husserl in his "Reversal of the Copernican Doctrine in the Interpretation of the Current Worldview" (1934).

the frenetic rise of a scientific and technological world can belie this profound truth. In summary, Ptolemaic world-building, which has been overcome in modern scientific thinking, not only corresponds to naïve conscious experience but also, essentially, to the natural, inherited organisation of man, whose *modus operandi* is mainly unconscious. One of the foundations of the Ptolemaic worldview is the conscious thinking and shaping of immediate experiences. However, at the basis of the important foundation of this immediate experience, there is the double historical-natural structure of man. This emerging structure, which, as a whole, is unknown to us, is inserted from the beginning in this geocentric environment (Portmann 1968g: 227).

Secondary World and "Copernican" Man: The Theoretical Function

Homo Nocturnal, Homo Diurnal

This primary world, born of Ptolemaic man, is now contrasted with a secondary, Copernican world, based on rationality. In the history of Western thought, this polarity has often been perceived as a real contrast of irreconcilable and alternative worlds, destined to be resolved with the ultimate superseding of the first man, of the first world, by the second. Thus, Portmann himself identifies a premise of such dualism in Linnaeus' taxonomic cataloguing, particularly in the distinction he introduces between *homo nocturnus*, of forests, and *homo diurnus* or *sapiens* (Portmann 1948: 131–132). The nocturnal man appears, in fact, to be a nocturnal caveman with a pale complexion and poor vision. More recently, this dualism is revived in the prelogical, magical mentality, as proposed by Lucien Lévy-Bruhl in *The 'Soul' of the Primitive* (1927, orig. L'âme primitive),[17] an alternative to the evolved and rational soul, but also in the "definitive dreamer" described by André Breton and recalled by Max-Pol Fouchet (Portmann 1948: 131). In all these cases, however, the primitive is constructed as a "counter-image" to civilised man, in relation to whom he must represent the premise and the antagonist.

[17] In the introduction to the Italian edition of the 1990 work, the ethnologist Ernesto De Martino clearly highlights how this approach still indulges the evolutionary image of a rectilinear progress; he also emphasises that prelogism does not exhaust the primitive, and that the hiatus between civilised and primitive man does not help, but hinders, the understanding of man (Lévy-Bruhl 1990: 13–17).

The World of Rationality

On the opposite side to the primitive and Ptolemaic man there is, therefore, the rational Copernican man. He is definitively based on a *gestalt* vision of the whole, from a cosmic immersion, thanks to the theoretical function – a function capable of distinguishing and selecting, of placing distance, of analysing and calculating what now appears before him as something objective. The Copernican man is the man governed by rationality, who distrusts his senses, who learns through experience, who knows how to distance himself from his own archaic, innate conditioning and is able to project himself into the invisible worlds of the infinitely small and the infinitely large. He is the mature, adult and conscious man, typically Western, who acts in a mediated and artificial way, who subordinates emotions and feelings to the prevailing forces of reason, the aesthetic and imaginative function to the theoretical one, who expels the mythical from the rational and, above all, who isolates himself from the world, from the environment, from the original context in which everything seems connected in profound unity. The Copernican man is the man of science and technology, called upon to reconstruct an image of the world, based upon a plurality of atomised entities.

Rationality, Technique and Coaction

The transition from the Ptolemaic to the Copernican type of man takes place in Europe through a *decision* taken, according to Portmann, at the time of the Renaissance (Portmann 1968c: 90–91, 1968i: 328–333). The direction taken by the West is not, therefore, its destiny, but the result of a choice in favour of an approach to the world based upon scientific rationality. The West opts for an understanding of reality based on centrality, on exalting the "I", which easily risks slipping into a sort of hypertrophy.[18] Its distancing, objectifying function leads us to dissect the world, to divide it into portions that lose all reference to totality; on the other hand, it presupposes an autonomous, disembodied and independent subject. Modernity presupposes and fuels a form of "hypertrophy of the ego" (*Hypertrophie der Ich-Position*) (Portmann 1968i: 324). Science and technique, therefore, become the emblems of an approach which carries within itself the risk of a deviation if we do not make some form of correction (Tolone 2012). Modern technology, in particular, with its "virtual infinity", is structurally exposed to the risk of "too much", of the "big", and brings with it a natural propensity to coaction: every new possibility that appears, thanks to science, in a circumscribed field, forces its application on a large scale, to the point of transforming this application into a permanent and lasting need (Portmann 1965: 7; Jonas 1982). In line with a series of reflections, which recall

[18] In this regard, Portmann speaks of *Irrtum des Okzidentes* ("the mistake of the Occident") (Portmann 1968i: 332).

impressively those of Hans Jonas, Portmann emphasises how the decision for the Copernican man, for science and technology, determines an unstoppable momentum, which Jonas exemplifies with the "law of coaction", according to which: "knowledge can lead to power, power can lead to doing and doing to a duty to do" (Jonas 2000: 128, orig. 1993). The ethologist Konrad Lorenz also supports something similar when he affirms that "technomorphic thinking has confused, over and above this, and in an almost compulsively neurotic manner, the mere existence of the possibility of realising a particular process technically with the obligation actually to carry the possibility through" (Lorenz 1987: 14).[19] This decision in favour of the Copernican man is therefore problematic, as while he appears to us completely natural today, he is the bearer of a vulnerability, of a pathology, which we must address.

Excess and Pathology of the Intellect: Portmann and Lorenz

Portmann, in fact, supports the idea – shared in different forms by Hans Jonas, Konrad Lorenz and Edgar Morin – that the hypertrophy of the self, connected to the exaltation of science and technology as the only valid approach in the world, has *pathological* implications. Contemporary man – even more centred on himself – is as sick as his relationship with the world, increasingly manipulative and functional (Portmann 1971a, b, c). It is not so much his manipulative attitude towards reality, typical of man as a cultural and artificial animal,[20] that is problematic, but the fact that this manipulative attitude is now exercised, without considering the primary world, in a sort of neurotic exaltation of the technical-intellectual function of man.

An ethological explanation of this possible intellectual extenuation can be found in a "dysfunction of intraspecific competition", as hypothesised by both Konrad Lorenz and Irenäus Eibl-Eibesfeldt (Lorenz 2002: 37–45). Lorenz, in particular, shows how intraspecific competition, unlike natural selection, often hinders the survival of the species. The case of the argus pheasant is emblematic. The exaggerated development of the nuptial livery results in its inability to fly; the impact with the extra-specific environment, however, prevents the species from proceeding on a path towards catastrophe.[21] The environment acts as a principle of reality and decides on the validity of such a path. In man, who, instead, has learned to control all environmental powers, the regulatory and healthy effect of the environment is

[19] "It has just about become a 'must' of the technocratic religion" (Lorenz 1987: 14). For Eibl-Eibesfeldt this is also linked to a kind of *Funktionlust*, to the pleasure of exercising a new, artificial function, developed by man (Eibl-Eibesfeldt and Hass 1985: 55).

[20] This attitude is actually considered a *Grundphänomen unseres Menschenseins* ("basic phenomenon of our human existence") (Portmann 1971a: 215).

[21] Konrad Lorenz discusses this, in particular, in *Civilised Man's Eight Deadly Sins* and in *The Waning of Humaneness* (1987).

missing, and he thus succumbs to the mercy of the diabolical consequences of intraspecific selection (Lorenz 1974: 24–30). Intellectual agonism, exasperated competition, and technical development risk becoming his end. As the ethologist Oskar Heinroth jokingly recalled, "next to the wings to of the argus pheasant, the hectic life of Western civilised man is the most stupid product of intra-specific selection!" (Lorenz 2002: 38).

In the view of Lorenz, this hypertrophic development seems to go hand in hand with the "atrophy" of a series of other typically human faculties, which puts man's humanity at risk. It even seems that – as in the case of symbionts and parasites – man has taken the road of "sacculinisation", that is, a "retrograde evolution" (Lorenz 1987);[22] parasites, symbionts and domesticated animals – relying structurally on another living form – lose a series of functions and vital information. In man this seems to be happening without the presence of an external species to depend on; in other words, the exoneration that the thought and technical apparatus guarantee to the individual makes man a kind of domesticated animal, dependent on technology and atrophied in its "Ptolemaic" functions.[23]

Although Portmann does not consider the process of self-domestication an adequate tool to explain man's intellectual evolution (Portmann 1954: 187–188), there remains the fear of a disharmonious development of man.

The Risk of a One-Dimensional Man: Portmann and Morin

The risk is that of a one-dimensional man, in whom the "rationalising illusion" distorts the dialogical dimension of reason. Its mythification, which arises from the disjunction from the primary world, determines the triumph of a false and abstract rationality. The effect of this "hyper-specialisation" of the theoretical function is the progressive affirmation of a technical culture – which divides and separates knowledge – the expulsion of global problems from research themes, the separation between humanities and the sciences, the weakening of global perception. Edgar Morin seems to share Portmann's conceptual ideas, making him one of the references of his complex thought. Since his work *Das Rätsel des Humanen* (The Enigma of Humanity, 1974), which not surprisingly in the German edition bears a preface by Portmann, he has embraced the intuition of our author: the "atrophy of the

[22] See the paragraph of the same name: "Retrograde Evolution, Or 'Sacculinisation'" (Lorenz 1987: 41–45).

[23] "Thought habits engendered by technology have become consolidated into doctrines of the technocratic system, which has isolated and protected itself through self-immunization [...] Forestalling the threatening apocalypse will devolve on their perceptions of value; their sensibilities of the beautiful and worthwhile must be aroused and renewed. And just these values are those being suppressed by scientism and technomorphic thinking. Educative measures begin with an exercising of *Gestalt* (form) perception, our only means for achieving a sensitivity to harmonies" (Lorenz 1987: 6–7).

natural mental disposition to contextualise and globalise" (Morin 1999: 17) causes the greatest human problems to disappear, to the advantage of specific, technical problems. The separation between aesthetic and theoretical, primary and secondary – as Portmann would say – takes on the appearance, for Morin, of the disjunction between *homo sapiens* and *demens*, *faber* and *ludens*, *empiricus* and *imaginarius*, *economicus* and *consumans*, *prosaicus* and *poeticus* (Morin 1974: 115–135, 1999: 27). This disjunction, for both Portmann and Morin, corresponds closely with the structure of the cerebral hemispheres. In man, the integration of the paleoencephalon and the midencephalon in the cortex – which seem to link the development of affectivity with long-term memory – appears complex. Primary and archaic impulses and affectivity are subject to a cortex, divided into two hemispheres – left and right – whose task, among other things, is to mediate between the three instances (impulse, heart and reason), between intelligence and affectivity, between the most animal and archaic part of man, and the most recent one (Kugler 1985). The close connection between affectivity and intelligence, highlighted by Damasio in his *Descartes' Error* (Damasio 1994), does not exclude an unstable and antagonistic relationship between different impulses.[24] As Morin maintains, perfectly in line with Portmann:

> The sapiens <-> demens dialogic was creative while also being destructive. Thought, science and the arts were irrigated by deep forces of affectivity, by dreams, anxiety, desires, fears and hopes. In human creations there is always the double command sapiens <-> demens [...] Creation bursts forth from the liaison between obscure psycho-affective depths and the bright flame of consciousness. (Morin 1999: 28)[25]

A "Spiritual Revolution"

What we need, then, is a "spiritual revolution", bringing an end to the tyranny of the theoretical function, to the atrophy of the primordial functions of man, and contributing to the integration of our faculties (Portmann 1968h: 314, 328).[26] It will certainly not be "the *homo divinans* who will save us" (Portmann 1948: 150) or a generic recovery of artistic or literary disciplines in school education: "Today, the young generation is oppressed by something very different: the upset of mental balance increased by rapid technical development, the lack of sure knowledge of the processes of complete spiritual formation, the healthy development of the complete

---

[24] One can see in Portmann an extraordinary attempt to harmonise man with himself, with his world, also involving the harmonisation of the two cerebral hemispheres. In the past, the connection between clustering techniques, proposed, in particular, by Gabriele Lusser Rico, and Portmann's approach has also been highlighted (Kugler 1985).

[25] We must avoid a Lévy-Bruhl-style "humanistic rationalism" – which attempts to reject the primitive in the sphere of an infantile weakness – as much as "neo-ethnologism", which sees modernity as a pure demolisher of archaic wisdom (Morin 1974: 133).

[26] The link between Portmann's thinking and Schiller's aesthetic education is important (Theunissen 1997).

human being" (Portmann 1968h: 319). Education in art, painting, poetry, music and creative activity is certainly one of the objectives that must be set by today's pedagogy (Nussbaum 1999) – and the contribution of Portmann's anthropology to educational reform in the global age has not gone completely unnoticed (Theunissen 1997; Retter 2000; Heykaus 2010), nor has it lost its innovative impetus. On the other hand, manipulation of technique and of education, is the fundamental instrument of his work of humanisation. Following in the footsteps of Benedetto Croce, Portmann sees in art, and in aesthetics in the broadest sense, "the root of all theoretical life" (Portmann 1948: 146). However, integrating the technical-cognitive dimension with the aesthetic one means, above all, fuelling a "familiarity with the wealth of forms", to recover morphological attention to the living beings around us (Portmann 1968h: 325; Conte 2013). Man, as a living being, needs to reactivate that original superindividual communication in which he has been immersed since the beginning, and which makes him an active part of a "symbiotic" system.[27] He needs to be part of that exchange of information and communications that reveal the play of all parts with the whole (Bleakley 2000: 50).[28] Each self-presentation (*Selbstdarstellung*), in fact, is part of an overall play of communications which grow together and interconnect, and on which every form of "'upwardly' oriented creative development" depends (Lorenz 1987: 176). Guaranteeing, above all in childhood, this sensitive understanding of nature, this perception of the harmonious connection of everything – every scent, every colour, every shape, every perceptible noise connects everything to everybody – replenishes the spirit of modern man and facilitates a *healthy* relationship with the world. Nature has the power to provide man with a homeland, to take him away from himself. The senses, so mistreated by philosophy, have the power to reveal meaning, to reveal the coherence of everything. For this reason, one of the great tasks of all human education is to strengthen the meaningful experience of nature, also in "the evaluation of the many stimulating possibilities of the mind, which slumber in this world of natural things and on which our soul so longingly waits in silence – even when it does not know the goal of such longing. The stimulating power of the things of nature around us is an almost undeveloped source of excitement, renewal and joyful spiritual movement, as well as a source of fertilisation of the most hidden creative powers in each of us" (Portmann 1968h: 324–325). If man's objective is not knowledge, but happiness, then the

[27] The holistic vision that underlies the integration of mental functions and worlds is incredibly similar to Bateson's "ecology", to the need, expressed in *Steps to an Ecology of Mind* (1972), to overcome the dichotomy that separates humanity from nature, unconscious and conscious, "internal mind" and "rational mind". The comparison between Portmann and Bateson seems worthy of further investigation.

[28] Unlike Kantian and idealistic philosophy – in which perception is always a human construction – the various forms of direct and indirect realism recognise different degrees of unmediated perception. "In such views (Adolf Portmann, James Gibson, James Hillman), what is often implied is that states of unmediated 'immediacy' are facilitated by an 'animalising' of the human psyche, as a pre-presentational state" (Bleakley 2000: 50). See also Zahavi and Gallagher (2012).

world of conscious thought, of separating rationality, must be integrated with that of the dream, of the unconscious.[29]

Biology and Depth Anthropology

The Creative

In Portmann's primary world every living person is in a condition of immediate relationship with other living beings and with the environment, being part of an original biosemiotic process. All organisms appear to be in tune with each other and share the same expressive grammar.[30] The starting point is an "original background", from which all living relationships emerge and grow in a mutually conditioning link, in an original contamination between the intentionality of different bodies. As in Weizsäcker's *Circle of Form* (Gestaltkreis 1997), the living being seems immersed from the beginning in a state of "pathic" and pre-reflexive circularity with the rest of the world, with which it shares a grammar of signals, a world of semiotic images and markers, which grow together in synchrony: a process of pre-reflexive and supra-individual interactions which generate meaning. At the root is life's innate desire to communicate, which in man is drastically implemented. This human awareness forces man to reconcile the meanings that come from the senses with the meanings that he *gives* to things. In reality, this cosmic communication process, rooted in genetic power and unconscious in animals, immediately becomes an *ethical-aesthetic* act in man. The "semantic organs" can interrupt their natural communication and redirect or overturn the meaning they bear. This means that the study of modes of communication and relationships in plants and animals (biosemiotic and zoosemiotic) in man assumes an original significance (Kleisner 2008). After all, Portmann's entire anthropology highlights the problematic relationship that, in humans, the exterior assumes towards the interior world (Coccia 2011). The possibility of lying, of dissimulation, is just one example of this potential divergence.[31]

The danger, which Portmann has in mind, is that the hyper-reflexivity of the modern subject is both the cause and the effect of a progressive "disembodiment",

[29] "Our spiritual life – and, in it, our art – can only find a new fullness if man sees himself as great in thought as in dream" (Portmann 1948: 152).

[30] For a comparison between Portmann, Scheler and biosemiotics see Cusinato (2018).

[31] This is evident in the interpretative difficulty of medical semiotics and in the issue of psychosomatic medicine. Quoting Marcel, Portmann reminds us: "We live in a paradoxical situation, so that we can neither detach ourselves from our bodily conduct nor identify ourselves within it" (Portmann 1968h, 301). In man, it is not in fact excluded that internal organs, hidden beneath the surface, can be placed in the service of representation and can go beyond their original function of response to simple stimuli. In this regard Portmann refers to the *anthropotherapy* of Alexander Mitscherlich. Cf. Thure von Uexküll (2014: 64).

a de-naturalisation of man: of a compromised perception of the original expressive unity.[32] Therefore, it must be recognised that the conscious dimension of man and life is only an important, but limited, part of existence, of interiority. In reality, the vital process seems to be directed by "a hidden 'Self'" (*verborgenes "Selbst"*) (Portmann 1968e: 166), by "shaping forces", by "dark hidden causes" (Portmann 1990a: 113) that represent the large shadowy zones of consciousness: our action takes place "in the mystery of a reality which has hidden within us and before us the Creative" (Portmann 1971b: 363). This does not mean admitting the existence of an *entelechia* or orthogenetic force, as for Teilhard de Chardin (Portmann 1960c), but recognising the dark half of our experience, "of the dream and of the night which the rational one disregards [...] the dark half of the language that not only thinks rationally in us, but dreams in us" (Portmann 1968d: 101, 116).[33] It is an "us", before the "I", which makes the individual a being already preordnained to the supra-individual, to cosmic sympathy (Portmann 1965a: 24).

With respect to a dominant biology of the invisible, he opts for a basal anthropology that investigates in depth the foundations, the anchorage, of man so that a new world experience can grow in him. In this sense, Portmann's anthropology emerges as an anthropology of the deep. It investigates the primary rootedness – which in man always has a double socio-genetic value – recognizing the role that this latent wisdom can still exercise for the development of an integral humanity.

Like the psychological unconscious, the socio-biological one is likely to escape full clarification; like psychoanalysis, it is unimaginable that happiness and health can be achieved by making the unconscious aware. Life rests on a dark background, on an *original game between the parts and the whole* on which the creative dimension of the living being is based. Every image of man must consider that potential of images and connections of which he is the unconscious repository: that which is *foreign* in ourselves (Portmann 1968c: 93). In this sense, Portmann's basal anthropology is therefore a "depth anthropology".

The Sleep of Reason Produces Demons

The "hidden sources of life" act within us and through us; a *demonic* force – at least as man has defined it over the centuries – continuously overtakes conscious action (Portmann 1965c: 100, 1968f: 208–209). Life appears for the most part hetero-directed, by demons that do not dissolve, even with technical progress.[34] Nor is it

[32] Portmann's reflection is part of a more recent debate that relates world-openness and *epoché*, the risk of a pathological separation between "being-body" and "having-body" (von Weizsäcker 1997; Gebsattel 2014; Kimura 2014; Blankenburg 2012).

[33] Separate attention should be given to a comparison between the "depth biology" of Portmann and the depth psychology of Jung. Cf. Vajdová, Chap. 13, in this volume.

[34] Portmann, by way of example, refers to the rabies virus. The patient's senseless self-injury – which in this way encourages the further spreading of the virus – becomes immediately under-

desirable for this to happen; on the contrary, only a regenerative sleep of rationality can allow demons to survive;[35] only an integration of the secondary world with the primary one can prevent man from suffocating in an artificial world, constructed exclusively by himself. This does not mean a return to myth, but to *enigma* (Portmann 1948: 128). Of course, the idea of keeping together the mythical and the rational, the primary and the secondary, the theoretical and the aesthetic, seems, in turn, a mythical idea. The basic intuition is to place a limit on man, for his own good, in the name of the unknown foundation of life: a foundation that is based upon *play*, in time and space, of which human consciousness is just one of the protagonists. Portmann's anthropology takes on the features of a true *eco-logical* thought, in which the protection of man takes place by recognising the existence of a "pathic" dimension: of an *oikos*, of an original homeland. Man's objective, in fact, is not "to know as much as possible" (Portmann 1968e: 173), but to give a meaning, to "redeem", the world (Portmann 1965a: 31), to build one of his own on the basis of an original wisdom that lies in the primary world. In fact,

> The representational value of world things, which determines their symbolic power living in our mind, will always remain rooted in our structure according to the geocentric sphere. It is the great expressive power for shaping our experience of the world. We have to recognize the importance of the task, which is, on the one hand, to know our structural anchoring in the geocentric, but on the other hand to recognize the need to grow out of this world experience into another. (Portmann 1956: 230)

Therefore, in our opinion, the first task posed by Portmann's anthropology is to protect the homeland's natural life, as a place of archaic biosemiotic communications that constitute an essential experience of man and that can make him happy. The second is to avoid the primary and natural world becoming only the "raw material" on which to exercise the creative expressiveness of the split man. Post-human cannot in any way mean post-natural.

References

Bateson, G. (1972). *Steps to an ecology of mind*. San Francisco: Chandler Publishing Company.

Bateson, G. (1979). *Mind and nature. A necessary unity*. New York: Dutton.

Blankenburg, W. (2012). *Der Verlust der natürlichen Selbstverständlichkeit: Ein Beitrag zur Psychopathologie symptomarmer Schizophrenien*. Berlin: Parodos.

Bleakley, A. (2000). *The animalizing imagination. Totemism, textuality and ecocriticism*. London: Macmillan.

Cobley, P. (2010). *Cultural implications of biosemiotics* (Vol. 2016). Dordrecht: Springer.

Coccia, E. (2011). *La vita sensibile*. Bologna: il Mulino.

standable if it is interpreted as subordination, to the simplest organisational states, of the most complex and advanced form of life: life preserves a logic that goes beyond the rational one of man (Portmann 1968f: 206–207).

[35] Portmann speaks of a real "praise of sleeping" (Portmann 1968i: 320).

Conte, P. (2013). Fanerologia. In A. Portmann, (Ed.), *La forma degli animali* (pp. IX–XXXIII). Milano: Cortina.
Cusinato, G. (2018). *Biosemiotica e psicopatologia dell'ordo amoris*. Milano: Franco Angeli.
Damasio, A. (1994). *Decartes' error. Emotion, reason and the human brain*. New York: Putnam.
DeSilva, J. M. (2016). Brains, birth, bipedalism, and the mosaic evolution of the helpless human infant. In W. R. Trevathan & K. R. Rosenberg (Eds.), *Costly and cute. Helpless infants and human evolution* (pp. 67–86). Albuquerque: School for Advanced Research Press.
Eibl-Eibesfeldt, I., & Hass, H. (1985). Sozialer Wohnbau und Umstrukturierung der Städte aus biologischer Sicht. In E. Eibesfeldt et al. (Eds.), *Stadt und Lebensqualität* (pp. 49–84). Stuttgart: DVA.
Fischer, J. (2008). *Philosophische Anthropologie. Eine Denkrichtung des 20. Jahrhunderts*. Freiburg/München: Karl Alber.
Görner, R. (2016). *Hölderlin und die Folgen*. Stuttgart: Metzler.
Heykaus, B. (2010). *Rythmisch-tänzerische Bewegung im Unterricht an weiterführenden Schulen. Eine pädagosich-anthropologischeFundiereung in der Nach-Pisa Zeit*. Bern: Peter Lang.
Hoffmeyer, J. (1996). *Signs of meaning in the universe*. Bloomington: Indiana University Press.
Jonas, H. (1982). Technology as a subject for ethics. *Social Research, XLIX*(4), 891–898.
Jonas, H. (1993). *Dem bösen Ende näher. Gespräche über das Verhältnis des Menschen zur Natur*. Frankfurt a. M.: Suhrkamp. [Italian translation: Jonas, H. (2000). Sull'orlo dell'abisso. Conversazioni sul rapporto tra uomo e natura. Torino, Einaudi (pp. 126–146)].
Kimura, B. (2014). *Tra. Per una fenomenologia dell'incontro*. Trapani: Il pozzo di Giacobbe.
Kleisner, K. (2008). The semantic morphology of Adolf Portmann: A starting point for the biosemiotics of organic form? *Biosemiotics, 1*, 207–219.
Koechlin, F. (2004). Don Quijote der Laboratorien. *WOZ, 24*(2004), 1–5.
Kugler, R. (1985). Die Spannung zwischen primärem und sekundärem Welterleben. In A Cesana & O. Rubitschon (Eds.), *Philosophische Tradition im Dialog mit dem Gegenwart* (pp. 193–198). Basel: Birkhäuser.
Lévy-Bruhl, L. (1927). *L'âme primitive*. Paris: Presses Universitaires de France. [Italian translation: Lévy-Bruhl, L. (1990). *L'anima primitiva*. Torino, Bollati Boringhieri.]
Lorenz, K. (1974). *Die acht Todsünden der zivilisierten Menschheit*. München: Piper.
Lorenz, K. (1987). *The waning of humaneness*. Boston/Toronto: Little, Brown and Company.
Lorenz, K. (2002). *On aggression*. London/New York: Routledge.
Morin, E. (1974). *Das Rätsel des Humanen. Grundfragen einer neuen Anthropologie*. München/Zürich: Piper.
Morin, E. (1999). *Seven complex lessons in education for the future*. Paris: UNESCO.
Nestor, M. (2017). Adolf Portmann: die natürliche Erziehbarkeit und Erziehungsbedürftigkeit des Menschen. Forum Naturrecht und Humanismus. https://naturrecht.ch/wp-content/uploads/03-Nestor-Portmann-14.4.17.pdf. Accessed 15 Mar 2019.
Nussbaum, M. (1999). *Cultivating humanity. A classical defense of reform in liberal education*. Cambridge: Harvard University Press.
Pinotti, A., & Tedesco, S. (2013). *Estetica e scienza della vita*. Milano: Cortina.
Plessner, P. (1975). *Die Stufen des Organischen und der Mensch*. Berlin/New York: de Gruyter.
Portmann, A. (1948). L'art dans la vie de l'homme. In J. Cassou, E. Ansermet, T. Maulnier et al. (Eds.), *Débat sur l'art contemporain* (pp. 126–152). Neuchâtel: edition de la baconnière.
Portmann, A. (1951). *Zoologie und das neue Bild des Menschen. Biologische Fragmente zu einer Lehre vom Menschen*. Hamburg: Rowohlt.
Portmann, A. (1954). Biologie auf neuen Wegen. In J. Moras & H. Paeschke (Eds.), *Deutscher Geist zwischen Gestern und Morgen* (pp. 172–188). Stuttgart: Deutsche Verlags-Anstalt.
Portmann, A. (1956). Ein Wegbereiter der neueren Biologie. In J. von Uexküll & G. Kriszat (Eds.), *Streifzüge durch die Umwelten von Tieren und Menschen: ein Bilderbuch unsichtbarer Welten* (pp. 7–17). Hamburg: Rowohlt.
Portmann, A. (1958). *Don Quijote und Sancho Pansa*. Basel: Reinhardt.

Portmann, A. (1960a). Ptolemäer und Kopernikaner. Eine biologische Studie. *Der Monat. Eine Internationale Zeitschrift, 12*(139), 23–30.
Portmann, A. (1960b). L'ontogenèse humaine et le problème de l'origine. *Dialectica, 14*(1), 37–52.
Portmann, A. (1960c). *Der Pfeil des Humanen. Über P. Teilhard de Chardin*. Freiburg/München, Alber.
Portmann, A. (1962). Preface to a science of man. *Diogenes, 10*(1), 1–26.
Portmann, A. (1964). *Sinnvolle Lebensführung*. Zürich: Rhein.
Portmann, A. (1965a). Das Lebendige als vorbereitete Beziehung. In *Aufbruch der Lebensforschung* (pp. 13–32). Zürich: Rhein.
Portmann, A. (1965b). *Sinndeutung als biologisches Problem* (pp. 58–88).
Portmann, A. (1965c). *Kampf und Frieden in Biologischer Sicht* (pp. 89–116).
Portmann, A. (1965d). *Der Biologische Beitrag zu einem neuen Bild des Menschen* (pp. 117–147).
Portmann, A. (1965e). *Die Ordnung des Lebendigen im Deutungsversuch der Biologie* (pp. 188–228).
Portmann, A. (1965f). The special position of man in the realm of living. *Commentary, 40*(5), 38–41.
Portmann, A. (1965g). *Die Tiergestalt. Studien über die Bedeutung der tierischen Erscheinung*. Freiburg/Basel/Wien: Herder.
Portmann, A. (1968a). Die Biologie und das Phänomen des Geistigen. In *Biologie und Geist* (pp. 9–49). Freiburg/Basel/Wien: Herder.
Portmann, A. (1968b). *Das Ursprungsproblem* (pp. 50–75).
Portmann, A. (1968c). *Der naturforschende Mensch* (pp. 76–99).
Portmann, A. (1968d). *Mythisches in der Naturforschung* (pp. 100–132).
Portmann, A. (1968e). *Die Zeit im Leben der Organismen* (pp. 150–173).
Portmann, A. (1968f). *Die Bedeutung der Bilder in der lebendigen Energiewandlung* (pp. 174–209).
Portmann, A. (1968g). *Die Erde als Heimat des Lebens* (pp. 210–232).
Portmann, A. (1968h). *Um eine basale Anthropologie* (pp. 293–308).
Portmann, A. (1968i). *Biologisches zur ästhetischen Erziehung* (pp. 309–333).
Portmann, A. (1971a). Manipulation des Menschen als Schiksal und Bedrohung. In *Entläßt die Natur den Menschen? Gesammelte Aufsätze zur Biologie und Anthropologie* (pp. 215–229). München: Piper.
Portmann, A. (1971b). *Der Mensch im Bereich der Planung* (pp. 343–363).
Portmann, A. (1971c). *Biologie als technische Weltmacht* (pp. 364–379).
Portmann, A. (1974). *An den Grenzen des Wissens. Vom Beitrag der Biologie zu einem neuen Weltbild*. Wien/Düsseldorf: Econ-Verlag.
Portmann, A. (1990a). *A zoologist looks at humankind*. (J. Schaefer, Trans.). New York: Columbia University Press. [German original: Portmann, A. (1969). *Biologische Fragmente zu einer Lehre vom Menschen*. Third, enlarged edition. Basel: Schwabe.]
Portmann, A. (1990b). On the uniqueness of biological research. *The Journal of Medicine and Philosophy, 15*, 457–472.
Retter, H. (2000). *Anthropologische und gesellschaftliche Vorausetzungen der Erziehung*. Brunschwig: Technische Universität Braunschwig.
Theunissen, G. (1997). *Basale Anthropologie and Ästhetische Erziehung*. Bad Heilbrunn: Julius Klinkhardt.
Tolone, O. (2012). Adolf Portmann: The expressive function of technique. *Ethics & Politics, XIV(1)*, 230–243.
Tolone, O. (2015). Helmuth Plessner und Adolf Portmann. Zur philosophischen Bestimmung des Menschen durch Exzentrizität und Frühgeburt. In C. Köchy & F. Michelini (Eds.), *Zwischen den Kulturen. Plessners "Stufen des Organischen" in zeithistorischen Kontext* (pp. 141–159). Alber: Freiburg/München.
Tomasello, M. (2014). *A natural history of human thinking*. Cambridge: Harvard University Press.
von Gebsattel, V. (2014). Zum personalen und anthropologischen Aspekt der Neurosen. *Psychologik, 9*, 52–108.

von Uexküll, T. (2014). Psychosomathische Medizin als ärtzliche Grundhaltung. *Psycho-logik, 9*, 200–214.

von Weizsäcker, V (1997). *Der Gestaltkreis. Theorie der Einheit von Wahrnehemen und Bewegen* (Gesammelte Schriften. Bd. 4). Frankfurt a.M.: Suhrkamp.

Wheeler, W. (2010). Delectable creatures and the fundamental reality of metaphor. Biosemiotics and animal mind. *Biosemiotics, 3*, 277–287.

Zahavi, D., & Gallagher, S. (2012). *The phenomenological mind*. London: Routledge.

Oreste Tolone has obtained the National Scientific Qualification (ASN) as Associate Professor in Moral Philosophy at the "G. d'Annunzio" University of Chieti-Pescara. His areas of specialisation are philosophical anthropology, medical anthropology and the philosophy of religion. His publications include: "La Posthistoire di Arnold Gehlen" (2007); "Adolf Portmann: The Expressive Function of Technique" (2012); "Plessner's Theory of Eccentricity. A Contribution to the Philosophy of Medicine" (2014); "Plessner und Portmann. Der Mensch zwischen Exzentrizität und Frühgeburt" (2015); *Alle Origini dell'Antropologia Medica. Il Pensiero di Viktor von Weizsäcker* (2016); "The 'Introduction of the Subject'. Anthropology, Medicine, Education" (2019).

Chapter 11
The "New Morphology" Between Biology and Philosophy: The Hermeneutic Dimension of Portmann's Thought

Jiří Klouda

Abstract Among the peculiarities of Portmann's biological philosophy is its contradictory reception. Though it was intended to reform a traditional biological discipline, morphology, in academic biology itself this original doctrine had surprisingly little impact. Instead, Portmann's theoretical biology remains a focus of interest for philosophical disciplines like phenomenology and aesthetics and enjoys the attention of lay readers as well. Based on this internal tension, this chapter reflects on the theoretical status of a field that was ignored as a scientific reform but is still studied as an unacknowledged philosophy. Through discussion of the central concepts of "form" and "authentic phenomenon," it demonstrates that these categories do not have a straightforwardly empirically objective, "scientific" character but, rather, require an approach more akin to philosophy. Portmann, however, did not want to address the philosophical aspects of his morphology as a philosopher and styled his project in a manner more empirical than its true character called for, which to a certain degree complicated the reception and influence of his original biological thought.

Introduction: Portmann's Theoretical Biology as a Literary Genre

Before introducing the main theoretical components of Portmann's biological project, it will be useful to dwell briefly on the literary aspects and rhetorical strategies of his relevant writings. The texts that can be classified as theoretical biology represent, from a quantitative perspective, a significant – perhaps the greatest – part of Portmann's published work. From roughly 1950–1965, Portmann published an almost bewildering series of articles, essays, and studies whose ambitious titles

Chapter translated by Andrew G. Christensen.

J. Klouda (✉)
University of Hradec Králové, Hradec Králové, Czech Republic
e-mail: jiri.kl@centrum.cz

F. Jaroš, J. Klouda (eds.), *Adolf Portmann*, Biosemiotics 23,
https://doi.org/10.1007/978-3-030-67810-4_11

reveal, often quite unequivocally, that they aim at general questions concerning the basic characteristics of living beings and thereby a fundamental reorientation of biology. Examples of such ambitious titles include: "A New Picture of the Organism" (Portmann 1953; Um ein neues Bild vom Organismus), "Biology on New Paths" (Portmann 1954; Biologie auf neuen Wegen), "A Path to a New Picture of the Organism" (Portmann 1958a; Unterwegs zu einem neuen Bild vom Organismus), "Toward of Philosophy of the Living" (Portmann 1959a; Zur Philosophie des Lebendigen), "Transformations of our Image of the Living Being" (Portmann 1960b; Wandlungen unseres Bildes vom Lebendigen), "New Horizons in Biology" (Portmann 1970b; Neue Fronten der biologischen Arbeit, first published 1958), and "A New Conception of the Organism" (Portmann 1965b; Die neue Auffassung vom Organismus). However, most of these studies focused on fundamentals were not addressed to professional, specialized circles; they were published in anthologies and magazines aimed at a wider, educated but strictly lay, public. The case is similar with three extensive monographs considered to be the acme of Portmann's output: *Animal Forms and Patterns* (Portmann 1965a; Die Tiergestalt; first edition 1947), *Animals as Social Beings* (Portmann 1964; Tier als soziales Wesen; first edition 1953) and *New Paths in Biology* (Portmann 1960a; Neue Wege der Biologie).

From a literary critical perspective, these texts are closest to the genre of popular science writing – they are not specialized monographs, manuals, or textbooks.[1] Portmann's texts differ from popular science in their content, however. It is not new methods or discoveries of biology that are popularized here, nor is the possible future significance of this or that discipline developed for readers. Rather, the author points out blind spots in established biological science in the name of a future discipline, a "new," "complex" morphology, whose creation is the "fundamental biological task of our times" (Portmann 1970d: 71).[2] Somewhat paradoxically, at first glance, a program for an as-yet unestablished discipline, whose factual acceptance could come only from the side of specialized academic readers, is thus popularized and defended for the lay public.

This does not exhaust the list of paradoxes that the reader of Portmann's texts encounters. Despite the very general scope that the titles of Portmann's books and studies suggest, the actual texts on the "new morphology" are surprisingly not very theoretical. Contrary to expectations, we find here few explicit theoretical constructions or definitions and a minimum of quantifying methods, and even the central conceptual tools remain in flux and seemingly consciously provisional, often changing from study to study. This applies even to the concept of inwardness (*Innerlichkeit*), which plays a key role in Portmann's thought yet by his own admission remains, rather than a firmly defined term, "a placeholder, a scientific symbol" (Portmann

[1] Portmann's only "standard" specialized monograph is the textbook *Introduction to the Comparative Morphology of Vertebrates* (Einführung in die vergleichende Morphologie der Wirbeltiere) (Portmann 1948; revised 1959a, 1965a).

[2] On the "new" or "complex" morphology, see also Portmann 1953: 225–226, 1958a: 34, 39, 1958b: 170–171.

1963b: 16).[3] Instead of clearly formulated theoretical models and diagrams, the reader of Portmann's theoretical biology primarily finds a continuously repeating set of examples and observations (color patterns in opisthobranch snails, the coloration of slime molds, Oudemans' phenomenon, *descensus testiculorum* in higher mammals, etc.) along with commentary on them and remarks about how such phenomena defy the explanatory methods of contemporary biology.

When we place this small degree of theoretical components next to the principally general focus and the "lay" orientation of these texts, we cannot avoid the impression that the entire project of the "new morphology" was destined to fail and that the wide popularity it gained among lay readers[4] is merely the flip side of the rejection and disregard from mainstream professional biology. In the following pages, I will attempt to demonstrate the limits of, and alternatives to, this perspective, which, in the end, was not entirely alien to Portmann himself or to many of his sympathizers.[5]

The Project of the "New Morphology"

Portmann's theoretical biology is often labeled as a phenomenology of living beings or "bio-phenomenology." It is rooted in a traditional biological field in which Portmann was long engaged pedagogically and with which he identified in defining his scientific endeavors: morphology. Since the end of the nineteenth century, this discipline had found itself increasingly in the shadow of the swiftly developing research aimed at cellular and eventually sub-cellular structures. These deep approaches necessarily led to attempts to elaborate the general constant units (first cells, then, from the mid-twentieth century, genes) that would form the baseline for the description of living beings. The great wealth of concrete knowledge that descriptive morphology had accumulated over the centuries in this situation finds itself sidelined in biological research. Forms and other observable characteristics of living beings acquire meaning only to the degree that they become observable indicators of physiologically and chemically defined inner processes. A second current that significantly affected the development of morphology was the Darwinian doctrine of natural selection, which saw selection as the primary motor of evolution. Thus from the diversity of forms and colors of actual living beings attention was focused only on that which could be established as having some selection value, which provided its bearer a presumptive advantage in the form of increased reproductive possibilities, while everything else lost its relevance. It was during the

[3] On the variable and metaphorical terminology of Portmann's biological and anthropological writings, cf. Illies 1981: 113–114.

[4] A number of Portmann's books (including *Animal Forms and Patterns*, *The Animal as a Social Being*, and a collection of essays, *Biology and Spirit* [*Biologie und Geist*]) were widely distributed in the 1960s thanks to licensed editions by German publishers Herder, Rowohlt, and Suhrkamp.

[5] This "melancholy perspective" can be seen, for example, in Koechlin 2004.

interwar period, when these two currents (the molecular/genetic and evolutionary) were merging together into so-called neo-Darwinism or the Modern Synthesis, that Portmann began work on his reform of morphology.[6]

In presenting the core of Portmann's "doctrine," we can begin with his observation that in lower animals whose surface is transparent, the inner organs are arranged in a symmetrical fashion or are concentrated in an opaque cover symmetrically situated in a transparent body. By contrast, organisms with a completely opaque body surface have asymmetrically arranged organs, even though symmetry is respected on the surface, including the placement of colored areas or patterns on the skin, fur, or feathers. Thus the visibly inaccessible parts of animal (including human) bodies are "constructed" according to a different plan than those that are visibly accessible (Portmann 1957, 1965a: 17–59). (It is similarly the case with plants when comparing roots and above-ground parts. As a zoologist, Portmann primarily works with observations from the animal kingdom, but he often points out that his findings apply in principle to plants and fungi as well). To him, this observation suggests that visual appearance (and sensory phenomena generally) is a fundamental component of the inherent plan of living beings. The reality that living beings "show" themselves is not just a secondary or self-evident property of their anatomical and physiological organization, but rather one of its *primary* attributes which, as the comparison of the structure of transparent and opaque animal bodies suggests, has an influence on the arrangement of even the most vitally important organs. Portmann calls this fundamental property of living beings "self-manifestation" or "self-presentation" (*Selbstdarstellung*). This should not be understood only as a behavioral or ethological quality; rather, we should keep in mind its morphological anchorage in the construction of the living body.[7] Portmann termed a being's widely apprehended, sensually perceptible manifestation (from external bodily structure to various behavioral displays) its "form" (*Gestalt*), as opposed to its anatomical and primarily unmanifest shape (*Form*); the new conception of morphology differs from its traditional version precisely in the shifted emphasis on form in this wider sense (Portmann 1965a: 34).[8]

If we thus include self-manifestation – "appearing" – among the foundations of the structure of the living body as a spontaneous function, then this "phenomenon" is a vital activity just like ontogenesis, the acquisition and assimilation of nutrients, and reproduction. Portmann is here understandably faced with the question of why such a basic property of living matter escaped the attention of natural scientists for

[6] While the first publications containing the foundational thoughts of Portmann's theoretical biology appear shortly after the Second World War, we know from the archival research of M. Ritter that the first of the preserved notebooks containing handwritten materials for the lecture "Tiergestalt" dates from 1938–39 (Ritter 2000: 210).

[7] Only in the lowest, fully transparent organisms do the anatomical forms and their visual self-presentation coincide. With the appearance of opaque external boundary layers, active differentiation emerges, where manifest parts are constructed differently from hidden ones.

[8] In a certain sense, it can be said that Portmann renews morphology by incorporating ethology, which was rapidly developing at the time.

so long. He sees a reason in the preference for unambiguous functional structures whose practical significance for preserving the individual or species can be postulated and experimentally verified. Such a focus, according to him, is also the blind spot of modern biology, which forcibly interprets a range of conspicuous features of living beings according to functional selection mechanisms, or else ignores or trivializes them.

When Portmann speaks of the self-manifestation or appearing of living beings, he has in mind their so-called "authentic phenomenon" (*eigentliche Erscheinung*), which differs from *phenomenon* in the wider sense of the word. Such generally perceived phenomena obviously include the "alien" phenomena provided to us by a body subjected to autopsy, vivisection, chemical analysis, or x-ray images. In contrast, the authentic phenomenon arises from the specific spontaneity of a living being and appeals directly to the senses of perception without the use of scientific instruments. Portmann calls this spontaneity of all living beings "inwardness" (*Innerlichkeit*). The inwardness of living beings is their ability to relate to the world by means of self-centralized acts, on the basis of their own ("inner") representation of the surrounding world (Portmann sometimes uses "centricity" or "centrality" as equivalent terms). Therefore a living being does not simply react to neutral stimuli, but operates according to its own (species-specific) image of the world.[9]

Portmann is convinced that there is a relationship between the fullness or complexity of this relationship to the world (i.e. the intensity of its experience), the inwardness, and the self-manifestation of a given being.[10] Beings whose experience of the surrounding world is more developed have a more distinctive, more explicit self-presentation. This relation can be substantiated by a range of empirical observations: in lower, primarily scent-oriented small mammals the head is not significantly set apart from the torso either anatomically or by fur coloration, whereas in higher mammals (in whom long-distance vision plays a decisive role) the head is clearly separated as an anatomical structure and very often also by a specific coloration pattern ("mask") distinct from the coloration on the rest of the body. The opposite, anal pole of the body of ungulates, carnivores, and primates is similarly accentuated. It generally applies, that the authentic phenomenon of higher animals is more distinctive and, on the species (and individual) level, less interchangeable, and this rule is repeated in the difference between young and mature individuals. (This can be seen, for example, when comparing inconspicuous, small mouse-like rodents, which are difficult for non-experts to distinguish, with the characteristic and easily distinguished forms, patterns, and colors of ungulates or large carnivores). In higher animals we find conspicuous, visually distinguishing structures not only in the form of coloration of body surfaces, but anatomically fixed elements (antlers and horns in ungulates, protuberances in various insect species, the striking forms of bird feathers).

[9] For more on inwardness, cf. Portmann 1959a: 418–422, 1960a: 47–65.

[10] On the basis of the intensity of inwardness and self-manifestation it is possible to distinguish the "rank" of living beings – their place in the hierarchy of species. See, for example, Portmann 1960a: 86–101. Generally, higher-ranking beings are more intuitively comprehensible to us.

Here we are already approaching the more difficult points of this theoretical-biological model. If we are going to regard the relationship to the world via inwardness and the resulting self-manifestation as the two most fundamental roles of living beings, then we are bound to regard the sensually perceptible form of a living being as an autonomous reality, not as a secondary product of other, apparently more basic functionally-defined processes. The theory of natural selection considers individual components of the forms of living beings according to their usefulness or the value a given element provides its bearer in terms of preserving the individual or increasing its chances of reproduction. According to it, living beings with features contradictory to their individual preservation and their ability to pass on their genetic material cannot exist in the long term. Portmann, in opposition to these selectionist assumptions, needs both to give evidence that self-presentation is not explicable only as a by-product of metabolic processes and to convince his readers that it is, in itself, practicably distinct from characteristics with demonstrable selection value.

The difficulty is that if Portmann wanted to definitively refute the selectionist assumption of the usefulness of various elements interpreted as the self-manifestation of living beings, he would have to positively substantiate self-manifestation as an empirical, scientifically describable category. Self-manifestation is the entirety of the sensory impact of a given living being, which biologists understand always on the basis of their own inwardness, but it cannot enter into scientific (always partial) argumentation.[11] His approach, therefore, is always one of negation, directed at concrete scientific explanations whose inadequacy he attempts to demonstrate. Thus in the case of the formation of regular, conspicuous color patterns on animal body surfaces, Portmann must show that adaptive processes cannot explain them, nor, by contrast, can they be interpreted in terms of the selection value of neutral epiphenomena of ontogenesis or metabolism. In this case, he argues that seemingly secondary effects of ontogenesis or metabolism are extracted and synthesized (which suggests some non-random "role" in the course of life) in a complex manner, often with a high energy demand. Regarding a selectionist interpretation of self-manifestation, Portmann shows that indisputably functional traits of animals like the phenomena of crypsis, aposematism, or features emphasizing the sexual attractiveness of the bearer *assume* self-manifestation (evolution can favor certain pre-existing elements) but cannot fully explain it.[12]

In this context, Portmann makes the somewhat counterintuitive claim that the authentic phenomenon of the living is primarily *unaddressed*. This means that the self-manifestation of living beings cannot be understood in terms of motives

---

[11] For more on this, see section 4 below. According to Portmann, "totality" and "purposefulness" form the background against which we *understand* a living being, but they can never scientifically *explain* anything concrete – they cannot legitimately be used as a predicate. Hence also Portmann's critique of biological vitalism and finalism (Portmann 1954: 180).

[12] Portmann acknowledges mutation, which neo-Darwinism regards as a productive moment of evolution, only to the degree it can be empirically verified, as in so-called micromutation. He regards the elevation of random mutation to the actual motor of evolution (without empirical evidence of the origin of new species) as a mythical element in modern natural science.

(cryptic, semantic, etc.) directly related to the perception of other living beings. Portmann speaks of the so-called "unaddressed authentic phenomenon," whereas an addressed phenomenon, explainable as an evolutionary modification linked to selection value, has a derived value and existence. Portmann is himself aware of the somewhat paradoxical nature of this claim about the authentic phenomenon, which is defined as presenting itself directly to the senses of observation but nevertheless is not intended to show itself to a concrete observer and thus stands outside the category of communication. In this sense, he speaks of a serious "epistemological problem" which he refers for consideration to philosophers.[13] Perhaps the "unaddressability" of the authentic phenomenon could be positively determined by its relation to light as an all-embracing medium in which the self-manifestation of living forms takes place. With the development of an opaque body surface and the shapes and colors that appear on it, a living being takes on a relation to the illuminated environment as a whole, whereby light finds itself in the position of the "addressee" of the unaddressed phenomenon.[14] The doctrine of the principally unaddressed nature of the authentic phenomenon and its relation to the medium of light is a sort of speculative culmination of Portmann's new morphology.

The Inner Tension of the Morphological Project

Even though Portmann's manifestos on reformed morphology were addressed primarily to a lay readership, it is nonetheless surprising how little impact this indisputably original concept had on academic biology. The few contemporary critical responses reveal a (not entirely incorrect) concern about erasing the line between science and philosophy.[15] Nor did any of Portmann's own students take on or develop (or get the chance to develop) the reform of morphology as a whole on an academic level. Eminent contemporary philosophers were significantly more obliging toward Portmann, discussing his works with appreciation. They were driven in this, however, by their own interests and did not devote great attention to the

[13] "ein schweres erkenntnistheoretisches Problem" (Portmann 1957: 367). In his theoretical texts, Portmann often directs problems connected with the authentic phenomenon to the attention of philosophers (see Portmann 1958a: 45, 1960b: 74, 1970d: 71).

[14] On this aspect of Portmann's teaching, see especially Portmann 1957: 367–369, 1960b: 73–74, 1965c, 1970c. Note that Portmann regards extreme cases of organisms living entirely without contact with light (e.g. endoparasitic, cave-dwelling, and deep-sea species) as evolutionary modifications of forms that originated in illuminated environments.

[15] In 1963, the president of the German Zoological Society, A. Remane, somewhat officially criticized Portmann's theories as "armchair metaphysical speculation" (according to Illies 1981: 28, 260, fn. 16). Not even his more favorably inclined colleagues could hide their puzzlement – the ethologist P. Leyhausen, for example, preserved (without wanting to identify with) another contemporary impression of Portmann's work: "Gestaltmystik" (Lorenz & Leyhausen 1973: 391). On the reception of Portmann's books translated to English in the 1960s, see Jaroš, Chap. 7, in this volume.

"epistemological problem" linked to the category of the unaddressed phenomenon and its relation to the medium of light, which Portmann had referred to them.[16]

If we are to regard Portmann's morphology as a standard scientific discipline whose primary aim is the cumulative production of positive empirical knowledge,[17] then it is not difficult to arrive at a fundamental critique of it. As previously indicated, a "form" and its self-manifestation resist a clear-cut identification as a holistic category. As we have seen, in his explications of self-manifestation as one of the main properties of living beings Portmann often refers to conspicuous structures emphasizing the appearance of organisms. At the same time, he emphasizes the failure of attempts to explain these structures in terms of other life functions. Such an approach can be criticized on a methodological level, as he demonstrates self-manifestation only negatively – i.e. through the insufficiency of regular functionalist explanations considering the given component to be "purposeful," providing the bearer a selective advantage. In this aspect, Portmann's project depends on an argument of the same sort used by the very science (and its contemporary standards) that he is trying to overcome.[18] On the level of empirical or scientific argument, then, the new morphology has a problem regarding the positive existence of its very subject.

Attempts at a positive theoretical understanding of self-manifestation face the difficulty of how to practicably distinguish it from other life functions. "A living being stands before us as a unit. Nobody can completely isolate the functions of preservation and self-presentation; every detachment from the whole is an artificial intervention" (Portmann 1960a: 222). Portmann proposes to solve the relation of self-manifestation and other functions hierarchically, in the sense that self-manifestation transcends other functions.[19] Because this highest function exceeds and, in its way, comprises or orients all other functions, it is itself a function in a different sense than the others.[20] However plausible this solution sounds (living beings manifest themselves in their specificity even when finding food or reproducing), it is problematic on the methodological side. If self-manifestation transcends life-preserving functions such that it incorporates and somehow elevates them to

[16] On interpretations of Portmann by M. Merleau-Ponty, H. Arendt a M. Grene, see Conte, Chap. 9, in this volume.

[17] Cf. Ritter 2000: 253. In the Czech sphere, the strained compatibility of Portmannian morphology with the model of "normal science" led to its interpretation as an alternative biological paradigm in opposition to neo-Darwinism. On this, see Komárek, Chap. 14, in this volume and Jaroš 2010.

[18] Cf. Portmann 1960a: 227: "morphology must consider phenomena that go beyond the current framework of functional understanding, which physiology generally – and with good reason – must define."

[19] See, for example, Portmann 1958a: 39, 1970d: 73.

[20] See Portmann 1970c: 54: "In earlier discussions on the self-presentation of organisms I spoke of this aspect as a "function." If this term is meant to designate a role in a self-preserving system, then self-presentation does not fall under this narrow concept of function." Also: "Above all functionality in the service of preserving the individual and species, in the service of social acts or in the defense against enemies – in front of all these functions and *first giving them meaning* is basic existence as self-presentation" (op. cit. 57; italics mine). An identical formulation is found in Portmann 1965a: 236.

moments of the whole manifesting the "form," then these functions also always have a self-manifesting value. In other words, in living beings we find no purely single-purpose functions (these exist rather in our interpretations of life and in monocausal laboratory experiments). Portmann himself states that in higher animals living in complex social structures we can observe utilization of the majority of observable external surfaces (coloration, posture) and a range of internal organs and physiological functions (excretion) for various means of communicating inner states to conspecifics and members of other species.[21] Hidden physiological processes thus enter the outer sphere of the self-presentation of a living form and, thanks to this acquired significance, cease to be *exclusively* physiological facts.[22] The flip side of this is the devaluation of all those arguments in which self-manifestation is interpreted as being *in opposition* to vital functions.

One might ask, with a degree of malice, whether Portmann's intended reform of biology does not in the end lead to a grandiose confusion: it is not just a critique of science but has itself a positive aim, namely a "new conception of the organism"; this, however, does not have an empirically and scientifically demonstrable existence, and therefore the desired new biology cannot be a positive science. Nor is it metaphysics, as its opponents charged, as it does not aim to transcend the realm of living nature, and its own statements are critically tied to the state of empirical science. Nonetheless, in the remaining pages I will attempt to demonstrate that this inner tension within Portmann's project, along with its problematic theoretical status, represents a possible key to a positive understanding of the "new morphology."

Cognition of Living Nature and a Typology of Stances Toward the World

Aside from theoretical-biological manifestos, Portmann wrote a series of texts dedicated to the analysis of human cognizance of nature and forms of relation to nature. These texts, despite their unsystematic nature, represent an indispensable basis for judging the scientific theoretical status of his investigation. In them, Portmann presents several dual oppositions whose components are related but not fully transferrable. Portmann never attempted to systemize them, and the following summary necessarily risks a certain schematization.

---

[21] See Portmann 1960a: 216, 1965a: 196.

[22] In Portmann's time this was also a classic theme of philosophical anthropology, where, with the aid of such models, the difference between culture (understood principally as the prerogative of humans) and nature was delineated. On this, see Halák and Klouda 2018. Portmann, who acknowledged culture in the strict sense as the domain of humans only, nevertheless referred to the moment of freedom (non-necessity) that self-appearing introduces into the realm of nature (Portmann 1958a: 33).

In his lecture "On the Uniqueness of Biological Research" (Portmann 1951; Über die Eigenart des biologischen Forschens), Portmann develops an analogy with the dual view of a theatrical performance. Basically, there are two views of a play (and the combination of these views): the view of the stage, observing the action of the play, and the view from backstage, taking in the hidden work of make-up artists and lighting technicians, the scenic design, sound effects, etc. Stagecraft can be examined in great detail (both technically and historically), yet its primary significance is in making possible the drama on the stage. We take a different stance toward it, and even as theater critics we cannot cease to be audience members interested in the action performed on-stage. Similarly, two fundamental views of nature are possible: analytical causal exploration (the backstage view) and the non-analytic, naïve spectator view of the stage of life. In the latter, we are not interested in actions because we know their physical or chemical causes, but because they have *meaning* for us. Portmann does not explicitly address the mutual relationship between the two perspectives and simply refers to the necessity of their cooperation, but he notes that exact research (the backstage view) can often produce findings that are physically or chemically correct but irrelevant to a knowledge of life (Portmann 1951: 84, 90). Thus the view of the stage (the exploration of meaning) is given a certain preference in the study of living beings.[23]

In another text, "How Man Examines Nature" (Portmann 1963c; Der naturforschende Mensch), Portmann distinguishes between the scientist and the naturalist. Natural scientists (*Naturwissenschaftler*) strive to break through sensory perceptions to the ultimate building blocks and general laws. Therefore they look at the normally imperceptible interior of living beings and try to master the processes that take place there. Their method is not naïve observation, but (more or less invasive) experimentation. Their interest reduces the variety of living forms to a relatively narrow selection of organisms exceptionally suitable for testing given hypotheses: laboratory animals. Modern science comprises knowledge-producing power, knowledge able to control an array of processes within living beings, but it can serve only as an "auxiliary means" in comprehending life as such (Portmann 1963c: 66).

An analysis of the opposite position, that of the naturalist (*Naturkundige*), brings a greater shift. Naturalists live in the "naïve and aesthetic world of primary, sensory life," working with unanalyzed sensory perceptions without methodically separating their emotive and imaginative connections. Portmann cumulatively labels this stance an "aesthetic function" in the wider sense of the word, which comes from a root meaning "perception" (*αἴσθησις*), though he does not exclude the more narrow sense of *aesthetics* as the theory of beauty. Naturalists are interested in the inexhaustible variety of perceptible shapes, colors, tones, and scents, and their subjective interest and passion for these things is part of their knowledge. Their truths belong to Pascal's *ordre du cœur*, as suggested by the characteristically passionate interest of observers and collectors, the tireless endeavors of classifiers and

[23] More on this see Wild, Chap. 8, in this volume.

systematists, and the sacrifice of explorers of the past and present ages. The deep significance that sensorily perceptible living forms have for naturalists connects them to artists: for them, a living being is not a mere exemplification of general laws and ideas; rather, it takes on a symbolic value, as the sensory fullness of perceived processes suggests to human awareness, through analogy and imagination, what life is and what, at the deepest level, it is about.[24]

In comparison with the theatrical metaphor, there is an apparent and greater degree of contextualization here. The dichotomy of scientist/naturalist clearly reflects changes in knowledge linked with the development of the West over the last few centuries. More important, however, is the connection of the naturalist with the "aesthetic stance" as a certain approach to the world in general – one which surpasses the realm of science as such. Portmann here demonstrates for the third time that a scientific stance alone is not sufficient for learning about life, that the naturalist (aesthetic) component plays an important role here, as it refers to prescientific and extra-scientific life. Thus, according to him, biology is the study of nature (*Naturforschung*), but he is hesitant to specifically label it natural science (*Naturwissenschaft*), though he never denies the essential importance of the scientific component.

Portmann's analysis of the primary and secondary worldviews, which he sometimes labelled the Ptolemaic and Copernican approaches, can be thought of as the third phase of his thought (Portmann 1959b, 1960c).[25] In these deliberations the focus of Portmann's interest shifts to the relation of the aesthetic and theoretic functions, changing in the course of the individual human life. Each of us is born into the world and grows up in a primary, Ptolemaic, aesthetic stance. We eventually lose this, to a greater or lesser degree, thanks to Copernican elements, particularly of modern Western civilization, but we never abandon it completely. It is the life work of every person to continually form and cultivate a harmony between both approaches, which together form the specifically human approach to the world (human "inwardness").

Portmann developed these reflections over a full decade (c. 1950–60) in which he also completed his most important studies and books dedicated to his morphological project. He never attempted to systematize them (though their connection is obvious), preferring rather to point out the limited applicability of these parables. More importantly, he formulated these reflections while writing books and theoretical studies on the new morphology that contain marginal references to this matter but do not explicitly mention it. If we were to summarize the results of Portmann's "methodological" investigations, we will see the complexity of learning about

[24] Cf. Portmann 1963c: 79: "Such an experience of the life of nature leads [...] to the most hidden experience of the mysterious ground at work in everything around us and in ourselves. It is precisely through the strangeness of natural things, through insight into ways of life that are alien to us, that the stimulating effect of all forms of life, which go beyond all understanding, is greatly strengthened and the soul is fortified to confront the abyss within us."

[25] For a detailed analysis of all aspects of Portmann's opposition of the Ptolemaic and Copernican worldviews, see Tolone, Chap. 10, in this volume.

nature. We always somehow understand life, somehow know what is happening in this or that act of the play on the stage. We understand the meaning (*Bedeutung*, *Sinn*) of these actions because we ourselves are living beings who autonomously relate to the world and integrate with it. In a certain way, we thus understand processes such as growth, reproduction, foraging, rivalry, and fleeing from danger; we understand other types of inwardness on the basis of our own inwardness. Causal natural science is able to look behind the scenes of these processes, but the identification of their meaning is left to the aesthetic stance. The latter, however, is not unproblematic – it does not consist simply in opening one's eyes and other senses. In the series of texts just discussed one finds a deepening of the "view of the stage" – through the naturalist, aesthetic, and primary stances. Unless we have this aspect of Portmann's thought in mind, it will be difficult to evaluate the import and scope of his reform of morphology, as its central concepts – the living form and its authentic phenomenon – are codetermined precisely by this (still problematic) aesthetic stance to the world.[26]

The Hermeneutic Character of Portmann's Thought

The fact that Portmann developed his reflections on cognizance of nature outside of his texts dedicated to the new morphology might arouse a conviction in his readers (and perhaps it did in himself) that his biology is essentially that of the standard empirical discipline, acquiring objective scientific facts.[27] If we agree that its goal is *not* the production of objective facts (and thus is not an empirical science in the usual sense), then all of the objections to it raised above lose their relevance. What remains is only the question of how to positively characterize its theoretical status.

We have repeatedly seen that causal physico-chemical science alone cannot comprehend life – its phenomena are not the authentic phenomena of living forms. Portmann does not deny the significance of modern natural science, nor does he believe that biology should revert to a sort of premodern *historia naturalis*. The basic precondition for the meaningful functioning of causal natural science, however, is the very perception of living beings, which biologists (and everyone else) are capable of as living beings, not solely or primarily as biologists. The fundamental requirement of biology (grasping nature) thus depends on human inwardness and its condition. Portmann's morphological project, if we respect the thematic continuity of all its aspects, goes beyond biology in the narrow sense in that it explicitly links concrete cognizance of life with the preconditions of this cognizance. Portmann

[26] See Portmann 1957: 370: "The distinction between authentic and inauthentic phenomenon in the first place separates two worlds of discovery and experience: the naïve aesthetic world of primary, sensory life and the theoretical world of human curiosity."

[27] Objectivity or factuality is understood here in a relative sense: (relatively) objective knowledge is that which we do not need to further problematize in a given context and can continue to utilize without need of belaboring its assumptions.

was well aware of this precondition of biological cognizance and did not exclude the very ground of cognizance – human inwardness, including its social dimension – from the subject of his (biological) research.

Portmann also demonstrates a continual awareness of the conditionality of knowledge in his texts on the new morphology. He presents the first publication of his project as an effort "to clearly glimpse the most common assumptions of current biological research" (Portmann 1963b: 15). Biology for him continues to remain a human product, and its concrete form is multiply conditioned and by no means obvious. Portmann does not consider this cultural, social, and historical (but also individual) conditionality as something irrelevant or counterproductive for his own scientific research: "all science, as a human product, is determined by pre-existing norms and worldviews. This fact is detrimental to real science only when these intellectual forces go undetected and proliferate, when the natural scientist does not pay attention to their origin and considers his own thinking to be completely objective but that of others to be biased" (Portmann 1970e: 115). Portmann's awareness of the productive role of the preconditions of scientific knowledge aptly figures in his interest in the history of biology, and his knowledge of previous research went considerably beyond routine familiarity.[28] The recurring idea that in each age there is a limit to what biology can discover (Portmann 1951: 76) attests how far Portmann was from naïve epistemological objectivism.

It is important to note, however, that Portmann did not understand the preconditions of biological knowledge in an isolated or static way. Particularly in his analyses of the naturalist and primary stances there is an evident attempt to demonstrate that cognizance of living forms is not only enabled by human inwardness, but that it, in turn, helps to cultivate inwardness and assists its harmonious formation, in which he sees an important role also for biology, on both an individual and societal level.[29] The circular structure or oscillation in which the preconditions of cognizance

[28] In Portmann's bibliography we also find a short study on the history of entomology (Portmann 1930), and his intellectual autobiography (Portmann 1974) is an essentially historical work. What most attracted his attention in this area was the development of evolutionary theory, particularly the selectionist paradigm. Aside from studies on the context of the genesis of Darwin's theory (Portmann 1965d, 1970f), he left behind several folders of notes and lecture materials related to the Darwinian centennial of 1959, which are held in the library archives at the University of Basel. M. Ritter presents in full an earlier fragment in which Portmann reflects on the aesthetic contexts that may have influenced the differential reception of evolutionary theory in Romance and Anglophone countries (Ritter 2000: 234). To this context also belong Portmann's considerations on the linking of biology with economic and social interests and the exercise of power in the modern state, and the influence of these on biological research (e.g. Portmann 1970g).

[29] See Portmann 1963c: 82–83: "As technical complexity and the need for extensive specialization continue to increase, a large number of extremely specialized natural scientists will inevitably emerge. As a counterbalance to this development, it is important to support the other method – that of the naturalist, as we have called it; this task deserves the attention of all who participate in intellectual life. The main goal to be achieved, however, is to shape the natural scientist into a harmonious awareness of his full role: to increase our knowledge of natural things and to convey comprehensive, rich images of larger areas of nature." There is a direct path from here to Portmann's ideas on reforming the education system and culture in general, which mainly concerned a renewal

are deliberately acknowledged and discussed differentiates Portmann's project from Kant's philosophical foundation of life science, which it is sometimes compared with (Kugler 1967: 13–15). According to Kant, we are able to consider existence as organic (living) only because as acting beings we have a concept of purposeful causality. This final causality, however, only has a regulative validity – it does not extend to the production of scientific knowledge with constitutive validity. In the same way, the human capability for purposive action, thanks to its function as a condition of biological cognition does not fundamentally change. Portmann, by contrast, assumes that cognizance of living nature fundamentally influences human inwardness, both in cognitive abilities, which in the cognizance of living nature are not limited to the category of mathematical natural science, and in relationship to itself.

As we know, the emphasis on the irrevocable presuppositionality of all knowledge and its circular structure was systematically developed especially in the twentieth century in hermeneutic philosophy.[30] Among its convictions is the idea that philosophy and empirical disciplines are not separated by an ontologically-based disjunction (e.g. empirical vs. transcendental), and it understands their relationship rather as two poles of a single continuum (Boehm 1978a: 9–10).

Both philosophy and science are products of human activity, and the difference between them is not exclusively rooted in the subject of knowledge, but proceeds from the attitude and always multifaceted praxis of the knowing subject: the difference between a scientific "empirical fact" and a "philosophical idea" lies not in things but in *interpretation* – in the behavior of the knower, which always implies some evaluation of a given situation. By way of example, we need only to consider the fundamentally philosophical relevance of great creative scientists or, by contrast, the frequent inclination of philosophy toward factual ("empirical") conceptions of its traditional themes. The perspective of a hermeneutic relativization of the relation of science and philosophy offers a possibility of understanding Portmann's morphology in a more positive way: not as bad, inherently contradictory science, but as a *hermeneutic dimension* of biological cognizance.

It thus programmatically always lies partly outside of (natural) science so as to regulate its research on living organisms and to reflect its constitutive ties to the prescientific world. Constant regard of its own assumptions does not orient it only to the past, before science, but conserves options for the further development of science, frees it from rigid forms of thought, and maintains lines of contact between biology and other areas of learning. This, in the end, is what is exemplified in Portmann's reassessment of morphology, for which he saw a transformation from the propaedeutic role that scientific discoveries in the twentieth century were

---

of "aesthetic education" through the cultivation of sensory experience. To him, therefore, reform of the Western system of education is directly linked to the reform of biology. On this, see Portmann 1963d. Here we also see a reason behind Portmann's unusual communication strategy: biology is in fact "a public matter."

[30] See, for example, Bollnow 1992.

pressuring it into, to a top-level synthesizing field and the heuristic vanguard of biology.[31]

The Living Form as a "Critique of Reason"

Following this sketch of the hermeneutic perspective, let us now look again at the epistemological problems related to the authentic phenomenon. We have already noted the shift that characterized Portmann's typology of attitudes toward the world; whereas in the theatrical metaphor, the view of the stage and the backstage view are easily distinguished, it is more difficult to separate the aesthetic and theoretic, the primary and secondary views of the world.

Portmann cautions that the primary aesthetic approach is not alogical, but functions "from the very beginning as a unit with the rational function of our mind and the creative power of our imagination" (Portmann 1960c: 26). The goal here is not to separate both approaches into static contrary alternatives but to create the greatest possible harmony between them. The primary aesthetic stance thus does not consist merely in setting aside scientific instruments and opening the eyes and other senses. Aesthetic perception is a task to be continually addressed, a "beautiful occupation" that we are continually learning (Portmann 1963c: 79). If the authentic phenomenon of a living form is related by definition to an aesthetic stance, it is not something that could appear automatically and without effort on our part. In a later study dedicated to the question of appearing, Portmann defines the authentic phenomenon as "that part of what is accessible to us in a living being that presents itself directly to the perceptive sense, speaking directly to the senses without any cunning of reason" (Portmann 1970d: 68). This obviously does not mean merely the absence of external equipment and methods of observation – the "tricks of reason" can be even the concepts and rational constructions schematizing our sensory experience. These are not easily suppressed in humans, as beings of reason and language, but, as we have just seen, this is not the goal. The primary aesthetic stance, and the related perception of the authentic phenomena of living forms, is not a realized alternative that one can simply opt for, but a process, an "event" (Portmann 1953: 224, *Geschehen der Selbstdarstellung*).

That this process must occur also within our cognition is a consequence that Portmann did not fully work through. Perhaps he was hampered by a conviction that the aesthetic stance is basically "naïve," which can suggest an image of carefree ingenuousness. If, however, the aesthetic stance is a task and emerges as the result of certain activity, then the desired artless naïveté is not an initial state but the objective of an exertion which reason itself takes part in, just as the construction of the new morphology was a rational enterprise. Though he clearly stated that "in our

[31] See, for example, Portmann 1965b: 257: "So, too, morphology is not some initial stage of further biological work [...] but one of the abiding, *ever renewing* tasks of life research" (italics mine).

human world, natural forms have symbolic value; they have sense and meaning, which, for us, makes the foreign and elusive visual self-presentation a parable for the content of our experience" (Portmann 1959b: 159), he did not apply these views in the theoretical contexts of his texts on morphology. And where he did engage more deeply with the epistemological question, he ended up confessing that "differentiation of the authentic and inauthentic phenomenon is the first big problem that research on self-manifestation must deal with (Portmann 1970d: 74).[32] If we proceed on the assumption that Portmann's thought progressed further in his texts on attitudes to the world than in his manifestos on the new morphology, where he maintained the voice of the empirical biologist, then the thesis of the symbolic value of self-manifestation (authentic phenomena) gives us a clue as to which perspective to take on these epistemological problems.

We know that life can be grasped only by life, inwardness only by inwardness. In the case of humans, which interests us most here, as we know biology only as a human epistemological project, we can distinguish two components of how this grasping takes place. The main role is played by the senses, by which other animal or plant forms are perceived. The second component is interpretation, which we know from Portmann's analysis as the determination of the meaning or significance of what life is about, what its main functions are. This type of interpretation, of course, is already present on the prescientific level of human existence (thus concepts and tropes like "energy imbalance," "preservation of species," "adaptation to the environment" and "selfish gene" are intuitively accessible and belong, in a way, to folk opinions. Nevertheless, the border between these two components is not clearly defined: the aesthetic stance does not exclude perception of its own emotional and imaginative context – it does not deal in abstract sensory data. On the contrary, it has its own analogous and imaginative connections; unanalyzed sensory perception is always figurative – in modern terms, it perceives "with an excess of meaning." The rational interpretive component analytically stabilizes these meanings and renders them unambiguous.[33] At the same time, these interpretations, thanks to intersubjectively and historically transmitted speech, have their own effect on sensory experience, which tends to see above all that which we already know.

The emphasis of naturalists on aesthetically experiencing the sensory fullness of living forms can critically regulate and relativize the conceptual categories through which we understand them – and, since we understand precisely because we are also a living form, it can also develop our own inwardness, our lives. In this lies the priority of sensory perception, as through it we encounter life in greater measure

[32] For a similar statement, cf. Portmann 1960a: 112. Both texts compare the difficulty of distinguishing the authentic and inauthentic phenomena to the problem of differentiating the conscious and unconscious components of the human psyche. It is worth noting the frankness with which Portmann was able to put at stake the empirical nature of the basis of his biology.

[33] Viewing the predominance of the intellectual dimension in Western civilization, Portmann often takes on the role of an advocate for sensory experience, though he emphasizes that he is concerned with "inspired" sensuality (Portmann 1963d: 247). Elsewhere he repeatedly points out that the aesthetic stance does not amount to irrational ravings (Portmann 1963c: 82, 1965a: 241).

than we ourselves realize.[34] Thus an aesthetic awareness of living forms continually modifies our rationality, including concepts and interpretive models that help to create the "trick of reason" which permanently threatens to outwit its own bearer. Aesthetic perception, by contrast, can be focused and intensified through awareness and dedication of will. The self-manifestation of living forms thus acts as a regulator and critique of our own cognizance, which it keeps functional, itself dynamic and alive.

The appearing of a living form thus links not only science with extra-scientific life, but also two philosophically traditional components of human cognition – rationality and sensuality. It forms a living, pulsating boundary inside cognizance, an epistemological land belonging to no one, where there is no difference between active thoughts and passive perceived facts, where thought is figurative and the senses have their own speech and reason.[35] Portmann evidently realized that self-manifestation takes place substantially in our cognition,[36] but he did not manage to utilize or meaningfully incorporate this view in the context of his planned reform of morphology. Apparently, it was the idea of cognition as something merely subjective that diverted him from this path – this despite the fact that he himself had convincingly demonstrated that the cognizance of a living form occurs as the meeting of the recognizer and the recognized, that in describing living forms we necessarily say something about ourselves also. Nor, in the end, did he utilize the idea that in its critical and regulative dimension, sensory cognition works toward a de-subjectification of the knower.[37] Instead, by circumventing the philosophical aspects of his morphology, Portmann mainly gives the impression that it is an empirical discipline whose central category is unproblematically objective in nature.[38] His introduction of the quasi-objective medium of light as a dimension of unaddressed phenomena points in the same questionable direction. Certainly, the appearing of a living form consists in movements that come from the inwardness of a given living being (whether it be the movement of ontogenesis, physiology, or behavior). Yet with the advent of non-transparent bodily boundaries, which serve as expressive optical surfaces, the act of self-manifestation is potentially directed outward, to

[34] In nearly all of Portmann's texts we find an emphasis on "mystery" and the reality that organisms are and always remain more than what the biological research of a certain era can say about them. See, for example, Portmann 1965b: 259, 1970e: 114.

[35] Cf. Boehm 1978b. In this context, mention should be made of Portmann's creative talents and artistic study, which prompted an extended decision on whether to follow the path of the painter or the biologist. See Illies 1981: 33–54 for examples of his drawings and sketches.

[36] Cf. Portmann 1953: 226: "Living forms are original texts which speak to us in a foreign, distant language. Every generation must continually attempt to translate it into the expressions of their time; it is an undertaking which never ends, and which every new discovery must serve to improve."

[37] A similar function is emphasized by B. Prévost (2013), who interprets the unaddressed appearance as "pre-subjective interiority."

[38] Z. Neubauer also saw the presumed objectivity of Portmannian categories as a significant problem of his project (Neubauer 2008: 80–81). G. Kühne-Bertram discusses the category of *Selbstdarstellung* as a typical example of a "hermeneutic concept" in the framework of the Continental philosophical tradition, to which she assigns Portmann (Kühne-Bertram 2015: 74–79).

another. And in this other, in "recognition," this movement then finds its realization, similarly as the recognition of a cognizant being is reliant on an appearing form, which draws the recognizer out of himself and realizes in this life activity (i.e. cognition), *his* inwardness. In a case of vision without light, neither of these would work, but this does not mean that the unaddressed component of the phenomenon relates directly to light: the perceiver experiences this component as that mysterious and unclassifiable surplus of meaning. In the end, introduction of the category of unaddressed authentic phenomenon is epistemologically paradoxical only if we understand a phenomenon as something isolated, standing outside these relationships, or if we understand our cognizance as a simple and static concordance of the recognizer and the recognized. If, however, each perception is an event involving and exceeding the perceiver, then every addressed phenomenon is at the same time always primarily unaddressed.

The fact that Portmann did not utilize in his original biological model the hermeneutic concept of knowledge, which had at the time already been formulated and which he presumably was familiar with,[39] is not the greatest deficiency to have impeded the reception of his work. A greater pitfall was his deliberate retreat from the philosophical questions that his work placed in his path – by both separating his methodological-epistemological research from his texts on the new morphology and attempting to refer these "serious epistemological problems" to philosophers, as if they were no longer relevant to his "empirical biology." Perhaps his circumvention of the philosophical aspects of the morphological project, in a time before the establishment of borderline disciplines like philosophy and didactics of biology, had socioprofessional motives, as his original theory must have appeared (also to Portmann himself) more as a challenge to mainstream academic science – one which would have meant an absolute overturning of institutional scientific structures. And even though the "new morphology" was not acknowledged by contemporary biology, it remains relevant thanks to a reality that Portmann somewhat underestimated: that great scientists are also covert philosophers.

References

Boehm, G. (1978a). Einleitung. In H.-G. Gadamer & G. Boehm (Eds.), *Seminar: Die Hermeneutik und die Wissenschaften* (pp. 7–60). Frankfurt am Main: Suhrkamp.

Boehm, G. (1978b). Zur einer Hermeneutik des Bildes. In H.-G. Gadamer & G. Boehm (Eds.), *Seminar: Die Hermeneutik und die Wissenschaften* (pp. 444–471). Frankfurt am Main: Suhrkamp.

Bollnow, O. (1992). Hermeneutische Philosophie. In K. Salamun (Ed.), *Was ist Philosophie? Neuere Texte zu ihrem Selbstverständnis* (pp. 92–107). Tübingen: J. C. B. Mohr (Paul Siebeck).

Halák, J., & Klouda, J. (2018). The Institution of Life in Gehlen and Merleau-Ponty: Searching for the common ground for the anthropological difference. *Human Studies, 41*(3), 371–394.

[39] His writings contain citations and references to works of philosophers such as M. Heidegger, W. Szilasi, O. Bollnow, and H. Plessner.

Illies, J. (1981). *Adolf Portmann. Ein Biologe vor dem Geheimnis des Lebendigen*. Herder: Freiburg/Basel/Wien.
Jaroš, F. (2010). Darwinismus a portmanismus: střetnutí nesouměřitelných biologických paradigmat? *Teorie vědy/Theory of science, XXXII(/3)*, 301–318.
Koechlin, F. (2004, June 10). Don Quijote der Laboratorien. *Die Wochenzeitung* (Basel), 1–5.
Kugler, R. (1967). *Philosophische Aspekte der Biologie Adolf Portmanns*. Zürich: EVZ-Verlag.
Kühne-Bertram, G. (2015). *Konzeptionen einer lebenshermeneutischen Theorie des Wissens*. Würzburg: Königshausen & Neumann.
Lorenz, K., & Leyhausen, P. (1973). *Antriebe tierischen und menschlichen Verhaltens*. München: Piper.
Neubauer, Z. (2008). Esse obiectivum – esse intentionale. Auf dem Wege zur phänomenologischen Biologie. In D. Pleštil & W. Schad (Eds.), *Naturwissenschaft heute im Ansatz Goethes* (pp. 70–88). Stuttgart/Berlin: Johannes M. Mayer.
Portmann, A. (1930). Zwei Naturforscher. Jan Swammerdam und François Huber. *Die Ernte, 11*, 97–113.
Portmann, A. (1948). *Einführung in die vergleichende Morphologie der Wirbeltiere*. Basel: Schwabe.
Portmann, A. (1951). Über die Eigenart des biologischen Forschens. In E. Grassi & T. v. Uexküll (Eds.), *Die Einheit unseres Wirklichkeitsbildes und die Grenzen der Einzelwissenschaften* (pp. 69–93). A. Francke Verlag: Bern.
Portmann, A. (1953). Um ein neues Bild vom Organismus. In K. Piper (Ed.), *Offener Horizont* (pp. 213–226). München: Piper.
Portmann, A. (1954). Biologie auf neuen Wegen. In J. Moras & H. Paeschke (Eds.), *Deutscher Geist zwischen gestern und morgen* (pp. 172–188). Stuttgart: Deutsche Verlagsanstalt.
Portmann, A. (1957). Transparente und opake Gestaltung. In *Rencontre/Encounter/Begegnung. Contributions à une psychologie humaine, dédiées au professeur F. J. J. Buytendijk* (pp. 355–370). Utrecht/Antwerpen: Het Spectrum.
Portmann, A. (1958a). Unterwegs zu einem neuen Bild vom Organismus. In *Die Welt in neuer Sicht* (pp. 24–46). München/Planegg: Otto Wilhelm Bart-Verlag.
Portmann, A. (1958b). Selbstdarstellung als Motiv der Lebendigen Formbildung. In *Geist und Werk. Aus der Werkstatt unserer Autoren zum 75. Geburtstag von Dr. Daniel Brody* (pp. 139–173). Zürich: Rhein-Verlag.
Portmann, A. (1959a). Zur Philosophie des Lebendigen. In F. Heinemann (Ed.), *Die Philosophie im XX. Jahrhundert. Eine enzyklopädische Darstellung ihrer Geschichte, Disziplinen und Aufgaben* (pp. 410–440). Stuttgart: Klett.
Portmann, A. (1959b). Die Welt der Augen. *Die Ernte, 40*, 131–159.
Portmann, A. (1960a). *Neue Wege der Biologie*. München: Piper.
Portmann, A. (1960b). Wandlungen unseres Bildes vom Lebendigen. In *Wege zur neuen Wirklichkeit* (pp. 53–74). Bern: Ilallwag.
Portmann, A. (1960c). Ptolemäer und Kopernikaner. Eine biologische Studie. *Der Monat, 12*, 23–30.
Portmann, A. (1963a). *Biologie und Geist*. Herder: Freiburg/Basel/Wien.
Portmann, A. (1963b). Die Biologie und das Phänomen des Geistigen. In A. Portmann (1963a), 13–45.
Portmann, A. (1963c). Der naturforschende Mensch. In A. Portmann (1963a), 65–83.
Portmann, A. (1963d). Biologisches zur ästhetischen Erziehung. In A. Portmann (1963a), 247–265.
Portmann, A. (1964). *Tier als soziales Wesen*. Herder: Freiburg/Basel/Wien.
Portmann, A. (1965a). *Die Tiergestalt*. Herder: Freiburg/Basel/Wien.
Portmann, A. (1965b). Die neue Auffassung vom Organismus. In H. W. Bähr (Ed.), *Die Naturwissenschaft heute* (pp. 255–259). Gütersloh: C. Bertelsmann Verlag.
Portmann, A. (1965c). Erleuchtung und Erscheinung im Lebendigen. In A. Portmann (Ed.), *Aufbruch der Lebensforschung* (pp. 33–57). Zürich: Rhein-Verlag.

Portmann, A. (1965d). Die Idee der Evolution als Schicksal von Charles Darwin. *Eranos-Jahrbuch, 33*, 365–403.
Portmann, A. (1970a). *Entläßt die Natur den Menschen? Gesammelte Aufsätze zur Biologie und Anthropologie*. Zürich: Ex Libris.
Portmann A. (1970b). Neue Fronten der biologischen Arbeit. In A. Portmann (1970a), 13–28.
Portmann, A. (1970c). Die Erscheinung der lebendigen Gestalten im Lichtfelde. In A. Portmann (1970a), 40–57.
Portmann, A. (1970d). Die Beurteilung der Erscheinung im Organischen. In A. Portmann (1970a), 58–75.
Portmann, A. (1970e). Das Problem des Lebendigen. In A. Portmann (1970a), 88–119.
Portmann, A. (1970f). Die entscheidende Wendung im Evolutionsdenken der Biologie. In A. Portmann (1970a), 168–185.
Portmann, A. (1970g). Die Biologie als technische Weltmacht. In A. Portmann (1970a), 364–379.
Portmann, A. (1974). *An den Grenzen des Wissens. Vom Beitrag der Biologie zu einem neuen Weltbild.* Wien/Düsseldorf: Econ.
Prévost, B. (2013). Interiorita, exteriorita, výraz podľa Adolfa Portmanna. *Filozofia*, *68*, 82–90 [translated by A. Záthurecký from French original Intériotité, extériorité, expression. A partir d'Adolf Portmann].
Ritter, M. (2000). Die Biologie Adolf Portmanns in zeitgeschichtlichem Kontext. *Basler Zeitschrift für Geschichte und Altertumskunde, 100*, 207–254.

Jiří Klouda is an assistant professor at the First Faculty of Medicine, Charles University in Prague. He is also working on a research project at the Department of Philosophy and Social Sciences, University Hradec Králové. His research focuses on Continental philosophy, philosophical anthropology, philosophy of life sciences and topics concerning ancient Greek science, philosophy and literature. Among his most recent publications are "Wilhelm Dilthey: Lived Experience and the Symbolical Productivity of the Body" (In J. Parry – P. Allison, eds., Experiential Learning and Outdoor Education. Traditions of Practice and Philosophical Perspectives, London – New York: Routledge 2020); "Institution of Life in Gehlen and Merleau-Ponty" (In *Human Studies*, 41, 2018, with J. Halák).

Part IV
Historical Context and Later Reception

Chapter 12
The Beauty of Organisms: Biological Aesthetics Between Darwin and Portmann

Karel Stibral

Abstract This chapter will deal with one of Adolf Portmann's most characteristic topics – the aesthetic dimension of nature. Throughout his work, Portmann constantly draws attention to our inability to explain the richness and beauty of living forms solely through their functionality, usefulness, and fitness, as neo-Darwinism does. However, the neo-Darwinian interpretation of aesthetic phenomena in nature is very different from Darwin's own thinking about the realm of organismal beauty. This chapter analyses Portmann's arguments in this area as well as some parallels to Darwin's while also analysing some misunderstandings, some of which were rooted in contemporary science. In some respects, Darwin was even more radical than Portmann, compared to neo-Darwinism. This chapter will show that, in the case of aesthetic phenomena in nature, there is a trilateral discussion between Portmann, (neo-) Darwinism and Darwin rather than a Portmann-Darwin opposition.

Introduction

The appearance of living organisms is definitely one of the most important topics in Portmann's work. Appearance, which is often one of the manifestations of living matter, of its abundance and beauty, was emphasized in most of Portmann's works. It is significant that, for example, in Czech Republic Portmann was usually characterized as a representative of an aesthetic approach (Neubauer 2002: 172–211), of a biological or morphological aesthetics, as an opponent to reductionist neo-Darwinism, which was solely focused on utility without any sense for the beautiful abundance of nature.

It is clear that Portmann denies that a living being's appearance is limited only to its function, as he repeatedly emphasizes its aesthetic effect. And in his work, we feel a constant need to interpret appearance from this point of view. Not surprisingly, Portmann defines himself in comparison to the Darwinian scientific mainstream.

K. Stibral (✉)
Masaryk University, Brno, Czech Republic
e-mail: karel.stibral@seznam.cz

F. Jaroš, J. Klouda (eds.), *Adolf Portmann*, Biosemiotics 23,
https://doi.org/10.1007/978-3-030-67810-4_12

He is absolutely right in this, though he does not always differentiate between Darwin and his followers. So, I would like to draw attention to perhaps the most fundamental difference between Darwin's views and what we understand as Darwinism and neo-Darwinism. Although we can now find a number of publications on the subject of aesthetic phenomena in nature from the perspective of contemporary (neo-) Darwinism (e.g. Dissayanake 1992; Dutton 2009; Rothenberg 2012; Davies 2012), with the exception of a few studies[1], only a little attention has been paid to Darwin himself in this field. This is unfortunate because Darwin himself was rather complicated in his interpretation of aesthetic phenomena in nature. In this he could have even been considered a "dissident" within the paradigm that he himself established.

Therefore, in this chapter I would like to offer a short comparison of both great naturalists. Perhaps this comparison can foster a better understanding of Darwin and how he differed from his followers, as well as the specifics of Portmann himself.

Although at first sight Portmann appears to be an opponent of both Darwinism and Darwin himself, upon closer inspection we might question whether it was more a misunderstanding than a disagreement. In some ways, as we shall see, the opposition is very real. At the same time, there are topics that Portmann originally dealt with, but his opinions are questionable both from the point of view of the already established classical ethology and from our current knowledge. It is even possible that Darwin's original approach might be more radical than Portmann's, when compared to (neo-) Darwinian biology. But it might also be worthwhile to consider whether a synthetic view on aesthetic phenomena would be possible, as the Czech biologist Karel Kleisner mentions in his essay, "Vote Portwin!" (Kleisner 2006).

To begin with, just a few methodological notes: this chapter will be a comparison of ideas, and I would rather focus less on the field of contemporary biology, even if I make an occasional exception.

We should be aware of the problems of terminology amongst frequently used terms such as *aesthetics*, *aesthetic*, *beauty*, *taste*, etc., which are sometimes mentioned. Of course, in this chapter, I will maintain a broader understanding of the concept of aesthetics, but I will primarily take into account the more recent narrower understanding of aesthetics rather than the broader and older one, rooted in the ancient Greek word *αἰσθάνομαι* (*aisthánomai*, "to perceive by senses").

Of course, this comparison of both naturalists has a number of pitfalls rooted in the different periods in which they lived, and therefore logically there is a difference between their cultural context, education and terminology. Portmann seems to be more educated in the humanities, in addition to his own experience of drawing (see e.g. Ilies 1976). He uses many examples from the history of ideas and of art, for

[1]The majority of works are focused on Darwinian or evolutionary interpretation of art and other aesthetic phenomena or some evolutionary topics of art (Larson and Brauer 2009). Only a few deal with the interpretation of Darwin's own aesthetic ideas, as in Wolfgang Welsch's article (2004) and some chapters of the book *Endless Forms* (Donald and Munro 2009) or *Darwin and Theories of Aesthetics and Cultural History* (Larson and Flach 2013), and of course, Cronin's book, from a historical point of view (Cronin 1994).

example from the works of Heinrich Wölfflin, Goethe, and from the problems of historical and contemporary art.

But Darwin was also not uneducated; after all, he had studied theology. If we look at his early *Notebooks* and *Letters* (Darwin 1903, 1985, 1987)[2], we find that he had a surprising range of cultural interests at a younger age, although he was not artistically talented (in fact, he was one of the few scientists of that time who could not draw anything), he listened to music, visited art galleries, read the works of Romantic writers, and even attended unspecified lectures on aesthetics at Cambridge (to John Morley, 24 March 1871, Darwin 1903: I, 324–327).

Especially in the period between returning from the Beagle voyage and his marriage, Darwin read several interesting books on aesthetics, most notably Burke's *Philosophical Enquiry into the Origin of Our Ideas of the Sublime and Beautiful* but also Lessing's *Laocoon*, the works of Archibald Alison, Dugald Stewart, Hermann Helmholtz, Joshua Reynolds, Herbert Mayo, Alexander von Humboldt, Wordsworth's poetry, etc., an altogether interesting combination of texts of empirically based British aesthetics of the second half of the eighteenth century, Romantic works and contemporary physiologically oriented theories (see especially Darwin's notebooks marked as M, N, OUN, written between 1838 and 1840, Darwin 1987: 517–629).

So, it is evident that Portmann and Darwin use very different terminology. From the point of view of philosophical aesthetics, both terminologies are too broad and sometimes vague, mostly in Portmann's works. But thanks to their experience and knowledge, both Portmann and Darwin are quite familiar with what they are talking about and their approaches are part of a wider, more thoughtful context in comparison with many contemporary scientists.

The Function and Appearance of Living Beings

Among the topics associated with the aesthetic dimension in the work of both naturalists, I would point out the problem of the usefulness and appearance of living beings. This theme is crucial for Portmann, who literally focuses on it throughout his entire work.[3] Because it will be mentioned in another chapter of this book, I would just re-emphasize that he sometimes associates this topic with the field of aesthetics. Maybe of the greatest significance is when he analyses our current thinking in the article "The Biological for Aesthetic Education" (Biologisches zur ästhetischen Erziehung), published in *Biology and Spirit* (Biologie und Geist; Portmann 1968: 309–333). Our thinking, despite Portmann, is mainly concerned with the

[2] The books are quoted, but Darwin's letters and notebooks are quite accessible online these days, see Wyhe 2002, or Darwin 2018a, b.

[3] See also the titles of some his works: "The Beauty of Butterflies" (*Falterschönheit*, Portmann 1935, English edition 1945), *About the Wonder of Birdlife* (*Vom Wunder des Vogellebens*, Portmann 1984).

functional and utilitarian structures and explanations arising from the overwhelmingly dominant theoretical function (of our mind) over the aesthetic function. Unfortunately, the "technical spirit", or "theoretical function" of our mind interprets only a very selective view of reality as meaningful:

> Because of such pervasive effects we too easily forget that many representatives of the same animal types do not possess such strikingly functional forms, yet they still exist.... When trying to understand life forms, we must step away from the least conspicuous forms and realize that most often we have only noticed extremes. (Portmann 1960: 84)[4]

Throughout his entire work we encounter a rejection of the reductive functional point of view and a focus on the richness of forms, the abundant traits in the appearance of a living being that lack any direct purpose. For Portmann, this is a crucial moment during a living being's formation. As he summarizes:

> [...] When we look at the diversity of appearance in a group of related animals, we realize that in addition to structures serving the needs of preservation, there are completely opposing possibilities of free appearance, whose meaning does not end with only survival. Looking at these forms, we find that the appearance of living beings can only partially be interpreted through the notion of a function that applies to the pattern of 'preservation'. If we were to deal with technically incomprehensible forms, we would not meet only a form, an appearance, whose technical form we could not grasp 'functionally' – which happens naturally – but also with an infinite number of phenomena that resist this understanding. (Portmann 1960: 84–85)

Portmann does not reject the functional explanation at all, of course, but he clearly rejects the classical neo-Darwinian approach of finding a functional explanation in all cases or phenomena. A lot of these phenomena could be, in the vocabulary of today's neo-Darwinian biology, described as a neutral mutation, but only theoretically in most cases. When consulting scientific journals, we see an omnipresent search for a functional explanation – each form and behaviour has to be explained as an adaptation.

In this, Portmann could be compared to the French writer and entomologist Roger Caillois, who also emphasized the inability to explain everything by function. Caillois points out that reduction of the objectives of nature based only on suitability, is in fact an unacceptable form of anthropocentrism. Caillois appeals to the possibility of aesthetic laws of nature, laws of spectacular abundance, whose goals we however do not know – which is why he can go on to compare, for instance, the wings of a butterfly to artists' creations (Caillois 1962). But Caillois speaks directly of the uncompromising aesthetic principle; for him, aesthetics is one of the principles ruling everything living, and is just as uncompromising as the law of the function or preservation of species.

In this respect, Portmann is perhaps more cautious, or rather, the aesthetic dimension of living beings is part of the wider *self-presentation* efforts of living beings

---

[4]I use my own translation, the English translation from the Harper & Row publishing house (Portmann 1964a) is unusable – it is full of inaccuracies and in many terms shifts the original meaning of Portmann's words (e.g. towards mainstream biology).

(*Selbstdarstellung*). These tend to *hypertely*[5], creating structures exaggerated beyond the necessity of a functional goal (Portmann 1960: 214). Portmann even goes so far in some places to assert that this *self-presentation* prevails over functionality – or that it is the actual sense of living beings. So, it is not just some neutral mutation or Gould and Lewontin's *spandrel* (Gould and Lewontin 1979), where within the rigid determinism and function there remains a small and marginal space for free creativity:

> It is […] increasingly obvious that the well-known forms, preferred by our technical way of thinking, must be considered as second-rate – that our attention has been imposed only by the easy comprehensibility of their appearance. We learn gradually that the unspeakable richness of the organisms belongs to another circuit, where we must not ask firstly about the function, technical suitability, about the role in the drama of survival. Here it is necessary to finally see and recognize pure forms of self-expression. (Portmann 1960: 147)

Portmann's explanation of the non-deliberate conception of a living being's appearance within the division of phenomena (addressed and unaddressed phenomena) is also crucial (e.g. Portmann 1960: 109–111, 147–149, 215).[6] Here, from the point of view of the aesthetic dimension, extra attention should be given to unaddressed phenomena, often complex and "aesthetic" structures, such as those of Radiolaria or the algae Desmidiales, whose forms are not determined by the perceiver's sensory organs. Of course, from the point of view of biological aesthetics, Portmann did not just answer the question of whether nature itself tends to create aesthetic, beautiful structures. And here we should mention not only Caillois, but more importantly, Ernst Haeckel, who argued that organisms are endowed with an "artistic" direction in their own natural "ornamental potential" (*ornamentale Potenz*) originating from the "feeling of symmetry" (*Symmetriegefühl*) (Haeckel 1899; Haeckel and Breitenbach 1913). We should also mention other notable figures of German biology, for example, Karl Möbius (1908; see also Kockerbeck 1997), or the founder of the morphology of plants in Bohemia, Josef Velenovský (1910: III, 945, see also 1921, 1930). At the turn of the twentieth century, when these scholars published their works, the aesthetic principle was accepted by them as one of the evolutionary principles.

Interestingly, despite all his enthusiasm for natural beauty, Portmann never declared a similar hypothesis – he supposedly preferred to think that an aesthetic property arises only in the perception of a living being, and that only humans can indulge in the appreciation of beauty. We cannot see any such belief in Darwin either; rather, he sees the origin of "aesthetic" structures through another mechanism, as we will see – and he emphasized the general ability of higher beings to perceive and appreciate them.

---

[5] From ancient Greek *hyper* (ὑπέρ, which means *over* or *above)* and *telos* (τέλος, *purpose*, *goal*, or even *end*) This term was mainly introduced in Swiss entomologist Karl Brunner von Wattenwyl's *Über Hypertelie in der Natur*, 1873. Portmann also uses this term and translated it into German as "übers Ziel hinaus".

[6] See also expositions of the topic in this volume by M. Wild, P. Conte and J. Klouda, Chaps. 8, 9, and 10.

We should also focus on how Darwin perceived the problem of functionality in relation to aesthetic phenomena. After all, Portmann gives the hypertrophied functional approach – and rightly so – in connection with neo-Darwinism or Darwinism. In this context, today's biologists mostly cite Darwin as the one who combines standard beauty with functionality, mostly in the sense that beauty is functional for males as regards sexual selection. A "beautiful" appearance is undoubtedly very useful for males and often interpreted in contemporary biology as a sign of fitness, adaptation, good genes, health, good ontogenesis, and so on. Beauty refers, to put it simply (as Xenophon's Socrates already suggested), to purposefulness, functionality – there is no beauty without purpose; beauty, function, and usefulness are convertible.

The connection of beauty and functionality in Darwin's work was questioned only by his contemporary critics, such as John Ruskin (e.g. Ruskin 1872: 178–179, 1888: 84; see also Prodger 2009). Even current critics from the humanities, such as Diana Donald (2009: 24), continue to attribute this approach to him.

In most examples, and at first glance, Darwin says that beauty really is connected with functionality: he often mentions the beauty of males and its function in sexual selection. But firstly – it is worth noting, as Wolfgang Welsch (2004) or later but more extensively Stanislav Komárek (2003, 2016: 87–135) pointed out, that Darwin is often sceptical towards the protective, warning or communicative functions of those vibrant colours in animals. He often considers that they evolved without this purpose:

> We can, in our ignorance of most of the lowest animals, only say that their bright tints result either from the chemical nature or the minute structure of their tissues, independently of any benefit thus derived. (Darwin 1874: 323)

These are not necessarily observed only in invertebrates; at the same time Darwin gives an example of beautiful red arterial blood, which, however, serves no purpose (ibid.). The alluring blush of a maiden's cheek is not only a side effect of their colouration, but also an effect of other causes; haemoglobin certainly has a different task than to attract or interest men.

The danger of explaining a "beautiful" appearance in the terms of functionality applies also to flowers, when, for example, Darwin writes in a letter to botanist Thomas Meehan:

> I am glad that you are attending to colours of diœcious flowers, but it is well to remember that their colours may be as unimportant to them as those of a gall, or, indeed, as the colour of an amethyst or ruby is to these gems. (Darwin to Thomas Meehan 9 Oct 1874, Darwin 2018a)

Similarly, he wrote a letter to Wallace regarding the colouration of a tiger, and he argued against any cryptic interpretation of it (Darwin to A. R. Wallace 12 and 13 Oct 1867, Darwin 2018b). Secondly, if we take a closer look at Darwin's views on aesthetic issues, we will see that (at least at a younger age) he consistently considered aesthetic delight to be independent of utility, as he directly expresses in his *Voyage of the Beagle* (Darwin 1839: 481). I believe that Wallace and others' criticism of Darwin for formulating the theory of sexual selection, including those ideas of female taste, was related precisely to Darwin's abandonment of functional explanations, and thus to the blurring of the entire evolutionary theory based on the survival of the fittest and best adapted.

Indeed, for Darwin sexual selection was an evolutionary principle that often went against natural selection – of course, not in every case. For example, fights between males among mammals often lead to the selection of the strongest males. But the structures formed within the framework of sexual selection, mostly in birds, could be disadvantageous in terms of natural selection. Certainly, it is obvious that these structures are useful for their bearers because they will allow them to reproduce. However, it is often forgotten that Darwin faced the question of why females do not choose more visibly fit males – those proficient in terms of strength, speed, and so on. Those specimens who would win the male duel, who can best protect their females, who would contribute to offspring with similar strength, (i.e. explicitly big, muscular males with long legs and long teeth). He questioned why females choose the "beautiful" one, however they define that. When given the option, why do not females choose the overly muscular males, such as the gorilla silverback, whose forms are suitable in natural selection or in sexual selection based on "tooth and claw", instead of being drawn to the only skin-deep bejewelled beauties such as the superb lyrebird? Where did this *taste* emerge? This taste evidently conspired in creating the structures we humans call beautiful, ornamental, decorative (to express it in Darwin's terms), though they are not very well suited for survival (see Fig. 12.1).

Darwinists, though not Darwin himself, try to translate beauty into fitness, but are not able to explain why females do not like self-explanatory fitness and why they choose a much more complicated situation where not only male fitness traits, but also female preferences have to be inherited and both traits have to co-evolve.

Even if Darwin in his scientific work combines aesthetic issues with the theory of sexual selection and clear "utility" or functionality, I do not believe that his approach should be reduced to such a concept. The well-known German aesthetician and philosopher Wolfgang Welsch (2004) argues in his article about Darwin's aesthetics in the same way: the idea that the "sense of beauty" arises *in a context of utility* without yet being *per se a sense of utility* or *reducible to utility* lies at the heart of Darwin's account of animal aesthetics. Darwin put it frankly already in *The Origin of Species*: "The effects of sexual selection, when displayed in beauty to charm the females, can be called useful only in rather a forced sense" (Darwin 1861: 219).

Darwin's ideas about beauty being the driving force behind the frame of sexual selection was formed as part of his quest for the different paths of evolution. The main evolutionary mechanism, according to him, was natural selection, yet sexual selection was also important, even if it sometimes goes *against* natural selection. Welsch also writes that in Darwin's point of view, "sexual selection is a strategy complementary and not reducible to natural selection", unlike in contemporary evolutionary biology, which conceives of it as a sub-region of natural selection (Welsch 2004). Darwin also, like other biologists of that time, accepted the inheritance of

Fig. 12.1 John and Elisabeth Gould: Lyrebird (*Menura novaehollandiae*). (Wikipedia commons https://commons.wikimedia.org/wiki/File:Lyre_bird.jpg)

acquired properties, the use and non-use of organs, a law of compensation or balancing of growth formulated by Geoffroy St. Hilaire and Goethe (e.g. Darwin 1861: 164–166). So, if I were to make a certain simplification, Darwin was much more "postmodern" than the majority of twentieth-century biology which tries to translate everything into only one principle. In any case, it might be useful to realize that while Darwin certainly did not go as far as Portmann in emphasizing the danger of translating function and purposiveness, he certainly did not completely turn away from this problem in its relation to aesthetic phenomena in nature.

Sexual Selection and the Definition of Beauty

The fundamental problem or contradiction between our two naturalists begins with their differing notions on the topic of sexual selection. As we have seen previously, Darwin combines the feeling of beauty with sexual selection, and he openly argues for the ability to perceive this beauty among animals. Although, of course:

> Obviously no animal would be capable of admiring such scenes as the heavens at night, a beautiful landscape, or refined music; but such high tastes are acquired through culture, and depend on complex associations; they are not enjoyed by barbarians or by uneducated persons. (Darwin 1874: I, 64)

But

> The taste for the beautiful [...] is not of a special nature in the human mind; for it differs widely in the different races of man, as will hereafter be shewn, and is not quite the same even in the different nations of the same race. Judging from the hideous ornaments and the equally hideous music admired by most savages, it might be urged that their æsthetic faculty was not so highly developed as in certain animals, for instance, in birds. (ibid.)

As illustrated by Darwin, there is no essential gap between animals and the uneducated members of the human race. Rather, it is a continuous growth in these abilities in the broader context of mental abilities, as we will see.

Darwin's explanation of aesthetic phenomena (respectively, with the idea that animals possess taste) has failed not only for scholars of the humanities. Even the proponents of evolutionary theory have looked at these ideas with scepticism. In fact, Darwin's explanation of the "aesthetic structures" in animals via female selection was rejected even by Darwinists. Female selection was considered the weakest of all Darwin's ideas. Yet, later reflection on this topic has undergone interesting development, both within Darwinian and non-Darwinian biology. As Helen Cronin pointed out:

> Paradoxically, much of the debate about sexual selection, from the 19th century to the present day, hasn't been about sexual selection at all but about natural selection. The issues raised by sexual selection, throughout the theory's history, have fallen into two categories. The first is the question of whether sexual selection is required to account for the phenomena at all or whether they can be explained instead by the standard forces of natural selection alone. For nearly century, the majority of Darwinians saw this as the major issue. They sought almost any alternative to sexual selection, and it was on natural selection that they relied above all. The second category of questions concerns mate choice – in particular, the reasons for making the choice and how, or even whether, Darwinian forces could allow them to evolve. These questions were raised from the first, but it is only relatively recently that the role of mate choice has become the main focus of attention. It is now a flourishing line of research – and an enormously fruitful line it has proved to be. (Cronin 1994: 119)

The truth is that Portmann was aware of this problem and did not want to reduce it to natural selection; but the author of this chapter is repeatedly surprised by how Portmann completely refuses Darwin's concept of animals having taste. He even explicitly rejects Darwin's hypothesis that peacocks direct their display at peahens, who must, therefore, be endowed with an aesthetic sense (Portmann 1960: 166). Portmann argues that there is no experimental evidence that females make such a

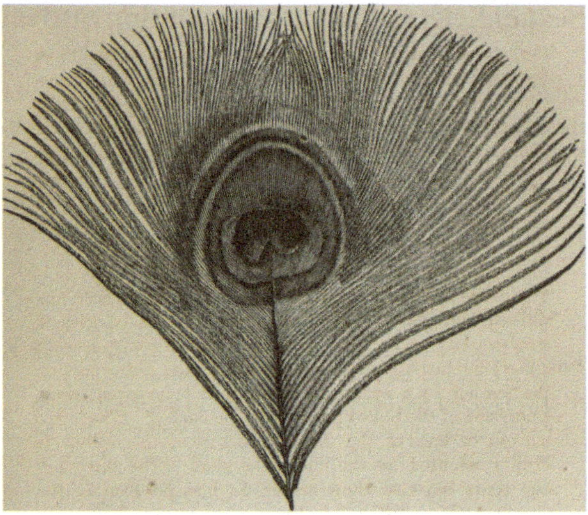

Fig. 12.2 Peacock feather from Darwin's *The descent of man, and selection in relation to sex* (Wikipedia commons https://commons.wikimedia.org/wiki/File:The_descent_of_man,_and_selection_in_relation_to_sex_(1873)_(20244858824).jpg)

choice, but rather, they look the other way and focus on the grain on the ground. According to him, the reason behind an Indian peafowl's (*Pavo cristatus*) ritual display is the release of hormones that cause sexual alertness in the female, or the "sex synchronization" of both partners (Portmann 1960: 170; see also Fig. 12.2).

Similarly, in *Animals as Social Beings* (Das Tier als soziale Wesen 1964b: 219–230) he dedicates a whole chapter to the lyrebird, enthusiastically describing the complexity and splendours of its musical expression and appearance, but his interpretation avoids any female choice. He writes:

> The singing fulfils a function in the lyrebird's family life, not merely as a constant sign of the mate's presence, but as a releaser of hormonal processes, as a constant stimulator of the moods the females need for carrying on their brooding and rearing business. (Portmann 1964b: 226)

The entire exaggerated tail of the peacock-pheasant, as well as the tail and song of the superb lyrebird, are further evidence to Portmann that such structures go beyond the requirements of sexual selection. For example, the eye pattern of the peacock's tail "is not needed for sexual attraction; it has a purely phenomenal significance" (Portmann 1964a: 118). We could wonder how he was limited here by the level of knowledge at that time. Matt Ridley later describes the paradoxical suppression of this viewpoint, which was not all so absurd after all:

> For a century after he proposed it, Darwin's theory of female choice was ignored while biologists tied themselves in furious knots to come up with other explanations [...] Any theory, it seemed, was preferred to the idea of female preference for male beauty [. . .] Yet it is impossible to watch peacocks displaying and not come away believing that the tail has something to do with the seduction of peahens. After all, that was how Darwin got the idea in the first place [. . .]. (Ridley 1993: 135–136)

It was not until the 1990s, after years of ignoring Darwin's theory, that biologists began to return to Darwin's original ideas[7], although there seems to be no general consensus explaining peacock behaviour (see Takahashi et al. 2008; Loyau et al. 2008).[8]

While the perception of beauty or taste could be discussed, it is evident that in many bird species (and mammals, too) there is a female who chooses. The complexity of courtship behaviours is fascinating, as is the complexity of structures combined with dance and speech sounds (for example, in bowerbirds, where the female's choice is clearly evident by how she carefully visits the individual bowers of males and does not look elsewhere).[9] Although Portmann declares in connection with the peacock that "a rich inner life (*ein reiches Innenleben*, Portmann 1960: 170) is expressed in the mating ritual", he in fact denied this in regards to the peahen.[10]

But in general, Portmann's argument is clear and, I suppose, correct. As Zdeněk Neubauer, the Czech microbiologist, philosopher, and follower of Portmann, stresses:

> The general biological law can be expressed: the beauty of living beings is usually somehow mysteriously combined with basic functions and especially with reproduction [...] But it also fulfils indirectly restrictions dictated by the necessity of existence. Wherever organs or organisms are relieved of the burden of satisfying the requirements of survival [...] the invention and creativity of living beings everywhere burst out into superabundance, not only in achievement but also in their variability; all life-energy, power, is "channeled" to unaddressed phenomena. (Neubauer 2002: 208)

What Is Beauty?

While I have mentioned aesthetic terms here, we have not yet dealt much with how both naturalists interpret them. I have not found a definition of beauty or aesthetics in Portmann – he uses the word "beauty" in reference to his personal interest rather than in connection with thinking about appearance. More often, he uses the term "aesthetic", or "aesthetics". He does not define it, but he obviously understands it

[7] On the neo-Darwinian approach towards beauty see issue of *Nature* in the year 1994 with the cover photo titled "Beauty is More Than Skin Deep" and 2015 as a special supplement, "Nature Outlook: Beauty", focused mainly on the human body and preferences (see Brody 2015).

[8] See also studies from the 1990s supporting a relationship between train features and mating success, primarily Petrie et al. 1991, but also Petrie and Halliday 1994, Yasmin and Yahya 1996.

[9] See a study about bowerbirds: Coleman et al. 2004.

[10] However, in 2018 an interesting study appeared that could support Portmann's view of peafowl behaviour as a general preparation for mating – the study shows that a peacock's tail feathers make low-frequency sounds that cause feathers on the head of females to quiver. It is not sure how the vibrations are connected with other parts of the mating display, but it seems that the female is not influenced only by the peacock´s visual appearance and show (Kane et al. 2018; see also Allen 2018). Perhaps Portmann (1960: 27) could be right, as he argues for an impact of the peacock´s behaviour on the brain and production of hormones in females. See also an interesting study about the perception of the peacock by predators (Kane et al. 2019).

quite broadly. Perhaps it is most used in the previously mentioned article that includes the adjective directly in its title "The Biological for Aesthetic Education" (Biologisches zur ästhetischen Erziehung, Portmann 1968). Here, the aesthetic function is understood as our ability to grasp the outer world through our senses, the perceptions of the qualities of forms and lines, colours, sounds, and tastes, in opposition to the theoretical function, founded on rational thought, quantification, etc. (especially 311–312).

In Darwin's work, the words "aesthetic" and "aesthetics" are infrequently used, but he deals with both the traditional concept of beauty[11] and taste (or sense of beauty), typical categories for the nineteenth century. To a lesser extent, he uses many other English words associated with aesthetic experience: *elegance*, *splendour*, *luxuriance*, *grandeur*, *harmony*, *decoration*, *monstrosity*, *deformity*, *mutation*, *pleasure*, *delight*, and the often used adjective *attractive*, though not only in the current biological meaning. In several places, he also tries to provide a definition of beauty:

> When an intense colour or two tints in harmony, or a recurrent and symmetrical figure please the eye, or a single sweet note pleases the ear, I call this a sense of beauty; and with this meaning I have spoken (though I now see in not a sufficiently guarded manner) of a taste for the beautiful being the same in mankind (for all savages admire bits of bright cloth, beads, plumes, etc.) and in the lower animals. (To John Morley, 24 March 1871, Darwin 1903: I, 324–327)

In short, Darwin sees beauty both in subjective reaction (the pleasure of our senses) and in the properties of an aesthetic subject: clear, bright colours, variety, symmetry, rhythm, characteristic features, a combination of novelty and uniformity. He sees the roots of aesthetic pleasure in physiology, in the properties of our senses.

Darwin's perhaps vague definition is an expected synthesis of the eighteenth century British tradition and the more positivist approaches of naturalists in the second half of the nineteenth century. Darwin defines beauty as pleasure, based on sensory perception, on objectively existing qualities of perceived objects (as opposed to the *associationism* of the eighteenth century, emphasizing the superiority of the train of associations in our mind over the qualities of objects); he also uses inspirations from physiology, as already Edmund Burke did, but also from Helmholtz and others (see Darwin 1987; also Stibral 2006, and 2014).

Both Darwin and Portmann agree about the emphasis on symmetry, which is, moreover, very characteristic of attempts to explore attractiveness and beauty in contemporary biology (see, e.g. Brown et al. 2005; Perett et al. 1994).[12] In this respect, the resignation of the argument that began in modern times, especially by Burke, is quite surprising, as it meant a relatively radical break-up of philosophical aesthetics with attempts to find a connection between beauty and symmetry,

---

[11] E.g. Amy Maxmen in a *Nature* supplement about beauty writes that the words "beauty" and "beautiful" appear in *The Descent of Man* 280 times (Maxmen 2015: S8). I was not so successful (only 196), but still. In, fact, as we see from the other frequent words connected with aesthetic experience, this work of Darwin´s is literally "full of aesthetics".

[12] Very rarely could we find a different point of view, as e.g. Swaddle and Cuthill (1995).

Fig. 12.3 An argus pheasant (*Argusianus argus*) – one of Darwin's favourite examples of aesthetic phenomena. An illustration from *The descent of man, and selection in relation to sex* by T. W. Wood (Wikipedia commons https://commons.wikimedia.org/wiki/File:Descent_of_Man_-_Burt_1874_-_Fig_52.png)

regularity, utility and perfection. Not only for Burke, but also for Schiller and Kant, these properties seemed to depend on reason, on what we would call with Portmann the theoretical function of our mind (Portmann 1968), not on our emotionality and thus on the feeling of beauty (see Fig. 12.3).

We could see the emergence of a paradox here – on the one hand, Portmann constantly talks about the appearance of living beings not subordinated to any function or adaptation, and at the same time avoids the argument similar to that of Kant, who used phrases such as "disinterested liking", "delight in purposiveness without purpose", etc. (Kant [1790] 1987). The more narrowly defined aesthetics of the modern era quite often considered symmetry as something irrelevant or even devaluing for an aesthetic experience. During the second half of the eighteenth century, when philosophical aesthetics developed, it was precisely nature – often imagined as wild landscapes such as mountains, waterfalls and forest – that was superior to human art because it was independent of regularity, symmetry, and so on. One can ask at this moment – what would be more appropriate to support Portmann's approach than Kant's purposiveness without *purpose*?

But perhaps this is because Portmann preferred to open the way to an even more radical concept of evolution, where the aesthetic dimension is no longer limited to the tastes of individual subjects or the usefulness of a "beautiful" appearance in the competition within species, even though it may result in dysfunction or disadvantage in natural selection. His concept of self-presentation is not only limited to aesthetic pleasure, but it aims at the radically independent creation and self-creation of living beings, where it is clearly visible in unaddressed phenomena.

The Hierarchy of Living Beings and the Opposition of Outside/Inside

Investigations into the aesthetic appreciation of visible organismal forms are closely related to the relationship between the *outside* and *inside* of a living being. It was one of the key issues for Portmann, as he often emphasized how the exterior is shaped by specific rules compared to the inside. While regarding anatomical structures the exterior is shaped with an emphasis on the phenomenon – often hypertrophied and symmetric structures; the interior, or the viscera, is purely functional and asymmetric – while two species of ducks can easily be distinguished by their external forms, we would hardly recognize them by their viscera and anatomy.

At the same time, Portmann introduces the concept of *Innerlichkeit* (inwardness, centricity), including not only physiological but also mental processes and their relationship to the outside. According to Portmann, the outside appears to be the expression of *Innerlichkeit* (Portmann 1960).

Darwin does not focus very much on this ambiguity, though to some extent he deals with the embodiment of inner states in one of his fundamental later works, *The Expression of the Emotions in Man and Animals* (1872).

There is also an interesting comparison of Portmann's and Darwin's views on the *hierarchy* of living beings, at the time a respectable theme following from the *scala naturae*. For Portmann, the problem of rank was absolutely essential (Portmann 1960: 86–101). While Darwin was aware of the hierarchy of organisms, the superiority of mental abilities over others, he was also a supporter of the lower forms, and expressly talks about the effort to raise or elevate lower forms into higher forms. Although his most famous and essential works contain thousands of examples from all possible levels of possible hierarchy, his attentive research was focused primarily on lower forms – see his crucial works about barnacles, experiments with earthworms, bees, his book on carnivorous plants, etc. Likewise, in the transition between humans and animals, he always emphasized the gradual transition and even the declining taste of the primitive man.

In any case, they both agree that the animal is not a mechanism, for obvious reasons, while neither of them approaches the later neo-Darwinian view of living organisms as "gene machines", where the genes themselves are the subject of natural selection and animals and humans are presented rather as a mechanism whose minds are some sort of black-box.

Both Portmann and Darwin emphasize the complexity of the animal psyche. Both agree that it is unthinkable to reduce higher animals to a machine, though there were differences in opinion between them. In fact, Darwin did not see any qualitative difference between the cognitive abilities of humans and animals: "As man possesses the same senses with the lower animals, his fundamental intuitions must be the same" (Darwin 1874: 36). "Nevertheless the difference in mind between man and the higher animals, great as it is, is certainly one of degree and not of kind" (Darwin 1874: 105). On the other hand, Portmann argues against the mechanistic paradigm in a completely conscious way:

> Biology today is no longer inclined to present a dull mechanical picture of animal life, beside which even the poorest idea of humanity would appear to advantage. On the contrary, a knowledge of the complexity of animal behaviour and experience, and its similarities with human behaviour makes it easier to see the uniqueness of Man. In this case, for instance, while recognising how far removed any animal imitation of sounds is from the least articulate human speech, we realise all that is involved in the mysterious complex of such animal imitation, which makes human speech seem a fortiori even more remarkable. (Portmann 1964b: 226)

For Portmann, the tone is different – humans are fundamentally different from animals. Although neither of the naturalists directly presented these problems of hierarchy in relation to aesthetics (maybe with the exception of Darwin's emphasis on the small difference between man and animals in intelligence and taste – the dividing line is ethics according to him) (see Darwin 1874: 70–104), in aesthetics as a discipline we find this problem mentioned. One of the few important books of philosophical aesthetics to significantly mention the aesthetics of nature (before the turn in the 1870s) is Friedrich Theodor Vischer's *Aesthetics, or the Science of Beauty* (*Aesthetik oder Wissenschaft des Schönen*, 1847). In the second part of this work we can find a systematically detailed aesthetics of nature divided hierarchically from minerals to mammals (Vischer 1847: 7–156). We find something similar in *General Aesthetics* (*Všeobecná aesthetika*, 1875), written by aesthetician and philosopher Josef Durdík, a representative of nineteenth-century Prague aesthetics and the leading promoter of Darwinism in the Czech lands (Durdík 1875: 407–450, see also Stibral 2013).

The last interesting topic I would like to mention is the reflections on art by both scientists. This topic itself could easily fill an entire chapter, but here I will only try to make a general comparison. For both, art is an important topic, though on relatively different levels. For Darwin, the theme of art appears in particular in the *Origin of Species*, where it is discussed in connection with animals as well as with humans.

In addition to his thoughts about aesthetic perception, Darwin considers the appreciation of simple art in animals, and also within the framework of sexual selection – males impress females not only by their beautiful forms, but also by their singing, dancing or, in exceptional cases, like by bowerbirds, by visual art and architecture. The female assesses the quality of these phenomena, not only the appearance of the male (see e.g. Darwin 1874: II, 51–70). On the other hand, Darwin underlines the continuity of these manifestations between human and animals – human art is a continuation of these manifestations, albeit more complex. And in some native peoples we find even a decline (for example, in comparison to the singing of birds) – according to him, it is not possible to talk about music among the African natives, although it is interesting how quickly they can learn music (ibid.: 334).

Even though Portmann grew up a century later, he shared similar problematic views, but surprisingly, he does not develop any theory about the biological basis of art. He does, however, often use excursions with the experience of art for argumentation in biology. With some exaggeration and reference to his previous drawing experience and skills, living forms are seen by him through the prism of artistic experience. For example, when he is quoting Wölfflin at the beginning of a section on hierarchy, we see how important it was to his thinking about the forms of living beings:

> In the natural sciences, the progress from the lowest living beings to the highest is characterized by the notions of differentiation and integration. It means that the individual members divide into contrasting forms and at the same time become increasingly integrated parts of the whole. The same could be said about the sense of development in art too. However, higher forms of organization are characterized by a higher degree of fullness of life [...]. Every form of life leads to an advancement of life. (Portmann 1960: 86)

He also appropriately shows in the example of functionalism that a purely functional explanation of architectural shapes and bodies of living beings does not hold up. It is not just humans who demand that buildings and flats are not just "machines for living", narrowed to purely functional targets, but animals exhibit similar demands (Portmann 1960: 83).

We can also find Portmann's educated and critical thinking about modern art, which, in his view, is based more on theory and intellectual components than on sensory perceptions (Portmann 1968) – what is more, he wrote it at a time before the enormous rise of conceptual art.

Conclusion

In conclusion, I think that in some respects we can find a surprising consensus between Portmann and Darwin: that the appearance of living beings cannot be reduced to the result of adaptations within natural selection. In this, both of them deviate from later neo-Darwinism. Nevertheless, it must be stressed that Portmann's opinion on the non-purposive origin of forms is definitely stronger. Likewise, both agree that we cannot reduce animals to mere mindless mechanisms. Darwin constantly emphasizes the animal psyche; it is no coincidence that he wrote an entire work on emotions and sees the transition between animals and humans as a continuum.

Portmann emphasizes a complex and complicated *inwardness* (*Innerlichkeit*), and while he draws attention to the richness of this inwardness, he quite clearly sees the difference between humans and animals. This is perhaps one of the reasons (or implications) for the vastly differing interpretation of aesthetic phenomena in nature. For Darwin, it is sexual selection above all in which such structures root, complemented by other evolutionary processes such as the law of compensation or balancing of growth, the co-evolution of insects and pollinated plants, or, of course, natural selection (semantic structures, cryptic colouration, etc.). In the context of sexual selection, it is a sense of beauty or taste inherent in females that influences the male form, and consequently its entire species, to evolve often in a direction going against the direction of natural selection.

In contrast, Portmann quite opposes this hypothesis and explains a number of phenomena of both behaviour and external structures as part of a ritualized "synchronization" of partners. He emphasizes that a substantial part of appearance and behaviour does not have any direct function; it is actually a *self-presentation* of the species (and individuals or sexes). Basically, what we would call an aesthetic structure is only one of the *facets* of this self-presentation.

However, both naturalists emphasize the symmetry of living beings as part of what leads to an aesthetic appreciation of their appearance – this is an explicit part of Darwin's definition of beauty, some kind of biological constant. With this, both are in agreement with contemporary neo-Darwinism.

Neither Portmann nor Darwin focuses on aesthetic preferences for landscape. Though Darwin was an enthusiastic devotee of landscape beauty, and there are numerous reflections on the subject in his travel writings, he did not include any hypotheses about it in his scientific work. These steps were taken by his later followers, such as Aldo Leopold, who combined the beauty of the landscape with biodiversity or Konrad Lorenz, who linked an increasing appreciation of landscapes and organisms with their growing complexity. The avalanche of works following Appleton's *The Experience of Landscape* (1975) combined the aesthetic pleasure of the landscape with the Darwinian roots of our responses to a specific habitat.

Both naturalists took an interest in art. Darwin involved this interest in his reflections on the biological origins of art among different animal species and man. Art, according to him, is an instinctive drive – an opinion we also find in a number of recent publications (Dissayanake 1992; Dutton 2009; Davies 2012). Portmann uses the examples from art to argue against the strictly utilitarian function.

Although they might disagree in some ways, I believe that their views, albeit sometimes from different angles, are opposed to later neo-Darwinism. In any case, both Portmann and Darwin attempt to "return to play" the aesthetic dimension of nature, to return to non-reductive explanations of it, which is still deserving of our attention.

References

Allen, M. (2018, November 28). Watch a peacock get a female's attention – By making her head vibrate. *Science*. https://www.sciencemag.org/news/2018/11/watch-peacock-get-female-s-attention-making-her-head-vibrate

Appleton, J. (1975). *The experience of landscape*. London/New York: J. Wiley.

Brody, H. (Ed.). (2015). Nature outlook: beauty. From aesthetics to animal attraction. *Nature, 526*, 1–17.

Brown, W. M., et al. (2005). Dance reveals symmetry especially in young men. *Nature, 438*, 22.
Caillois, R. (1962). *Esthétique généralisée. Impression en deux couleurs*. Paris: Gallimard.
Coleman, S. W., Patricelli, G. L., & Borgia, G. (2004, April 15). Variable female preferences drive complex male displays. *Nature, 428*(6984), 742–746.
Cronin, H. (1994). *The ant and the peacock*. Cambridge: Cambridge University Press.
Darwin, C. (1839). *Narrative of the surveying voyages of his majesty's ships Adventure and Beagle between the years 1826 and 1836, vol. III*. London: H. Colburn.
Darwin, C. (1861). *On the origin of species by means of natural selection, or the preservation of favoured races in the struggle for life* (3rd ed.). London: J. Murray.
Darwin, C. (1874 [1871]). *The descent of man, and selection in relation to sex*. 2 vols (1st ed.). London: J. Murray.
Darwin, C. (1903). *More letters of Charles Darwin*. 2 vols. (eds. F. Darwin & A. C. Seward), London: J. Murray, letter n. 241, s. I/324-327 on: http://pages.britishlibrary.net/charles.darwin/texts/more_letters/mletters1_05.html
Darwin, C. (1985). *The Correspondence of Charles Darwin I*. Cambridge/New York: Cambridge University Press.
Darwin, C. (1987). *Charles Darwin's notebooks 1836–1844*. Cambridge: British Museum-Cambridge University Press.
Darwin, C. (2018a). A letter to Thomas Meehan, *Darwin Correspondence project*, Cambridge University Library. https://www.darwinproject.ac.uk/letter/?docId=letters/DCP-LETT-9672.xml;query=thomas%20meehan;brand=default
Darwin, C. (2018b). A letter to Alfred R. Wallace, *Darwin Correspondence project*, Cambridge University Library. https://www.darwinproject.ac.uk/letter/?docId=letters/DCP-LETT-5648.xml;query=tiger%20wallace;brand=default
Davies, S. (2012). *The artful species: Aesthetics, art, and evolution*. Oxford: Oxford University Press.
Dissayanake, H. (1992). *Homo aestheticus: Where art comes from and why*. New York: Free Press.
Donald, D. (2009). Introduction. In D. Donald & J. Munro (Eds.), *Endless forms: Charles Darwin, natural science and the visual arts* (pp. 1–25). New Haven: Yale University Press.
Donald, D., & Munro, J. (2009). *Endless forms: Charles Darwin, natural science and the visual arts*. New Haven: Yale University Press.
Durdík, J. (1875). *Všeobecná aesthetika*. Praha: I. L. Kober.
Dutton, D. (2009). *The art instinct. Beauty, pleasure, & human evolution*. Oxford: Oxford University Press.
Gould, S. J. & Lewontin, R. C. (1979). The spandrels of San Marco and the Panglossian paradigm: A critique of the adaptationist programme. *Proceedings of Royal society B. Biology, 205*(1161).
Haeckel, E. (1899). *Kunstformen der Natur. Erste Sammlung*. Leipzig/Wien: Bibliographisches Institut.
Haeckel, E., & Breitenbach, W. (1913). *Die Natur als Künstlerin*. Berlin: Charlottenburg Vita.
Illies, J. (1976). *Das Geheimnis des Lebendigen*. München: Kindler.
Kane, S. A., Van Beveren, D., & Dakin, R. (2018). Biomechanics of the peafowl's crest reveals frequencies tuned to social displays. *PLOS ONE, 28*, 2018.
Kane, S. A., Wang, Y., et al. (2019). How conspicuous are peacock eyespots and other colorful feathers in the eyes of mammalian predators? *PLOS ONE, 24*, 2019.
Kant, I. (1987). *Critique of judgement*. Transl. & introd. by Werner S. Pluhar. Indianapolis/Cambridge: Hackett.
Kleisner, K. (2006). Biologické omalovánky aneb "Volte Portwina" (The biological colouring book, or "Vote Portwin"). In K. Stibral (Ed.), *Darwin a estetika. Ke kontextu estetických názorů Charlese Darwina* (*Darwin and aesthetics. Towards the context of Darwin's aesthetic ideas*) (pp. 153–157). Červený Kostelec: P. Mervart.
Kockerbeck, C. (1997). *Die Schönheit des Lebendigen. Ästhetische Naturwahrnehmung im 19. Jahrhundert*. Wien/Köln/Weimar: Böhlau.
Komárek, S. (2003). *Mimicry, aposematism and related phenomena. Mimetism in nature and the history of its study*. München: Lincom Europa.

Komárek, S. (2016). *Mimikry a příbuzné jevy. Dějiny vnímání a výkladu vnějšího zhledu živých organismů*. Praha: Academia.
Larson, B., & Brauer, F. (Eds.). (2009). *The art of evolution. Darwin, Darwinism and visual culture*. Hannover/London: Dartmouth College Press.
Larson, B., & Flach, S. (Eds.). (2013). *Darwin and theories of aesthetics and cultural history*. Surrey: Ashgate Publishing, Ltd..
Loyau, A., Petrie, M., et al. (2008). Do peahens not prefer peacocks with more elaborate trains? *Animal behaviour, 76*(5), 5–9.
Maxmen, A. (2015). Come mate with me. *Nature, 526*, 8–10.
Möbius, K. (1908). *Ästhetik der Tierwelt*. Jena: G. Fischer.
Neubauer, Z. (2002). *Biomoc*. Praha: Malvern.
Perett, D. I., May, K. A., & Yoshikawa, S. (1994). Facial shape and judgements of female attractiveness. *Nature, 368*, 239–242.
Petrie, M., & Halliday, T. (1994). Experimental and natural changes in the peacock's (*Pavo cristatus*) train can affect mating success. *Behavioral Ecology And Sociobiology, 35*(3), 213–217.
Petrie, M., Halliday, T., & Sanders, C. (1991). Peahens prefer peacocks with elaborate trains. *Animal Behaviour, 41*(2), 323–331.
Portmann, A. (1935). *Falterschönheit. Exotische Schmetterlinge in farbigen Naturaufnahmen*. Introduction by H. Hesse. Bern: Iris Books.
Portmann, A. (1945). *The beauty of butterflies. Introduction by J. Huxley*. New York/Toronto: Iris Books.
Portmann, A. (1960). *Neue Wege der Biologie*. München: R. Piper.
Portmann, A. (1964a). *New path of biology*. New York: Harper & Row.
Portmann, A. (1964b). *Animals as social beings. New York*: Harper & Row.
Portmann, A. (1968). Biologisches zur ästhetischen Erziehung. In *Biologie und Geist* (pp. 309–333). Frankfurt am Main: Suhrkamp.
Portmann, A. (1984). *Vom Wunder des Vogellebens*. München/Zürich: R. Piper.
Prodger, P. (2009). Ugly disagreements. Darwin and Ruskin discuss sex and beauty. In Barbara Larson and Fae Brauer (eds.). *The Art of Evolution: Darwin, Darwinisms and Visual Cultures*. Hannover/London: Dartmouth College Press, 40–58.
Ridley, M. (1993). *The Red Queen: Sex and the Evolution of Human Nature*. New York: Viking Books.
Rothenberg, D. (2012). *Survival of the beautiful. Art, science and evolution*. London: Bloomsbury.
Ruskin, J. (1872). *The eagle's nest. The works of John Ruskin* (Vol. IV). London: Smith, G. Allen.
Ruskin, J. (1888). *Proserpina*. New York: J. Wiley.
Stibral, K. (2006). *Darwin a estetika. Ke kontextu estetických názorů Charlese Darwina* (*Darwin and aesthetics. Towards the context of Darwin's aesthetic ideas*). Červený Kostelec: P. Mervart.
Stibral, K. (2013). Darwin and prague aesthetics. Towards the acceptance of Darwinism in Central Europe. *Estetika. The Central European journal of aesthetics, 6*(1), 81–100.
Stibral, K. (2014). Darwin o kráse aneb o spletitých cestách od estetiky k biologii (Darwin on beauty or the knotty paths from aesthetics to biology and back). In O. Dadejík, F. Jaroš, & M. Kaplický (Eds.), *Krása a zvíře* (pp. 245–280). Praha: Dokořán.
Swaddle, J. P., & Cuthill, I. C. (1995). Asymmetry and human facial attractiveness: Symmetry may not always be beautiful. *Proceedings of the Royal Society B. London, 261*, 111–116.
Takahashi, M., Arita, H., Hiraiwa-Hasegawa, M., & Hasegawa, T. (2008). Peahens do not prefer peacocks with more elaborate trains. *Animal behaviour, 75*, 1209–1219.
Velenovský, J. (1910). *Všeobecná botanika: Srovnávací morfologie III*. Praha: Česká akademie F. Josefa.
Velenovský, J. (1921). *Přírodní filozofie*. Praha: L. Souček.
Velenovský, J. (1930). *Obrázky*. Praha: L. Souček.
Vischer, F. Th. (1847). *Aesthetik oder Wissenschaft des Schönen*, 2nd vol. Reutlingen: C. Mäcken.
Welsch, W. (2004). *Animal aesthetic. Contemporary aesthetics*, 2. https://www.contempaesthetics.org/newvolume/pages/article.php?articleID=243

van Wyhe, J. (Ed.) (2002). Darwin's notebooks and reading lists. In *The complete work of Charles Darwin*. Online. http://darwinonline.org.uk/EditorialIntroductions/vanWyhe_notebooks.html

Yasmin, S., & Yahya, H. S. A. (1996). Correlates of mating success in Indian Peafowl. *The Auk. Ornithological Advances, 113*(2), 490–492. https://doi.org/10.2307/4088915.

Karel Stibral is an associate professor and head of the Department of Environmental Studies at Masaryk University, Brno, Czech Republic. He studied aesthetics and history of art and obtained his Ph.D. in the field of philosophy and history of sciences at Charles University in Prague. He also studied at the Jagellonian University, Cracow. He is focused on the history of the aesthetic perception of nature in the eighteenth to twentieth century, connections between biology and art, darwinism and aesthetics, and the history of nature conservation. He has published several books about these topics (in Czech, *Why is Nature Beautiful?*, 2005; *The Picturesque. The Aesthetics of Nature between Garden and Wilderness*, 2011; *John Ruskin and Nature*, 2011; *Aesthetics of Nature*, 2019, etc.)

Chapter 13
On the Brink of the Expressible: Adolf Portmann Meets Carl Gustav Jung on Eranos Ground

Ivana Ryška Vajdová

Abstract This chapter focuses on Carl Gustav Jung (1875–1961) and Adolf Portmann (1897–1982), two prominent figures in the intellectual discussion club called Eranos. I will outline how the nature of the Eranos platform broadened from the humanistic and religious studies introduced by its "*spitirus rector*," Jung, to incorporate natural sciences and adopted a more scientific approach in the process. This historical shift began with the arrival of Portmann in 1946 and resulted in his appointment as chairman of the group (1962–1977). Portmann and Jung were prominent figures in their respective disciplines, and their theories became the subject of a mutual critique on Eranos ground. The aim of this chapter is to describe the relation between Jung and Portmann from different perspectives, mainly through their discussion about the subject of archetypes and their relation to instinct. The chapter also indicates how the Eranos spirit shifted under the influence of Portmann.

Introduction

On the Swiss border with Italy, along Lake Maggiore, the Via Moscia runs from the town of Ascona. Right before one of its bends, where nobody expects anything but palms and pine trees between the road and the river, a deep stone stairway descends to Casa Gabriella. This three-storey villa belongs to Eranos, an organization that became a meeting point of some of the world's most prominent thinkers. In this chapter we give a brief overview of the history of Eranos and demonstrate how its peculiar spirit was reflected in the work of Adolf Portmann. We will see how Portmann's appreciation of a larger context when studying the animal and the human led to discussions with another notable Eranos member, Carl Gustav Jung.[1]

[1] Although Portmann and Jung met for the first time on Eranos ground, as young men they both studied under professor Friedrich Zschokke (1860–1936) at the University of Basel. Both wit-

I. R. Vajdová (✉)
University of Hradec Králové, Hradec Králové, Czech Republic
e-mail: ivanavajdova@gmail.com

F. Jaroš, J. Klouda (eds.), *Adolf Portmann*, Biosemiotics 23,
https://doi.org/10.1007/978-3-030-67810-4_13

The subject that became the center of their intellectual exchange was archetypes and instincts, and we show how both scholars clashed in their approach. We also show how despite the confrontation, both thinkers shared fundamental similarities; in placing particular knowledge into a more general, humanized context, they reached the very limits of their respective disciplines, and even language itself, in the process.

The Evolution of Eranos

From its very beginning, the Eranos gathering was conceived by its founders as a forum for the interchange between Eastern and Western religion and spirituality. Its general idea was to bridge cultures, epochs and disciplines not only by an intellectual understanding but also, as one of its participants puts it, "a knowing through direct experience [...] the continuing juxtaposition of the primordial and the modern, the individual and the universal, the scientific and the mythological" (Progoff 1966: 312). Carl Gustav Jung, who shaped early Eranos dramatically, described it as "the only place in Europe where scholars and interested lay participants could come together and exchange ideas, unrestricted by academic boundaries" (Hakl 2013: 7). It was also the place where Jung and Adolf Portmann engaged in an intense intellectual exchange.

The idea of Eranos was originated in the German city of Marburg in 1932 by patron, scholar and esotericist Olga Fröbe-Kapteyn[2] (1881–1962) and Rudolf Otto (1869–1937), the theologian, philosopher, and comparative religionist. It was Otto who gave Eranos its name, the Greek *ἔρανος* meaning "a shared feast" to which every attendant brings a small gift (Kerenyi 1955). For the Eranos gatherings, this gift took the shape of a lecture, the main goal of which was to bring together

nessed during the lectures very convincing demonstrations about the layers of an organism's evolution, which inevitably influenced Jung's own thoughts about the possible nature of the collective unconscious as a layered experience of the human species "with remnants of phylogenetic functions of perception and adaptation" (Jung 1916, appendix). Portmann recalled later in an article that he was convinced that these images of hidden reality in the context of the human species must have fascinated Jung (Portmann 1976).

[2] Olga Fröbe-Kapteyn (1881–1962) was born in London of Dutch parents. Her father, Albert Kapteyn, was a photographer, and her mother was a philosophical anarchist, a writer on social questions, and a friend of playwright George Bernard Shaw and anarchist Peter Kropotkin. Olga Fröbe-Kapteyn studied applied art in Zurich, but her great interest was spiritual research. She had a lively interest in searching for and collecting artefacts which she used to illustrate the topics of each year's Eranos meetings. In 1935, she systematically began to collect pictures that exemplified archetypal themes. She traveled around Europe trying to find and purchase photographs of ancient frescoes, paintings, sculptures, manuscript illustrations and primitive folk art. She then classified them according archetypal themes in what became known as the "Eranos Archive." After the war, in 1946, she started to send her photographs to London, and in 1956 the Warburg Institute accepted her entire collection. In 1960, the archive was renamed the Archive for Research in Archetypal Symbolism (ARAS), consisting of more than 25,000 representations (Gronning et al. 2007).

intellectuals from various areas contemplating the importance of spirituality in human culture and to establish a common ground on which the religious philosophies of the Orient and Western civilization could meet (Progoff 1966).

The very first symposium took place in 1933, a year after Fröbe-Kapteyn and Otto met, under the title "Yoga and Meditation in East and West." Lectures included "Meditation and Contemplation in the Roman Catholic Church" by Ernesto Buonaiuti, "On the Meaning of Indian Tantra Yoga" by Heinrich Zimmer, "Contemplation in Christian Mysticism" by Friedrich Heiler, and "A Study in the Process of Individuation" by Carl Gustav Jung.

This annual lecture program, called "Eranos Meetings" (*Eranos Tagungen*), continued with a few breaks until today. Jung himself gave 14 lectures between 1933 and 1951 (Vitolo 2015). Every conference usually lasted for a week, during which an open discussion between participants took place. Lectures were later published in the *Eranos Yearbook* (*Eranos Jahrbuch*), published by Olga Fröbe-Kapteyn from 1933 to 1961 – the year before her death. Six volumes of selected Eranos lectures, translated into English and edited by Joseph Campbell, were published between 1954 and 1968 as part of the Bollingen Series (Gronning et al. 2007: 248).

Jung and Portmann were not the only notable names among Eranos members. The circle was also joined by Gershom Scholem (Jewish mysticism), Gilles Quispel (Gnostic studies), Henry Corbin (Islamic religion), Mircea Eliade (history of religion), Erwin Schrödinger (physics), Heinrich Zimmer (Indian philology and art), Max Knoll (physics), Herbert Read (art history), Joseph Campbell (comparative mythology) and Erich Neumann (analytical psychology), to name a few.

But despite the abundant presence of prominent intellectuals, Eranos never gained wider recognition. As Thomas Hakl argued, the esoteric themes discussed during these symposia and their pronounced sympathy to esotericism and mysticism led the outside intellectual community to accuse the meetings of irrationalism (Wasserstrom 1999; Noll 1994). Indeed, it is hard to position Eranos on an intellectual map, as it found itself in a certain border zone between esotericism and science (Miller 2006). Many of its participants were adventurous in their respective disciplines and were even labeled as "an avant-garde in the field" (Hillman and Shamdasani 2013: 146), but one facet of the Eranos spirit is particularity interesting to this day. This is the joint effort of different personalities to reach the limits of scientific knowledge and to respect the fact that what lies beyond our capacity of articulation is equally as real as any scientifically accessible actuality. Portmann expressed this in his speech dedicated to the 80th birthday of Fröbe-Kapteyn in October 1961:

> Our thoughts aim to explore the hidden origins from which all greatness arises. All origins are hidden in obscurity. To conceive a mystery of living spirit with a vigilant mind, to express what is expressible and at the same time an awareness of the inexpressible is always present in the work of Eranos. (Portmann 1974: 226)

And it was in this spirit, through Eranos, that Portmann and Jung tried to formulate their complex biological and psychological theories with respect to the inscrutable background transcending any discipline.

The Change of the Guard

For two decades after its foundation, Jung had a major impact on the Eranos spirit. Jung's approach was adopted by Fröbe-Kapteyn herself (Portmann 1974: 224). Jung was considered a "spiritus rector", as denoted by Eliade, and a "volcano which brings to light a mysterious rock from the core of the Earth, an exceptional material that can be treated only by a volcanologist" (Portmann 1974: 226–227). And although the symposia were not "Jungian," their core subjects were archetypes, the collective unconscious and its subsequent topics treated from various pespectives (Gronning et al. 2007: 248; von Franz 1972: 156; Progoff 1966: 310). For Jung, the Eranos meetings were a testing ground for his individual thought processes, offering an opportunity for discussions with other creative thinkers. He presented his concept of individuation at Eranos in 1933, and at the next Eranos symposium he outlined his concept of archetype.[3]

In 1946, 13 years after C. G. Jung's first Eranos appearance, the Eranos circle was joined by a Swiss biologist, Adolf Portmann, who greatly influenced the future direction of the meetings. Not only was Portmann an academic but he was also known as a public intellectual. He gained notoriety with his radio shows, newspaper articles and public lectures, which he used as tools to introduce science to the general public. It was for him also an opportunity to agitate against political materialism and social Darwinism (Rieppel 2016: 100).

Adolf Portmann was also occupied by the idea of the unity of living organisms in both a psycho-biological sense and in relation to environment. His lectures caught Fröbe-Kapteyn's attention (Ritsema 1982), and she encouraged him to visit the Eranos meetings. On the one hand, Portmann perceived the invitation as an opportunity to enrich discussion and perception of complex topics, appreciating the Eranos ethos of not being limited to a particular scientific discipline, but on the other hand he also saw a need to introduce modern scientific approaches into the Eranos debates. That is why his introduction in 1946 signified a major turnaround in the spirit and content of Eranos. The title of Portmann's first Eranos meeting was "Spirit and Nature" (*Geist und Natur*) and it was the first time that natural scientists were also invited.

The aim of the first Eranos symposium which Portmann attended was to inspect the boundaries between the natural sciences and humanities. Jung spoke about "The Spirit of Psychology," while Nobel Prize holder Erwin Schrödinger spoke about "The Spirit of Science." The field of biology was represented by Portmann, whose first Eranos lecture held the title "Biology and the Phenomenon of the Spiritual" (Die Biologie und das Phänomen des Geistigen, 1946). From that moment, Portmann

[3] Jung's last lecture at Eranos (at the 1951 session on "Man and Time") was another example of his use of the meetings to test ideas. He called his lecture "On Synchronicity" (Über Synchronizität), and it dealt with a frontier subject between psychology and physics. Jung was indecisive whether to present it at all and had to be encouraged by Wolfgang Pauli, who studied the phenomenon from the perspective of quantum mechanics (Gieser 2005).

never missed a single meeting and gave more than 30 lectures – more than any other speaker before him. His lectures are available in Eranos journals, and most of them were also published in his books *Biology and Spirit* (Biologie und Geist, 1956) and *Awakening of Life Science* (Aufbruch der Lebensforschung 1965). Portmann's friendship with Fröbe-Kapteyn deepened, and eventually she entrusted him with the Eranos chairmanship. Portmann maintained this role for 15 years (1962–1977), partially with the help of Dutch sinologist Rudolf Ritsema and his wife, Catherine. Ritsema also chaired Eranos after Portmann's death (Ritsema 1982; Lochman et al 1982).

"The Original World Experience"

With the accord of Fröbe-Kapteyn, Portmann and other scientists eventually shifted the original Eranos spirit established by Jung. The thread of East-West spirituality[4] which ran through Eranos meetings for more than a decade was replaced by particular subjects of natural science, cultural criticism, sociology and theology. However, Portmann did not just negate the original conception of the symposium but displayed a candid interest in it. His aim, similarly to Jung and other Eranos members, was to place empirical knowledge into a wider frame. As he drew from a tradition of idealist morphology, he transcended the boundaries of his discipline for the sake of multidisciplinary dialogue. He opposed the instrumentalization of science in the attempt to dominate nature and to engineer life. His deep appreciation for art infused his approach to biology (Rieppel 2016), which understood biology as the research of life, which he argued is ungraspable by purely scientific methods, not only because of the obscurity of the concept of life and the currently insufficient level of knowledge, but also by its complexity:

> Life is more complex than the science devoted to its exploration. The reality which encompasses us and which we are part of is greater than what is possible to discover with the methods of research of a specific time. The boundaries which are set by these actualities are not marked by barbed wire or less treacherous obstacles, but the researcher indicates them.
>
> And wherever he senses the boundary, he will try to determine what is scientifically expressible with greater vigilance – but he will also understand that man, with the power of his mind, strives for a bigger picture of his being in the world. He who stands at a frontier is also looking across barriers to a different land. (Portmann 1969: 60)

For Portmann, strict analytical reason should be enriched by other modes of thinking, thus making a researcher better suited to perceive highly complex subjects. He explained that Eranos was a place where he could experience this way of thinking:

> My biological work has given me a strong sense of our original world experience. The Eranos meetings made the confrontation with archaic thinking particularly impressive; it was truly the central point of our work. (Portmann 1974: 225)

---

[4] As Portmann's biographer once wrote ironically: "Portmann and Eastern mysticism, that is not a rational connection, that is at most karmic sympathy!" (Illies 1981: 232).

To grasp the atmosphere built around Eranos by its "*spiritus rector*," Carl Gustav Jung — something that Portmann explicitly appreciated — we might consider Jung's *Archetypes and the Collective Unconscious* (1934/1959), where he analyzes the difference between the mind of primitive and modern man. The former has an "irresistible urge" to assimilate outer experiences into inner psychic events, hence conceiving of any natural happening as a mythological story, while the latter is able to detach himself to some extent from outer objects (Jung 1934: 6).

According to Jung, what really changed in human history was not the nature of the unconscious — that would contradict Jung's definition — but the rise of the conscious. The unconscious will always require us to assimilate oncoming experiences, and by doing so we gradually discover our position in the world and establish a relative balance between our psychological components. This is the cornerstone of the individuation process, the ultimate goal of Jungian psychoanalysis. In his *Symbols of Transformation* (1952), Jung equates "the original state of unconsciousness" to "containment in the primal mother," a containment which has been *sacrificed* (Jung's own term) by becoming conscious (Jung CW 5: §652). From this perspective we can better understand Jung's most general statement: "the collective unconscious is an image of the world,"[5] or, as I would rather reformulate it for the purposes of this chapter, "a totality of our connections to the world" – a definition which, to my best knowledge, would "survive" all of Jung's variations in his Collected Works.

Awareness of our connection to a larger whole that is continuous, potentially infinite and only very marginally illuminated by consciousness — this is the archaic mode of thinking rediscovered by modern humans and promoted by Jung in Eranos lectures. And although Portmann was critical of some aspects of Jung's thought, as we will see later, he himself recognized a role for archaic thinking in the work of a scientist:

> In the moment when our dreaming overcomes all possibility of reason, the imagination assumes its original role which is to *integrate us into a larger whole*, something larger than a world of elementary functions of survival. And in those moments the imagination prompts the thinker to write big words like *élan vital* and allows him to realize his deeds and creations [...] When reason is not fruitful anymore, where rational theories cannot illuminate anything, in a dreaming we glimpse great images of "Nature,""Mother The Creator," "Magna Mater," the eternal myth of the maternal foundation, the uterus and origin of all things. (Portmann 1950b: 202)

Portmann's description of an integration into a larger whole was not just a momentary poetic vision but a serious philosophical question linked to various aspects of his biological work, most notably to his inquiry into biological epistemology and his concept of living forms. We encounter them across his rich lectures and publications, in his Eranos lectures, and, most densely formulated, in his *New Paths in Biology* (1964). There he claims that all fundamental biological questions lead us

[5] The passage continues: "the collective unconscious is an image of the world that has taken aeons to form. In this image certain features, the archetypes or dominants, have crystallized out in the course of time. They are the ruling powers" (Jung CW 7: § 151).

beyond the limits of scientific statements. Life always means something more than what a given era with its scientific methods can testify, and if we want to explain the most characteristic processes of life, we reach the paradox that "Life is older than living beings"[6] (Portmann 1964: 6).

If we move our attention from the general notion of life to concrete living beings, we encounter the problem of wholeness again. An individual, whether animal or human, is indivisible and non-reducible to its parts.[7] It relates itself to the world and creates and preserves its own inner world. It is only in these ways that it is genuine and complete (Portmann 1964: 10). For Portmann, every life form is an immense reality which does not contradict a rational analysis, though it is not embraced by it (Portmann 1949). This leads him to the expression of the "unsolvable unity of the organism" (Portmann 2000: 126).[8]

The term which Portmann propagated and which was supposed to denote the complex connection between an organism and the environment was "inwardness" (*Innerlichkeit*). Portmann established the notion of inwardness as a heuristic instrument and at the same time as a task for future researchers to formalize:

> We have penetrated into two invisible zones - into the darkness which begins below the threshold of our optical instruments, and into the other darkness which hides our subjective experience […]. Nature comprises every aspect of life – subjective experience no less than structure. Biologists ignore this fact at their peril. (Portmann 1964: 34–37)

According to Portmann, if we want to perceive a living organism adequately, it is necessary to enrich the intellect by imaginative thinking. Both capacities are "important elements of the totality of a human." Only when employed together can we understand not only the diversity of living things, but also the "kind of experience that integrates them into a new world of meanings" (Portmann 2000: 93–94).

The experiencing of fullness, appreciation of greatness, and the encompassing of wholeness that Jung strived to cultivate during Eranos symposia can be appreciated by a scientist but do not suffice when it comes to solving practical problems. Portmann touched on this subject in his Eranos lecture "Mythology in Natural Research" (Mythisches in Naturforschung), the first of two Eranos lectures dedicated to reflection on Jungian subjects. Portmann made a kind of personal confession about the role of imaginative thinking in his biological work but also denoted its limits. When it comes to technical progress, the task of a scientist is opposite to

[6] Portmann refers here to John Desmond Bernal, a pioneer in molecular biology who studied the formation of organic compounds from inorganic matter, concluding that life appeared earlier than living organisms. The idea is formulated in his *The Origin of Life* (Bernal 1967).

[7] In the English preface of the book *Animal Forms and Patterns* (1967), the translator Hella Czech explains that the wholeness of an organism is better denoted by the German term "Gestalt" than the English "form": "The word *Gestaltung* implies both the process by which such forms have been produced and its result; organic form production, production of animal form, have seemed to be the most suitable equivalents" (Czech 1967: 7).

[8] Portmann does not imply here any vitalistic principle. He himself was an opponent of vitalism; as he wrote in the foreword to Jakob von Uexküll's *Bedeutungslehre*, we cannot suppose any secret agent of life which can enter as an all-explaining factor (Portmann 1956: 7).

what the archaic mind does. Science "disenchants the world," as Portmann demonstrates with the example of the human desire to fly. Only after we got rid of the mythological images of flying souls and Icarus' wings and started to construct unsightly but functional assemblages of wooden planes were we able to actually soar into the air. On the other hand, it was imaginative thinking that produced the desire to fly in the first place.

The imagination is for Portmann a source of inspiration but not the solution. And yet it was the imagination that gave shape to many biological theories. According to Portmann it was especially the period between the seventeenth and nineteenth century when imagination and symbolic thinking obstructed a factually eager mind. Portmann was very well aware that it was during the period in question that theories of the human unconscious and animal instincts started to develop, and that philosophers like Kant, Schopenhauer, Carus and Hartmann created the bedrock for Jung's theories. Also, in the nineteenth century a theory of organic memory as a possible technical explanation for how instincts are transmitted between generations was born, and Portmann knew that Jung's thoughts about the nature of the unconscious were strongly influenced by it.[9]

Instincts and Archetypes

Questions about the nature of instincts and archetypes and their mutual relation were crucial for Jung's inquiries about the human psyche and inevitably caught the attention of biologists.[10]

Jung himself did not elaborate on instinct as an isolated phenomenon, mainly using it as a border notion whose main purpose was to define the psyche. According to him, the psyche is delimited "from the bottom" by instincts and "from above" by spirit. The psyche is an emancipation from the compulsiveness of instinct. Human consciousness and will function in the zone between the instinctive and spiritual spheres. This liberation from biological determinism makes the development of psychic quality possible. Jung occasionally names the supposed characteristics of instinct as contrasting with the psychic sphere, mainly its compulsiveness and rigidity. When touching the necessity of defining instinct itself, he claims that "it is

[9] Theories of organic memory promoted by Butler, Hering, Semon, Ribot, Lazarus, Steinhal and Wundt were based on two basic conceptions: Lamarck's theory of the inheritance of acquired characteristics and Haeckel's biogenetic law that ontogeny recapitulates phylogeny (Otis 1994).

[10] In an article from 1937 called "The Effectiveness of Archetypes in the Instinctive Actions of Animals" (Die Wirksamkeit von Archetypen in den Instinkthandlungen der Tiere), Friedrich Alverdes described a biology without psychology as incomplete. He also pointed to Jung's psychology as helpful in the research of animal minds. Alverdes adopted Jung's concept of collective unconscious in the sense of a latent disposition to certain identical reactions and behavior. According him, not only humans but also animals have an unconscious. (Alverdes 1937) Adolf Portmann even uses the term "psyche" while describing the navigation ability of migratory birds (Portmann 1964: 28–29).

uncommonly difficult not only to define the instincts conceptually, but even to establish their number and their limitations" (Jung 1947: §374).

As well as instinct setting bounds for our minds "from the bottom," Jung describes archetypes as taking effect "from above," existing beyond the reach of our intellect. Although these two directions are in an apparent clash, somehow biological dynamism correlates with the spiritual. Jung even states that "psychologically [...] the archetype as an image of instinct is a spiritual goal toward which the whole nature of man strives" and that, "the archetype is a formative principle of instinctual power" (Jung 1947: §84). When Jung continues in this direction of thought, he quickly encounters the question of whether archetypes are not simply the instincts of the human species:

> To the extent that the archetypes intervene in the shaping of conscious contents by regulating, modifying, and motivating them, they act like the instincts. It is therefore very natural to suppose that these factors are connected with the instincts and to inquire whether the typical situational patterns which these collective form-principles apparently represent are not in the end identical with the instinctual patterns, namely, with the patterns of behaviour. I must admit that up to the present I have not laid hold of any argument that would finally refute this possibility. (Jung 1947: §404)

Jung indeed tried to contextualize archetypes biologically. In his *Psychological Types* (1921) he identifies the collective unconscious with Semon's phylogenetic mneme[11]:

> The psychic structure is the same as what Semon calls "mneme" and what I call the "collective unconscious." The individual self is a portion or segment or representative of something present in all living creatures, an exponent of the specific mode of psychological behaviour, which varies from species to species and is inborn in each of its members. The inborn mode of acting has long been known as instinct, and for the inborn mode of psychic apprehension I have proposed the term *archetype*. I may assume that what is understood by instinct is familiar to everyone. It is another matter with the archetype. (Jung 1921: 376)

It must be said, though, that Jung found Semon's theory insufficient.[12] Even in his time, Semon's theory was criticized as being based on Lamarck's theories of the inheritance of acquired characteristics.[13] Jung distanced himself from the Lamarckian

[11] In his fundamental paper "Mneme" (*Die Mneme*) (1904), Richard Wolfgang Semon (1859–1918) came up with a mnemic principle. It is based on the presupposition that agitation leaves traces. Repetition of the same stimuli leads to renewal of those traces, and the process becomes hereditary. All the organized matter has the ability to retain the traces so they can be evoked later. Semon called this ability "mneme" and the trace "engram."(Semon 1904, 1921). Among his notable friends was psychiatrist Auguste Forel (1848–1931), a leading figure of the Burghölzli Asylum and Eugen Bleuler (1857–1939), whose student was C.G. Jung (Schacter 2001).

[12] "We are forced to assume that the given structure of the brain does not owe its peculiar nature merely to the influence of surrounding conditions, but also and just as much to the peculiar and autonomous quality of living matter, i.e. to a law inherent in life itself" (Jung 1921: §748).

[13] Theodule Ribot considered Semon's book a metaphysical work (Ribot 1912: 267); it was also criticized by Hans Driesch (1908: 218–219) and August Weismann, whose 27-page critique of Die Mneme appeared as the lead paper in the 1906 volume of the *Archiv für Rassen- und Gesellschaftsbiologie* (Schacter 2001: 130).

position already in 1918 when he claimed that there cannot be anything like inherited ideas, but that we could think about inherited possibilities of ideas, "apriori conditions for fantasy-production" which are in a way similar to Kantian categories (Jung 1918: §14). Nevertheless, the fact that Jung tried to look for parallels between the unconscious and mneme encouraged criticism from Portmann, who considered it a central point of outdated thinking and pure speculation.

Jung was curious about Portmann's view on this matter and asked him to present it at Eranos. The resulting lecture, "The Problem of the Primordial Image in Biological Perspective" (Das Problem der Urbilder in biologisher Sicht) in 1950, was not only Portmann's reflection of instincts and archetypes from a biological perspective but also played a part in the Eranos commemoration of Jung's 75th birthday.

Early in the lecture, Portmann joked a little when he said that biologists reached for the term *instinct* instinctively. For Portmann, the manner in which biology explained patterns of unconscious behaviors was unsatisfactory. While trying to find the relation between archetypes and instincts, states Portmann, Jung worked on biological presumptions from the nineteenth century, when Jean-Henri Fabre established a concept of instinct as a rigid set of inborn, automatically triggered reactions. That, according Portmann, cannot fully explain the connection which animals and humans have with their environment and suggested that such biological phenomena are more akin to taxis or tropism.[14]

After Fabre, the concept of instinct underwent dramatic changes, namely through the work of scientists such as the Austrian zoologist Konrad Lorenz (1903–1989) and Dutch biologist and ornithologist Nikolaas Tinbergen (1907–1988), both Nobel Prize holders for physiology and medicine. Lorenz found middle ground between the early twentieth century behaviorist conception of instincts as mechanical, inborn reactions to specific stimuli (reflex theories) and a vitalist conception of instinct as an irreducible purpose which guides mental activity (Lorenz 1950: 232; Brigandt 2005). Of course, not everything is inherited, stated Lorenz. Some behavior is learned, and some is the product of insight (Lorenz 1970: 116). In the middle of the century, this was already perceived by many as a central theory of the new ethology, also thanks to the popularizing activities of Lorenz.

Adolf Portmann absorbed all these new impulses from ethology. In *Animals as Social Beings* (Das Tier als soziales Wesen), Portmann builds on the research on bees by Frisch, Roesch, Lindauer and others, which indicated that even in the behavior of such relatively "lower" creatures, we can trace phenomena like invention, nonconformity, and work towards a purpose, but also indolence (Portmann 1961: 99–104). For these inner states that we can only guess from the outside, Portmann chose the name "tuning" (*Stimmung*), taking from the musical sense, but in contrast with the physically clear basis of musical tones, the biological structures that underlie these affective-cognitive states are largely unknown (Portmann 1960:

[14] Fabre considered instinct to be any behavior which did not require cognition or consciousness to perform (Portmann 1961: 98–119).

58). During his own research, Portmann witnessed an animal's ability to orient itself in such complicated and variable situations that instead of naming this capacity "instinct," he inclined to Wolfgang Köhler's term "isomorphy," which denotes a correlation between a structure of the world and a structure of an animal and which does not suggest a strictly mechanical connection between inborn structure and behavior. Where in Portmann's debate about instincts is there a place for Jung's archetypes? Are archetypes a typical human feature or are they just instincts typical for the human species? The archetype, in the context of various Jungian descriptions of it, could be interpreted as a process of achieving a human meaning in accordance with biological processes[15] and with the world in general or, translated into Portmann's terms, as a human isomorphy, the specific human experience of the world.

Although it is a hidden structure, for Portmann the human unconscious can be studied by means of biology. It is not a chaos inaccessible by reason. Compared to an animal's mind, it might even be more accessible, as Portmann suggests, because with higher organizations of life the amount of hereditary structures grows. It is very probable that there are more such structures in humans than in any other species, and so we must presuppose an "enormous pool of preexisting forms of experience" (Portmann 2000: 113). This biological fundament, says Portmann, should be a starting point for complex psychology in its study of the structure of archetypes. It is also a task of biology to substantiate the phenomena which Jung called archetypal.

Rather than "archetype," Portmann chose in his lecture the term "archetypal structure" as more suitable for a biologist, presumably avoiding possible ontological connotations. He suggested three types or groups of these structures specific to humankind. The first type is essentially hereditary. Such structures have a fixed character from the beginning and correspond to the "triggers" found in animals. An example might be the recognition of a human face by a baby. At the same time, Portmann expresses his uncertainty as to what extent the archetypes of "woman", "man" or "father and mother" are hereditary.

The second archetypal group includes such structures in which heredity takes part only in a vague and open way. These structures allow our individual formation. Here Portmann places sexual preference but also the subjective experience of "home" and the resulting special relation to this place as examples.

The third group is formed by the practice and adoption of customs, traditions and goods. Everything that Gaston Bachelard marked as *complexes de culture* belongs here according to Portmann. In this case, Portmann concedes a possible merit of the Lamarckian conception of heredity, but only on a cultural level.[16]

[15] Jung also wrote: "We may say that the image represents the meaning of the instinct " (Jung 1947: §398).

[16] Bachelard established the notion of cultural complex as a part of his literary theory. He defines it as an unreflected attitude which affects the very process of reflection and imagination. While we believe that we are individually evolving, in fact we just cultivate a cultural complex. According Bachelard, a "realist" just chooses his reality the same as a historian chooses his version of history.

Jung himself was very dissatisfied with the outcomes of Portmann's Eranos lectures, and ever since, his animosity towards Portmann grew. According to Jung it is hopeless to explain the nature of archetypes to people who do not have direct experience with psychology. He wished that natural scientists would not enter a field they knew nothing about (Shamdasani 2003). According to Jung, only representations of archetypes in the form of symbols and myths, and not the archetypes themselves, could be empirically studied because by becoming conscious an archetype is always altered. For the same reason, Jung would also object to Portmann's notion of "archetypal structure" as a clear reference to the physiological structure of the brain and the senses. Jungian archetypes do not describe a structure of our mind, but the fact that the mind is capable of representing them: "'Archetype' is an explanatory paraphrase of the Platonic *εἶδος*" (Jung 1934: 4–5).This emphasis on the psychological and philosophical aspects of the archetype is probably the main reason for discord between the two scholars. Nevertheless, Portmann remained diplomatic and evaluated the whole situation as a partial success. After many years he wrote in a letter to Zwi Werblowsky:

> My treatment (of Jung's archetypes) was a part of C. G. Jung's birthday party and so naturally a polemic moment has dissolved and my respect for Jung's endeavour prevailed in my study. My opinion today is after all the same as before, that Jung's research contributed to the broadening of the discussion about the archetype. I discussed with Jung several times how his presentation implies a cultural meaning of the archetype stressed by Bachelard and not the inherited one.[17]

It must be said that in their studies of archetypes, Portmann and Jung did not share the same terminology. Portmann dedicated his Eranos lectures almost exclusively to examples of animal behavior and animal morphology. When he decided to proceed to the question of archetypes as a human feature, he limited his inquiry to early stages of human life (up to 4–5 years), thus the period of apparent cultural formation of an individual but still very rudimentary in comparison with the fully developed adult human,[18] and he very briefly mentioned Bachelard's culture complexes.

Both are connecting to a certain tradition. Bachelard's cultural complexes are dependent on historical, cultural, literary and personal contingencies but at the same time have an inherent and shared foundation (Bachelard 1999).

[17] July 11, 1973, Portmann archive (Portmann 1973).

[18] Portmann refers to the research of E. Kaila (1932) R. A. Spitz, and K. M. Wolf (1946) on smiling as one of the earliest emotional and social patterns of infant behavior. Kaila observed three-month-old children reacting to other people's faces. Children's reaction or smiling response was reduced and their anxiety increased when exposed to a motionless face compared to a vivid one (Kaila 1932). Spitz (1887–1974) discovered during his observations of children from three- to six-months-old that it is not a certain expression of the face that triggers the smile of the child, but the general shape of the face: domed forehead, two symmetrical eyes, nose. Various experiments with defigurations of the face presented to a child proved the effect of this specific configuration (Spitz, Wolf 1946). Here, claims Portmann, we clearly encounter the phenomenon described by C.G. Jung as the archetype, a reflection of a hidden psychic structure, an inborn readiness for action which is triggered when an appropriate configuration is reached.

Portmann was looking at human behavior and its unconscious patterns in a similar fashion as an ethologist studies the behavior of animals, i.e. as processes significant in themselves, without apparent transcendence to broader meaning, ethical dimension or personal development — topics that came together in Jung's notion of individuation, the lifelong path of a human along which archetypes serve as signposts. Compared to this Jungian picture of humans' life drama, Portmann's ambition in the archetype department was relatively modest. He wanted to demonstrate how biology can contribute to the discussion, mostly in a critical way, as the biological findings prove many hereditary speculations unfounded (particularly Lamarckianism), and that caution is needed when we try to link archetypes with instincts.

Conclusion

Portmann was the most notable biologist of the twentieth century to have evaluated Jung's study of archetype. It was Eranos that made this intellectual exchange possible. While having a great respect for Jung, Portmann found it necessary to critically evaluate Jung's conception of archetypes and instincts, seeing both subjects as apparently related to each other. But similar to Jung, when he stepped to the edge of the obscure world where instincts and archetypes resided, Portmann refrained from making definitive statements: "We are reaching the edge of the zone where any certainty of scientific claims exists." The proper research of this field, according Portmann, belonged to natural scientists who are able to perceive the human dimension to a much greater extent and can appreciate the richness of human experience. It belonged also to the researchers of the psyche who would inquire into biological structures more than before, with emphasis on the early stages of human development (Portmann 1950a).

Portmann's effort to conceptualize archetypes as a biological structure clearly irritated Jung, but despite all their differences we find fundamental similarities in both thinkers. Both of them reached the very limits of what is expressible not only by means of their disciplines but by human language itself. Both of them tried to formulate the most complex notions, be it the unconscious or life itself, and to place the knowledge of their respective disciplines into a more general, humanized context.

From a values perspective, Portmann and Jung were humanists who stood on the same ground against reductionist and mechanistic tendencies in science. Instead of an approach that saw every being as rigidly formed by the pressure of its surroundings or innate drives, both conceived as something foreign to its own nature, they perceived the environment or the world in general as an opportunity for development. To them, a being is not just formed by its surroundings, but forms itself and expresses itself. This revelation of possibilities is also one of the effects of the Jungian unconscious. More than anything, Jung's archetypes are a source of possibility and meaning. Similarly, Portmann conceives of an organism as a peculiar opportunity of expression and experience of the world. It was only through interdisciplinary meetings such as Eranos that such an exchange could have taken place. In Portmann's words:

> Collaboration in the circle of Eranos had an immense influence on the widening of my spiritual world. Above all it provided nutrition for my deep desire to cross the borders of my specialization and to put together results of my own work with results of other fields of knowledge – and at the same time to present those parts of biological research where I found deeper meaning to a wider circle of participants [...]. Eranos symposia also allowed me to grasp more clearly many questions which a lay person would like to address to a biologist. (in Ritsema 1982: 12)

Jung and Portmann met on Eranos ground and did their best as psychologist and biologist to show a better approach to humans and the world. Although they collided in their opinions on the specific subject of archetypes, the encounter itself became a valuable contribution to discussion about life and the unconscious. The fact that both of these great minds of the twentieth century struggled to answer the most fundamental questions of their respective disciplines and to translate them into the framework of others only proves how complex and inexhaustible the topics are.

References

Alverdes, F. (1937). Die Wirksamkeit von Archetypen in den Instinkthandlungen der Tiere. *Zoologischer Anzeiger, 119*(9/10), 225–236.

Bachelard, G. (1999). *Water and dreams: An essay of the imagination of matter.* Dallas: Dallas Institute of Humanities and Culture. [Bachelard, G. (1942). *Eau et les rêves*. Paris: Corti.]

Bernal, D. J. (1967). *The origin of life*. Cleveland: World Pub.

Brigandt, I. (2005). The instinct concept of the early Konrad Lorenz. *Journal of the History of Biology, 38*, 571–608.

Czech, H. (1967). Translator's preface. In A. Portmann, *Animal forms and patterns* (2nd ed., pp. 7–8). New York: Schocken Books.

Driesch, H. (1908). *The science and philosophy of the organism*. Aberdeen: University Press.

Gieser, S. (2005). *The innermost kernel: Depth psychology and quantum mechanics. Wolfgang Pauli's dialogue with C.G. Jung*. Berlin/Heidelberg: Springer.

Gronning, T., Sohl, P., & Singer, T. (2007). ARAS: Archetypal symbolism and images. *Visual Resources: An International Journal of Documentation, 23*(3), 245–267.

Hakl, H. T. (2013). *Eranos: An alternative intellectual history of the twentieth century*. Montreal: McGill-Queen's University Press.

Hillman, J., & Shamdasani, S. (2013). *Lament of the dead: Psychology after Jung's Red Book*. New York: W. W. Norton.

Illies, J. (1981). *Adolf Portmann: Ein Biologie vor dem Geheimnis des Lebendigen*. München: Herderbücherei.

Jung, C. G. (1916). *The structure of the unconscious*. Collected works of C. G. Jung (Vol. 7).

Jung, C. G. (1918). *The role of the unconscious*. Collected works of C. G. Jung (Vol. 10).

Jung, C. G. (1921 [1946]). *Psychological types or the psychology of individuation* (H. G. Baynes, Trans.). London: Kegan Paul, Trench, Trubner, & Co.

Jung, C. G. (1934 [1954]). *Archetypes of the collective unconscious*. Collected works of C. G. Jung (vol. 9/I), §§1–86.

Jung, C. G. (1947 [1954]). *On the nature of the psyche*. Collected works of C. G. Jung (Vol. 8).

Kaila, E. (1932). Die Reaktion des Säuglings auf das menschliche Gesicht. *Annales Universitatis Aboensis, 17* (Set. B), 1–114.

Kerenyi, K. (1955). Was bedeutet der Name Eranos? *Du, 15*(4), 39–40.

Lochman, J. M., Ritsema, R., Mislin, H., & Thürkauf, M. (1982). Gedenkfeier für Herrn Prof. Dr. Adolf Portmann. Freitag, 9. Juli 1982, in der Peterskirche, Basel. Privately Published.

Lorenz, K. (1950). The comparative method in studying innate behavior patterns. *Society for Experimental Biology, Physiological Mechanisms in Animal Behavior. Society's symposium* IV, pp. 221–268.

Lorenz, K. (1970). *Studies in animal and human behaviour* (R. Martin, Trans.). London: Methuen.

Miller, D. L. (2006). At the edges of the round table: Jung, religion and Eranos. In L. Cowan (Ed.), *Edges of Experience: Memory and emergence*. Proceedings of the Sixteenth International Congress for Analytical Psychology (pp. 129–144). Einsiedeln: Daimon.

Noll, R. (1994). *The Jung cult*. Princeton: Princeton University Press.

Otis, L. (1994). *Organic memory*. Lincoln: University of Nebraska Press.

Portmann, A. (1949). Mythisches in der Naturforschung. *Eranos Jahrbuch - Der Mensch und die mythische Welt*, Bd., *17*, 475–514.

Portmann, A. (1950a). Das Problem der Urbilder in biologisher Sicht. Sonderband für C. G. Jung zum 75. Geburtstag. *Eranos Jahrbuch - Aus der Welt der Urbilder,* Bd. *18*, 413–432.

Portmann, A. (1950b). Imagination et biologie. *Revue de théologie et de philosophie, 38*(156), 187–202.

Portmann, A. (1956). Ein Wegbereiter der Neuen Biologie. In J. Uexküll, & G. Kriszat (Eds.), *Streifzüge durch die Umwelten von Tieren und Menschen. Ein Bilderbuch unsichtbarer Welten* (pp. 7–18). Hamburg: Rowohlt.

Portmann, A. (1960). *Neue Wege der Biologie*. München: Piper. [Portmann, A. (1964). *New paths in biology (*A. J. Pomerans, Trans.). University of Wisconsin: Harper & Row.]

Portmann, A. (1961). *Animals as social beings* (O. Coburn, Trans.). New York: Viking.

Portmann, A. (1965). *Aufbruch der Lebensforschung*. Suhrkamp Verlag: Frankfurt am Main.

Portmann, A. (1969). Grenzen der Biologie. In L. Reinisch (Hg.), *Grenzen der Erkenntnis* (Herder-Bücherei Nr. 357, pp. 45–60). Freiburg: Herder-Verlag.

Portmann, A. (1973). Letter to Zwi Werblowsky, 11 July 1973, Portmann's personal correspondence. Archive of University, Basel.

Portmann, A. (1974). *An den Grenzen des Wissens*. Zurich: Ex Libris.

Portmann, A. (1976). Jung's biology professor: Some recollections. *Spring*, 149–154.

Portmann, A. (2000). *Biologie und Geist*. Göttingen: Burgdorf.

Progoff, I. (1966). Idea of eranos. *Journal of Religion and Health, 5*(4), 307–313.

Quispel, G. (1980). Gnosis and psychology. In B. Layton (Ed.), *The rediscovery of Gnosticism* (pp. 17–31). Leiden: Brill.

Ribot, T. (1912). Le rôle latent des images motrices. *Revue Philosophique de la France et de L'étranger, 73*, 248.

Rieppel, O. (2016). *Phylogenetic systematics. Haeckel to Hennig*. Boca Raton: CRC Press.

Ritsema, R. (1982). Adolf Portmann 1897–1982. *Eranos Jahrbuch, 51*, 1–6.

Schacter, D. L. (2001). *Forgotten ideas, neglected pioneers*. Philadelphia: Psychology Press.

Semon, R. (1921). *The mneme*. London: George Allen & Unwin. [Semon, R. (1904). *Die Mneme als erhaltendes Prinzip im Wechsel des organischen Geschehens*. Leipzig: W. Engelman.]

Shamdasani, S. (2003). *Jung and the making of modern psychology*. Cambridge: Cambridge University Press.

Spitz, R. A., & Wolf, K. M. (1946). The smiling response: A contribution to the ontogenesis of social relations. *Genetic Psychology Monographs, 34*, 57–125.

Vitolo, A. (2015). The analytical leitmotif of the Eranos conferences. In R. Bernardini & N. Cater (Eds.), *Eranos: Its magical past and alluring future: The spirit of a wondrous place* (pp. 107–124). New Orleans: Spring Journal.

von Franz, M. L. (1972). *C. G. Jung: Sein Mythos in unserer Zeit*. Frauenfeld: Verlag Huber.

Wasserstrom, S. (1999). *Religion after religion: Gershom Scholem, Mircea Eliade, and Henry Corbin at Eranos*. Princeton: Princeton University Press.

Ivana Ryška Vajdová studied history and philosophy at the University of Presov, Slovakia. Currently she is a PhD student at Charles University in Prague, Faculty of Science, Department of Philosophy and History of Science, under the supervision of Prof. Stanislav Komárek. She specializes in Jungian studies in the context of history of science, history of depth psychology, psychology of religion and philosophical anthropology. She has written several papers on these topics including "Jung's Psychological Analysis of Imago Dei," "How Deep is Jung's Love for Plato?", and "Three Dimensions of Living Forms – A Brief Look into Parallels between Adolf Portmann and C.G. Jung."

Chapter 14
The Czech Reception of Portmann

Stanislav Komárek

Abstract The University of Prague and its graduates, unlike universities in Anglophone countries, for example, had a certain "preadaptation" to the reception of Portmann, in that it was a "German type" university with a tradition similar to that which produced Portmann himself. Two Czechs studied under Portmann, but wider intellectual interest in him, however, came from Protestant theology circles. At the semi-legal "ecumenical seminar" at the Protestant Theological Faculty in Prague, a paper on Portmann's anthropology was given in the 1963–64 winter semester by Jan Miřejovský and Jan Milíč Lochman. Philosopher and psychologist Jiří Němec (a Catholic) came to know Portmann's work through these seminars, and, at his encouragement, biologist Zdeněk Neubauer and philosopher Václav Benda translated *Neue Wege der Biologie* into Czech and "published" it in typewritten samizdat form in 1979. In the same year, zoologist Stanislav Komárek was introduced to Portmann's thought by Zdeněk Neubauer at a seminar at the Faculty of Natural Sciences; since his return from exile in Austria in 1990, he has lectured regularly on Portmann, and his doctoral students Karel Stibral, Karel Kleisner, Filip Jaroš, and Ivana Ryška Vajdová belong to the new generation of Portmannian researchers in the Czech Republic.

The University of Prague as a German University; "Preadaptation" for the Acceptance of Portmann's Ideas

The University of Prague was in some way predisposed to the acceptance of Portmann's ideas from early times – a similar phenomenon in biology is called "preadaptation" (fish had four fins long before it turned out that these could evolve into legs after their ascent from water). Its tradition is historically that of the German universities of Central Europe and of "German" biology. Essentially all Continental biology outside of Romance countries can be considered "German" biology; in

---

The chapter was translated by Andrew G. Christensen.

S. Komárek (✉)
Charles University, Prague, Czech Republic
e-mail: stanislav.komarek@email.cz

F. Jaroš, J. Klouda (eds.), *Adolf Portmann*, Biosemiotics 23,
https://doi.org/10.1007/978-3-030-67810-4_14

addition to Germany and its historical territories, and other German-speaking countries, this included especially Holland, Bohemia, the Baltic region, and St. Petersburg in Russia. While most work in biology from this region between the years of 1750–1950 was in German, by no means does this necessarily correspond to the native language of the authors but rather to their affiliation with a single cultural sphere. In the present day, when global knowledge of German is greatly in decline and English has taken on the role of lingua franca, this extensive literature is becoming practically inaccessible and, from divergent paradigmatic starting points, difficult to understand even in German-speaking countries. It seems that this nearly "lost" world of ideas will, in a relatively short time, be rediscovered with wonder, similarly as was the world of Alexandrian natural science. One can only hope that this will occur before its literary monuments are reduced to ruins, like those of Alexandria. Thanks to centuries of direct geographical proximity to the German language and the resulting similarities in semantics, phraseology, and an overall similar style of thinking, Czech has been a more viable medium for the translation and interpretation of Portmann's thought than English, in which the art of expressing Portmann's ideas can, with a little exaggeration, be compared to the art of playing Mozart's *Eine kleine Nachtmusik* on, say, a Turkish drum.[1]

Just like German philosophy, German biology (in its capacity as one of its special applications) is characterized by an uncommon intricacy of thought and an ability to render problems in their full complexity. This property of German thinking makes it a powerful tool for learning about and operating with the world, but it also harbors many dangers arising from the possibility of exaggerating certain intellectual concepts (and eventually actions) ad absurdum. This particularity led not only to the great blossoming of German thought in the long nineteenth century but also to the unprecedented horrors of the Second World War. Following the war, the heritage of the past was abandoned across German thought, especially in biology (an analogous process took place in Japan), and replaced with Anglophone models considerably more schematic and flat, which certainly hold substantially lower risk of explosive destruction from "misuse or mishandling" – working with brisant ideas is understandably dangerous, akin to tossing bottles of nitroglycerin. Pockets of "German" thought in biology survived the aftermath of the war mainly in countries which did not feel responsibility for its instigation (Switzerland, Czechoslovakia, The Netherlands, Estonia) nor, therefore, the need to radically abandon their intellectual traditions for those of the victors; at present, such pockets, apart from universities in Prague and Tartu, are in further retreat. There are resonances of Portmann's impact still present in Swiss cantonal identity in Basel, where he spent the majority of his fruitful career.

---

[1] Portmann's biological thought has also been received in other major European languages, especially French and Italian. In French, this primarily relates to thought on man and his relation to animals (cf. the edited volume *Animalité et humanité autour d'Adolphe Portmann*, 1999); the Italian tradition has commented mainly on Portmann's aesthetic morphology (see Conte in this volume). Portmann's thinking on humans is reflected in the field of philosophical anthropology (cf. Wulf 2013).

From its beginnings, German biological thought expressed a greater skepticism than other schools toward the mechanistic view of living organisms and always stressed their specificity. Another typical aspect of German biological thought was its heightened sense of reflection on the changeability of living beings in time (the idea of transformation was also an important theme of German philosophy – e.g. Hegel, Marx, and Klages). This led on one hand to an interest in paleontology and on the other to the intensive study of the ontogenetic development of living organisms. For German biology, this intensive interest in individual ontogenesis, especially in its early stage, embryology, was wholly typical, fundamental, and constitutive, characterizing it from the second half of the eighteenth century well into the twentieth century.

From the eighteenth century, German biology was almost exclusively tied to the university and therefore generously supported, as there was always a significant number of universities which were distributed over a large area (in contrast to the centralization of French science in Paris, for example). A professor of zoology or botany was essentially a senior civil servant enjoying significant prestige and authority, ruling over the affairs of his department almost like a monarch, often for several decades, making a deep impression on the local style of research (Charles University in Prague was this sort of typical "German" university, even in its Czech-language branch from 1882). German thought, having previously held a strong religious charge, transmitted this ethos to the realm of science as "secular religion" in the nineteenth century, and scientific findings played the role of quasi-religious truth. Practical applications began to be expected from German biology only from the second half of the nineteenth century, when there was a general shift away from Romanticism in Europe and particularly from *Naturphilosophie* in Germany (this term is customarily used to refer to German natural philosophy in the first half of the nineteenth century). Even so, the desire for practical gains from science came relatively slowly, arriving only with the consolidation of Bismarck's ideal of a functional, efficient, and unsentimental Germanness. The German "national spirit" in the first half of the nineteenth century had a rather different form than in later periods, with a predominance of fantasy that was lively yet seldom practical. In a politically fractured and rather restrictive Germany, among the few escapes from the drab and philistine provincial daily life were science (including biology) and art, which frequently intertwined (an interest among artists in natural science and the artistic, especially literary, ambitions of many scientists were typical phenomena of the Romantic era).

The intimate connection between German biology, the universities, and the philosophy that was taught there gave German biological concepts a particularly philosophical character. Certain ideas from late Antiquity were revived in new forms in German *Naturphilosophie*, particularly an abundant use of analogies. A nice example of this is the derivation of the oral appendages of crustaceans or insects through morphological transformation from the walking legs, or the mandibular apparatus and middle ear bones from the branchial basket of vertebrates. A further range of concepts derived from this intellectual tradition can be found in botany or psychology – for example, Hofmeister's concept of alternation of sexual and asexual generations in plants, Freud's homology between childhood activity and adult sexuality, or Lorenz's homologization of various types of behavior. Although the expression of change in the living world was a fundamental theme of *Naturphilosophie*, and though creationist biases were almost nonexistent in Germany, concepts of natural history led rather to the revival of certain mythopoetic themes of origin, extinction, and metamorphoses into new and unusual forms on the basis of new "material" than to evolutionary theories of the Lamarckian-Darwinian type.

Between the wars, Continental Europe saw the formation of the so-called "German autonomist school" (Komárek 2004), emphasizing the autonomy and spontaneity of living organisms and the autonomous origin of their structures, especially through the example of animal coloration, which is subject to selection only to a limited degree and expresses the so-called "inwardness," *Innerlichkeit*, of living organisms. To this school belonged three scientists who studied the ontogenetic and phylogenetic development of organisms, especially butterfly patterns:

Fritz Süffert (1891–1945) from Freiburg im Breisgau, Karl Henke (1895–1956), working in Göttingen, and Boris N. Schwanwitsch (1889–1957), working in St. Petersburg. After the war, the only one from this group able to continue on without radical upheaval or the need to revise his standpoint in the face of external factors induced by "great historical events" was Adolf Portmann (1897–1982).[2]

It can be said that already in the 1920s, Portmannian tendencies were being propagated in the Czech sphere by professor of botany Josef Velenovský (1858–1949), who in his book on natural philosophy (Velenovský 1922) introduces the term "principle of ornamentalism" for the creative abilities and characteristics of plants and fungi; due to his unusual character, however, and the subsequent war, Velenovský's ideas did not see wider application in the local intellectual environment (Stibral et al. 2004; Stibral 2009a, b, c).

Czech Protestants, Jiří Němec, and "Bringing Portmann to Bohemia"

The contact of Czech dissidents and academics with Portmann's ideas dates to the 1960s. Already before the Second World War, however, Portmann's article *Evoluční problémy ptačího vývoje* [The Ontogenesis of Birds as an Evolutionary Problem] was published in the medical journal *Měsíčník Ciba* (which had strong ties to Switzerland) in the issue *On Normal and Abnormal Development* (Portmann 1937). Czech was therefore among the first languages in which Portmann's work was introduced. The earliest known paper on Portmann delivered on Czech soil was by Jan Mířejovský (1911–1984) in the context of the ecumenical seminar in "V Jirchářích"

[2] Portmann also continued the work of Jakob von Uexküll and his doctrine of the *Umwelt*; see his foreword to the complete edition of von Uexküll's *Streifzüge durch die Umwelten von Tieren und Menschen a Bedeutungslehre* (Portmann 1956). Some authors (Monastra 2000) assign Portmann to the tradition of idealistic morphology.

street in the winter semester of the 1963–1964 academic year. His co-contributor was his brother-in-law Jan Milíč Lochman, who would later become a famous theologian at the University of Basel. The ecumenical seminar in "V Jirchářích," organized at the Comenius Protestant Theological Faculty in Prague, was an important center of spiritual dialogue and contributed to the renewal of intellectual life of the 1960s and early 1970s. J. Miřejovský worked as a secretary for the World Council of Churches in Geneva from the years 1948–50. Jan Milíč Lochman (1922–2004) and Jan Heller (1925–2008) matriculated as theology students at Basel in 1947 when Portmann was rector and were later examined by him in anthropology. The interest of Protestants in Portmann's anthropology is evidently closely related to his holistic framework and his unwillingness to reduce man to a purely biological dimension; Portmann often points out the limited possibilities of the scientific method when it comes to evolution. J. Miřejovský returned in 1950 to Czechoslovakia, where he worked as a pastor; he was later followed by J. M. Lochman and J. Heller. J. M. Lochman returned after 1968 to Basel, where, as rector, it eventually fell to him to preside over and speak at Portmann's funeral. Portmann's biology had an even greater impact on the Czech sphere. Here, first mention must go to the contribution of the philosopher and psychologist Jiří Němec (1932–2001), one of the key figures of Czech intellectual dissent from the 1960s to the 1980s and of modern Czech intellectual life in general. This man came to renown not so much for a great body of work as for an uncommon erudition and unimaginable overview of the academic literature – I had the opportunity frequently to admire this aspect of his, including his ability to accurately quote from memory by the page, when we led the local branch of the Czechoslovak Society of Arts and Sciences together while in exile in Vienna in the 1980s. It was J. Němec, a regular participant of the ecumenical seminar in "V Jirchářích" street, who noticeably mediated Portmann's ideas for a wider intellectual circle of Czech dissent. Additionally, two Czech zoologists studied briefly under Portmann in the 1960s – Zdeněk Veselovský (1928–2006), an ethologist and director of the Prague Zoo, for a shorter time, and Leo Sigmund (b. 1931), a vertebrate anatomist and morphologist, for a longer time.[3]

Jiří Němec encouraged Zdeněk Neubauer and Václav Benda in their translation into Czech of *Neue Wege der Biologie* (1960), which was "released" under the title *O nové horizonty v biologii* [On New Horizons in Biology] in 1979 in a few typewritten copies (rare collector's items today) as Volume 73 of the Dispatch Editions (*Edice Expedice*) under Olga Havlová (1933–1996), wife of the future president. This samizdat, which came into my hands for one night in 1980, along with one two-hour lecture from Zdeněk Neubauer, was enough to completely change the trajectory of my life.[4] The idea that this sort of biology, which the communists in their stupidity had banned, was also pursued in the West kept me awake at night

[3] The author thanks the volume editors for findings and consultations on the introduction of Portmann's thought into Bohemia.

[4] In the 1980s a specialized work on encephalization in birds based on Portmann's research in comparative anatomy also came out—see Mlíkovský (1985).

from that time forward – I did not suspect, though, that this was an almost singular phenomenon, the final enchanting flowering of a fading intellectual order.

Zdeněk Neubauer as the Most Eminent Czech Biological Thinker and Cornerstone of Portmannism in The Czech Republic

Zdeněk Neubauer (1942–2016) was undoubtedly the most interesting Czech-language philosophico-biological thinker of all. He was the son of Zdeněk Neubauer (1901–1956), a law professor at Masaryk University in Brno, and Štěpánka (née Makowski), also a lawyer, brother of professor Kamila Bendová, who lectured on logic in the Faculty of Arts at Charles University, and brother-in-law to the philosopher and dissident Václav Benda (1946–1999). He was also a close friend of Václav Havel. He studied biology at the Faculty of Natural Sciences at Charles University, changing his original focus on mycology (the bizarre world of mushrooms always fascinated him) to microbiology and graduating in 1965. From 1967–70 he worked in the famed International Laboratory of Genetics and Biophysics in Naples, where he focused primarily on bacteriophage lambda, a bacterial virus of exceptional theoretical importance. After his return he completed a degree in philosophy from the Charles University Faculty of Arts in 1971. At this time, he began publishing works on philosophy and the interface of philosophy and biology, mostly in samizdat or abroad, sometimes under the pseudonym Sidonius. He left the university due to his nonconformist attitudes in 1982 and supported himself as a programmer for the company ČKD (Českomoravská Kolben-Daněk), where many other dissidents were working at the time. In 1990 he founded the Department of Philosophy and History of Science at the Faculty of Natural Sciences at Charles University, which he led for many years, from 1991 as associate professor and from 1992 as professor, and where he continued to work after his official retirement. He authored or edited approximately 25 books and through his lectures and publications deeply influenced an entire Czech intellectual generation. He studied philosophical problems in biology and physics, the epistemological significance of form and formality, and the role of mythic and symbolic motives in cognition. His other work concerned culture, the mind, and the body; consciousness and unconsciousness; religious studies, hermeneutics, and depth psychology.

Fundamental to the propagation of Portmann's ideas in the Czech sphere was Neubauer's comprehensive work summarizing Portmann's opinions on the significance and meaning of the external appearance of living organisms (Neubauer 1985), first printed in a compilation by the Czechoslovak Scientific and Technical Society, a hectographed publication somewhere on the border between a samizdat and regular book (the work was published again together with the similarly-themed *Biomoc* [Biopower] in Neubauer [2002]). In spite of its nondescript and inferior form of publication, this treatise was widely read in pre-revolutionary Czechoslovakia and was intensively promoted by the author himself in numerous public lectures. Neubauer and Benda's translation was published again in 1997 under the title *Nové cesty biologie* [New Paths of Biology], this time from a proper press, on the occasion of the centenary of Portmann's birth. It appeared as volumes 7 and 8 of the journal *Scientia et Philosophia* from the publisher Jůza and Jůzová in Prague as a subsidized, not-for-sale publication which, though widely distributed, did not find its way into the book market, which, for that matter, had virtually collapsed as a delayed result of revolutionary events. Volume 8 also includes Neubauer's comprehensive essay on phenomenological (eidetic) biology, "*Esse obiectivum – esse intentionale*" (Neubauer 1997); as an introduction see Komárek's "Kapitoly o Portmannovi" [Chapters on Portmann] on Portmann's life, intellectual roots, and beliefs in Volume 7 (Komárek 1997a). Also interesting is Neubauer's detailed work on the "Portmannian" aspects of the world of fungi, which Portmann himself, though he did look at slime mold, completely passed over. Among Neubauer's doctoral students, Tomáš Daněk (b. 1976, Ph.D. 2011), assistant professor in the Department of Development and Environmental Studies at Palacký University Olomouc, published a Portmann-tinged study on the life of bumblebees (Daněk and Markoš 2005).

The opportunity of cofounding the Department of Philosophy and History of Science at the Faculty of Natural Sciences at Charles University with Zdeněk Neubauer was one of the most significant experiences of my life. Our first discussion in this direction took place on December 6, 1989, during my first visit to Prague after the revolution, and we both saw Portmannian themes as fundamental to its purpose and direction. The Department was formally established in April of 1990 and from the beginning of winter semester in October 1990 commenced a regular schedule of lectures and Friday seminars. After the revolution of 1989, it seemed that the newly founded department would become a center from which Portmannian biology would spread, if not across the planet, at least across Central Europe, certainly. While such an idea might seem outlandish today, it certainly did not at the time: the global collapse of communism with its unimaginable accumulation of power and armaments lent support to a wide array of expectations, of which the decline or demise of neo-Darwinism of the Dawkinsian-Wilsonian type was a relatively sober and marginal example: if a bear can be brought down, surely a squirrel can be caught. Amidst the overall "end-of-an-era" atmosphere and the contemporary slogans about the victory of "truth and love over lies and hatred," a new beginning was expected – in everything, as it were, across Europe and perhaps even the world. While in the first years Portmannian biology aroused exceptional interest among a wide array of students and was for a number of years a widespread and well-known alternative locally, with the connection of society to informational post-industrial Western civilization, it has been relegated to a marginal and historicist position.

Stanislav Komárek and His Students

The author of this contribution (b. 1958, study in zoology at Charles University Faculty of Natural Sciences 1977–1982, exile in Austria 1983–1990: Natural History Museum, Ministry of Agriculture, Zoological Institute of the University of Vienna), following his return to the Faculty of Natural Sciences at Charles University in 1990 (associate professor 1993, professor 2001), has worked on biological mimicry, domestication, nature-culture relationships, and human ethology, as well as the Portmannian topics of the inwardness and proper phenomenon of living organisms and the interpretation of their external appearance. He has been regularly acquainting students with Portmann's views on natural aesthetics and biological anthropology in his lectures since the winter semester of the 1990–91 academic year and has directed several bachelor's and master's theses and doctoral dissertations on these topics. In 1997, on the centenary of Portmann's birth, he published two works on his life and work—one mentioned above (Komárek 1997a); the second, a shorter essay in the magazine *Vesmír* (Komárek 1997b), which was reprinted in book form. Another article in *Vesmír* (Komárek 2003) concerns Major Richard Hingston (1887–1966), a half-forgotten, Irish amateur naturalist who, in his noteworthy book *The Meaning of Animal Colour and Adornment* (1933) anticipates practically all of Portmann's views on the self-representation of organisms, perhaps with the difference that he understands this as a threat and not as a phenomenon *an sich* (Hingston apparently was unaware of Portmann). A number of chapters in the author's book on mimetic phenomena are dedicated to Portmann, the German autonomist school, and further questions regarding the interpretation of the external appearance of living organisms (Komárek 2016, first edition 2000, second printing 2004, English translation Komárek 2003, expanded third Czech edition 2016). Similarly, there is a detailed chapter on Portmann's scientific contributions, as well as other chapters on natural aesthetics and the evolution of artefacts, in the author's *Příroda a kultura* [Nature and Culture] (first edition 2000, revised second edition Komárek 2008a, English translation Komárek 2009). An extensive essay on Portmann, especially on his biologico-anthropological views, is in the book *Obraz člověka a přírody v zrcadle biologie* [The Image of Man and Nature in the Mirror of Biology] (Komárek 2008b); the essay, with minor alterations, also appeared in the Portmann-dedicated volume *Biologie ve službách zjevu* [Biology in the Service of Appearance] (Kleisner 2008b) in the same year (Komárek 2008c). Together with the Dutch bioethicist Henk Verhoog (b. 1938) of Leiden University, he wrote the entry on Portmann in Achterhuis's dictionary of thinkers, *Kritisch Denkerslexicon* (Komárek and Verhoog 1994). With his Ph.D. student Roman Figura, whose dissertation (in progress) is devoted to a formal analysis of bird coloration based on Portmannian foundations, he published a shorter study on the interpretation of the creativity of living organisms and the Cartesian prejudices usually associated with this (Figura and Komárek 2014).

From the author's graduate students, another four are engaged with Portmann to a great extent. Only those of their works which directly relate to the reception and interpretation of Portmann and his concepts are mentioned here, and not those on the interpretation of the external appearance of living organisms in a wider sense. First mention goes to Karel Stibral (b. 1971, Ph.D. 2000): originally studied aesthetics, from 2011 associate professor in the Department of Environmental Studies, Faculty of Social Sciences, at Masaryk University in Brno. He discusses questions of natural aesthetics, primarily from a Darwinian perspective but also including Adolf Portmann, in two of his books and one lengthy essay – *Proč je příroda krásná?* [Why is Nature Beautiful?] (Stibral 2005), *Darwin a estetika* [Darwin and Aesthetics] (Stibral 2006), and "Darwin o kráse" [Darwin on Beauty] (Stibral 2014). With his colleagues, he published the edited volume *Česká estetika přírody ve středoevropském kontextu* [Czech Aesthetics of Nature in the Context of Central Europe] (Stibral et al. 2009), in which he has three extensive contributions: on Czech Portmannism, especially in the Department of Philosophy and History of Science at Charles University (Stibral 2009a); on Portmann's place in the Central-European biological and cultural context (Stibral 2009b); and the above-cited work on Josef Velenovský as an intellectual predecessor of Portmann (Stibral 2009c). At the Department of Environmental Studies, Stibral has also directed a master's thesis on the history of the reception of Portmann's intellectual heritage in the Czech Republic (Náhlíková 2011) and regularly includes Portmann's concepts in his lectures, also from the point of view of species conservation and environmental ethics.

Another eminent expert on Portmannian themes in a broad sense is Karel Kleisner (b. 1979, Ph.D. 2006): originally studied biology, from 2015 associate professor in the Department of Philosophy and History of Science at Charles University. Among other things, he has edited a collection on Portmannian themes, *Biologie ve službách zjevu: k teoreticko-biologickým myšlenkám Adolfa Portmanna* [Biology in the Service of Appearance: on the Theoretic Biological Ideas of Adolf Portmann] (ed. Kleisner 2008b), in which he has a comprehensive study on Portmann's theoretical biological views (Kleisner 2008c). The edited volume includes, among other things, a Czech translation of five of Portmann's essays from the book *Entlässt die Natur den Menschen?* [Will nature release man?] (Portmann 1970) by Marco Stella. Stella (2008) is also the author of a Portmann-inspired chapter on the phenomenal aspect of domestication published in this volume. Kleisner (2006) deals with a comparison of the Darwinian and Portmannian view of the living world; two further papers concern biosemiotic aspects of Portmann's views (Kleisner 2008a, 2011); and another is on phenomenological and functional aspects of the scrotum in mammals (Kleisner et al. 2010). Other works are dedicated to the complementarity of appearance and optical perception in animals (Kleisner et al. 2014; Brejcha et al. 2018); the significance of eyespots and other superficial structures of living organisms (Kleisner and Kočnar 2009); evolution and analogy in the elytral ornaments in ground beetles of the genus *Carabus* (Kleisner et al. 2012); and finally, the application of Portmann's and Uexküll's principles to visual communication in animals (Kleisner and Maran 2014; Timo Maran, b. 1973, is currently head of the Department of Semiotics, University of Tartu, Estonia, and an expert in biosemiotics, particularly mimetic phenomena in nature; he works closely with the Department of Philosophy and History of Science in Prague).

Another of the author's notable students is Filip Jaroš (b. 1982, Ph.D. 2012): originally studied mathematics, from 2013 assistant professor in the Department of Philosophy and Social Sciences, University of Hradec Králové. In collaboration with his colleagues, he published the edited volume *Krása a zvíře: Studie o vztahu estetických a etických hodnot zvířat* [Beauty and the Beast: Essays on Aesthetic and Ethical Values of Animals] (Dadejík et al. 2014), in which appears his essay on aesthetic and functional-morphological aspects of coloration in wild cats (Jaroš 2014), drawing on his dissertation (Jaroš 2012) and his earlier work (Jaroš 2009). His work dealing with theory of science is also based on Portmannian foundations (Jaroš 2008, 2010, 2018). Another of the author's graduate students, Ivana Ryška Vajdová (b. 1985), in her dissertation in progress, is working on a comparison of Jung's and Portmann's theoretical concepts (for an example of her published output, see Vajdová 2015).

References

Animalité et humanité autour d'Adolphe Portmann: actes du XVe colloque annuel du Groupe d'étude Pratiques sociales et theories. (1999). *Revue européenne des sciences sociales, 37*(115).

Brejcha, J., Pecháček, P., & Kleisner, K. (2018). Complementarity of seeing and appearing. *Cognitive Architectures, 94*, 13–30.

Dadejík, O., Jaroš, F., & Kaplický, M. (Eds.). (2014). *Krása a zvíře. Studie o vztahu estetických a etických hodnot zvířat* [Beauty and the beast: Essays on aesthetic and ethical values of animals]. Praha: Dokořán.

Daněk, T., & Markoš, A. (2005). *Život čmeláků. Koláž o pobývání v různých světech* [The life of bumblebees: A collage on living in different worlds]. Červený Kostelec: Pavel Mervart.

Figura, R., & Komárek, S. (2014). O ptácích, lidech a umění [On birds, people, and art]. In O. Dadejík, M. Kaplický, & K. Stibral (Eds.), *Krása a zvíře* (pp. 203–218). Praha: Dokořán.

Jaroš, F. (2008). Náhodnost v biologické evoluci (Proč má úspěch kreacionistická reakce?) [Chance in biological evolution (why is the creationism movement successful?)]. In J. Nosek, V. Havlík (Eds.), *Evoluce a věda*. (pp. 261–280). Nymburk: OPS.

Jaroš, F. (2009). Adaptivní význam zbarvení kočkovitých šelem. Empirie, fikce a vykročení k estetice [The adaptive function of felid coat patterns: Experience, fiction, and a step toward aesthetics]. In K. Stibral, B. Binka, & O. Dadejík (Eds.), *Krása, krajina, příroda II* [Beauty, landscape, nature II] (pp. 124–134). Brno: Masarykova univerzita.

Jaroš, F. (2010). Darwinismus a portmannismus: střetnutí nesouměřitelných biologických paradigmat? [Darwinism and Portmannism: A clash of incomparable biological paradigms?]. *Teorie vědy, 32*(3), 301–318.

Jaroš, F. (2012). *The ecological and ethological significance of felid coat patterns* (Carnivora, Felidae). PhD thesis. Faculty of Natural Sciences, Charles University in Prague.

Jaroš, F. (2014). Zbarvení kočkovitých šelem ve světle morfologické a estetické funkce. [The coloration of wild cats in the light of morphology and aesthetics]. In O. Dadejík, F. Jaroš, & M. Kaplický (Eds.), *Krása a zvíře: studie o vztahu estetických a etických hodnot zvířat* (pp. 217–244). Praha: Dokořán.

Jaroš, F. (2018). Současné teorie sociálního rozumění na pozadí sporu o antropologickou diferenci [Current theories of social cognition and their relation to a dispute about anthropological difference]. *Teorie vědy, 40*(1), 3–27.

Kleisner, K. (2006). Biologické omalovánky aneb, "Volte Portwina!" [Biological coloring books, or, "vote Portwin!"]. In K. Stibral (Ed.), *Darwin a estetika* (pp. 153–157). Praha: Pavel Mervart.

Kleisner, K. (2008a). The semantic morphology of Adolf Portmann: A starting point for the biosemiotics of organic form? *Biosemiotics, 1*(2), 207–219.

Kleisner, K. (Ed.). (2008b). *Biologie ve službách zjevu: k teoreticko-biologickým myšlenkám Adolfa Portmanna* [Biology in the service of appearance: On the theoretic biological ideas of Adolf Portmann]. Červený Kostelec: Pavel Mervart.

Kleisner, K. (2008c). Mezi formou a funkcí aneb Po stopách knížete niternosti: K základům Portmannovy teoretické biologie [Between form and function, or, in the footsteps of the prince of inwardness: On the foundations of Portmann's theoretical biology]. In K. Kleisner (Ed.), *Biologie ve službách zjevu: k teoreticko-biologickým myšlenkám Adolfa Portmanna* (pp. 109–135). Červený Kostelec: Pavel Mervart.

Kleisner, K. (2011). Perceive, co-opt, modify, and live! Organism as a centre of experience. *Biosemiotics, 4*(2), 223–241.

Kleisner, K., & Kočnar, T. (2009). Oko za oko: Struktura a význam v evoluci ornamentury organismálních povrchů [An eye for an eye: Structure and significance in the evolution of ornamentation in organismal surfaces]. In K. Stibral & O. Dadejík (Eds.), *Krása, krajina, příroda* (pp. 254–264). Praha: Dokořán.

Kleisner, K., & Maran, T. (2014). Visual communication in animals: Applying Portmannian and Uexküllian biosemiotic approach. In D. Machin (Ed.), *Visual communication* (pp. 659–676). Berlin: DeGruyter Mouton.

Kleisner, K., Ivell, R., & Flegr, J. (2010). The evolutionary history of testicular externalization and the origin of the scrotum. *Journal of Biosciences, 35*(1), 27–37.

Kleisner, K., Keil, P., & Jaroš, F. (2012). Biogeography of elytral ornaments in Palearctic genus *Carabus*: Disentangling the effects of space, evolution and environment at a continental scale. *Evolutionary Ecology, 26*(4), 1025–1040.

Kleisner, K., Brejcha, J., & Pecháček, P. (2014). Evoluce a zdroje proměnlivosti sémantických orgánů [Evolution and sources of variation of semantic organs]. In O. Dadejík, F. Jaroš, & M. Kaplický (Eds.), *Krása a zvíře. Studie o vztahu estetických a etických hodnot zvířat* (pp. 103–160). Dokořán: Praha.

Komárek, S. (1997a). Kapitoly o Portmannovi [Chapters on Portmann]. In *K stému výročí narození Adolfa Portmanna (1897–1982)* [On the centenary of the birth of Adolf Portmann] (Scientia & Philosophia, Vol. 7, pp. 3–37). Praha: Jůza a Jůzová.

Komárek, S. (1997b). Adolf Portmann (1897–1982) – k stému výročí narození. *Vesmír 76*(12), 676–678. Reprinted in S. Komárek, (2006). *Hlavou dolů. Eseje o přírodě a společnosti* (pp. 77–82). [Upside down: essays on nature and society]. Praha: Dokořán.

Komárek, S. (2003). Co spatřil major Hingston [What Major Hingston saw]. *Vesmír, 83*(7), 405–409.

Komárek, S. (2004). *Mimikry, aposematismus a příbuzné jevy: mimetismus v přírodě a vývoj jeho poznání.* Praha: Dokořán. [English version: Komárek, S. (2003). *Mimicry, aposematism and related phenomena: Mimetism in nature and the history of its study.* Lincom: Münich.]

Komárek, S. (2008a). *Příroda a kultura. Svět jevů a svět interpretací.* Praha: Academia. [English version: Komárek, S. (2009). *Nature and culture: The world of phenomena and the world of interpretations.* Lincom: Münich.]

Komárek, S. (2008b). Adolf Portmann. In *Obraz člověka a přírody v zrcadle biologie* (pp. 188–218). [The image of man and nature in the mirror of biology]. Praha: Academia.

Komárek, S. (2008c). Adolf Portmann (1897–1982): Myšlenky, kořeny, názory [Thoughts, roots, opinions]. In K. Kleisner (Ed.), *Biologie ve službách zjevu* (pp. 79–108). Červený Kostelec: Pavel Mervart.

Komárek, S. (2016). *Mimikry a příbuzné jevy: Dějiny poznávání a výkladu vnějšího vzhledu živých organismů* [Mimicry and related phenomena: A history of the study and interpretation of the external appearance of living organisms]. Praha: Academia.

Komárek, S., & Verhoog, H. (1994). Adolf Portmann. In H. J. Achterhuis (Ed.), *Kritisch Denkerslexicon* (pp. 1–15). Houten, Zaventem: Bohn, Stafleu, Van Loghum.

Mlíkovský, J. (1985). Portmann's rule of precedence and the evolution of encephalization in birds (Aves). In J. Mlíkovský & V. J. A. Novák (Eds.), *Evolution and morphogenesis* (pp. 559–566). Praha: Academia.

Monastra, G. (2000). Adolf Portmann, Wilhelm Troll: due biologi della morfologia idealistica. *Rivista di Biologia, 93*(2), 1000–1004.

Náhlíková, T. (2011). *Alternativní směry v české biologii – žáci Adolfa Portmanna* [Alternative tendencies in Czech biology: The followers of Adof Portmann]. Master's thesis. Faculty of Social Sciences, Masaryk University in Brno.

Neubauer, Z. (1985): *Střetnutí paradigmat v současné biologii* [The clash of paradigms in contemporary biology]. Sborník přednášek cyklu Kybernetické problémy přírodovědy [Lectures on cybernetic problems in natural science]. Praha: Československá vědeckotechnická společnost.

Neubauer, Z. (1997). Esse obiectivum – esse intentionale. Cestou k fenomenologické biologii. [Toward a phenomenological biology]. In *K 100. výročí narození Adolfa Portmanna (1897–1982)* [On the centenary of the birth of Adolf Portmann]. *Scientia et Philosohia* 8: 113–160. Praha: Jůza a Jůzová.

Neubauer, Z. (2002). *Biomoc* [Biopower]. Praha: Malvern.

Portmann, A. (1937). Evoluční problémy ptačího vývoje [The ontogenesis of birds as an evolutionary problem]. Měsíčník Ciba, *4, O normálním a porušeném vývoji*, 1287–1290. (German orig.: Portmann, A. (1936). Evolutionsprobleme der Vogelentwicklung. Ciba Zeitschrift, *40*, 1387–390.)

Portmann, A. (1956). Ein Wegbereiter der neuen Biologie. In J. von Uexküll, G. Kriszat. *Streifzüge durch die Umwelten von Tieren und Menschen* and *Bedeutungslehre* (pp. 7–17). Hamburg: Rowohlt.

Portmann, A. (1970). *Entlässt die Natur den Menschen?* Munich: Piper.

Portmann, A. (1997). Nové cesty biologie. (Trans. V. Benda and Z. Neubauer.) In *K 100. výročí narození Adolfa Portmanna (1897–1982). Scientia et Philosophia* 7, 8 (Vol. 7, pp. 51–161), (Vol. 8, pp. 6–111). Praha: Jůza a Jůzová.

Stella, M. (2008). Domus a podoba: poznámky k jevové stránce a historii výzkumu domestikace [Domus and similarity: Notes on the phenomal aspect and the history of the study of domestication]. In K. Kleisner (Ed.), *Biologie ve službách zjevu. K teoreticko-biologickým myšlenkám Adolfa Portmanna* (pp. 139–170). Pavel Mervart: Červený Kostelec.

Stibral, K. (2005). *Proč je příroda krásná? Estetické vnímání přírody v novověku* [Why is nature beautiful? The aesthetic perception of nature in the modern age]. (See "Estetické otázky v biologii 19. a 20. století" [Aesthetic questions in biology in the 19th and 20th centuries)] (pp. 158–167). Praha: Dokořán.

Stibral, K. (2006). *Darwin a estetika. Ke kontextu estetických názorů Charlese Darwina* [Darwin and aesthetics: On the context of the aesthetic concepts of Charles Darwin] (pp. 130–133). Červený Kostelec: Pavel Mervart.

Stibral, K. (2009a). Český portmannismus – "Mystici z Viničné" [Czech Portmannism – "the mystics from Viničná street"]. In K. Stibral, O. Dadejík, & V. Zuska (Eds.), *Česká estetika přírody ve středoevropském kontextu* [Czech Aesthetics of Nature in the Context of Central Europe] (pp. 250–264). Praha: Dokořán.

Stibral, K. (2009b). Přírodovědná linie s akcentem na vývoj ve střední Evropě [The line of natural science emphasizing development in Central Europe]. In K. Stibral, O. Dadejík, V. Zuska (Eds.), *Česká estetika přírody ve středoevropském kontextu* [Czech aesthetics of nature in the context of Central Europe] (pp. 185–216). Praha: Dokořán.

Stibral, K. (2009c). Josef Velenovský – princip ornamentalismu jako evoluční princip [Josef Velenovský – The principle of Ornamentalism as an evolutionary principle]. In K. Stibral, O. Dadejík, & V. Zuska (Eds.), *Česká estetika přírody ve středoevropském kontextu* [Czech aesthetics of nature in the context of Central Europe] (pp. 217–225). Praha: Dokořán.

Stibral, K. (2014). Darwin o kráse aneb o spletitých cestách od estetiky k biologii [Darwin on beauty, or, on the tangled paths from aesthetics to biology]. In O. Dadejík, F. Jaroš, & M. Kaplický (Eds.), *Krása a zvíře* (pp. 243–280). Praha: Dokořán.

Stibral, K., Komárek, S., & Neustupa, J. (2004). J.: Přírodní estetika Josefa Velenovského [The natural aesthetics of Josef Velenovský]. *Estetika, 2004*(1–2), 117–126.
Stibral, K., Dadejík, O., & Zuska, V. (Eds.). (2009). *Česká estetika přírody ve středoevropském kontextu* [Czech aesthetics of nature in the context of Central Europe]. Praha: Dokořán.
Vajdová, R. I. (2015). Three dimensions of living forms – Brief looks into parallels between A. Portmann and C.G. Jung. *Humanum, 18*(3), 119–126.
Velenovský, J. (1922). *Přírodní filosofie I* [Natural Philosophy I]. Praha: Dokořán.
Wulf, C. (2013). *Anthropology: A continental perspective.* Chicago/London: The University of Chicago Press.

Stanislav Komárek studied zoology at the Faculty of Natural Sciences at Charles University in Prague (1982) and afterward emigrated to Austria, where he worked at the Natural History Museum, the Ministry of Agriculture, and the Zoological Institute of the University of Vienna. In 1990 he returned to his alma mater, where, with Zdeněk Neubauer, he helped found the Department of Philosophy and History of Science, in which he continues to work (from 1993 as associate professor, from 2001 as professor). He is interested in Portmannian themes in relation to animals and humans, human ethology, the relationships of nature and culture, the phenomenon of domestication, mimicry, and the interpretation of the external appearance of living organisms. In 2006 he was awarded the Tom Stoppard Prize for non-fiction.

Index

F. Jaroš, J. Klouda (eds.), *Adolf Portmann*, Biosemiotics 23,
https://doi.org/10.1007/978-3-030-67810-4

The manufacturer's authorised representative in the EU is Springer Nature Customer Service Centre GmbH, Europaplatz 3, 69115 Heidelberg, Germany. If you have any concerns regarding our products, please contact ProductSafety@springernature.com

Printed and bound by CPI Group (UK) Ltd, Croydon, CR0 4YY
30/09/2026
02225289-0005